EXPLOSIVE NUCLEOSYNTHESIS

EXPLOSIVE NUCLEOSYNTHESIS

Proceedings of the Conference on Explosive Nucleosynthesis
Held in Austin, Texas, on April 2-3, 1973

Edited by DAVID N. SCHRAMM
and W. DAVID ARNETT

UNIVERSITY OF TEXAS PRESS, AUSTIN & LONDON

International Standard Book Number 0-292-72006-8 (cloth);
   0-292-72007-6 (paper)
Library of Congress Catalog Card Number 73-9185
Copyright © 1973 by the University of Texas Press
All Rights Reserved
Printed in the United States of America

TABLE OF CONTENTS

Preface	vii
List of Conference Participants	xi
Chapter I: Abundances and Chemical Evolution of Galaxies	1
A New Table of Abundances, A.G.W. Cameron*	3
On the Origin and Evolution of the Light Elements, Beatrice M. Tinsley	22
Chemical Evolution of the Galaxy: Coefficients from Stellar Evolution and Alternative Solutions to the Problem of Few Metal-Poor Stars, R. J. Talbot, Jr.	34
Chapter II: Explosive Processes	45
The Hot CNO Process, Jean Audouze	47
Explosive Carbon Burning, W. Michael Howard	60
Nucleosynthesis During Explosive Oxygen and Silicon Burning, S. E. Woosley	70
The Dynamic r-Process, David N. Schramm	84
The p-Process, J. W. Truran	102
Chapter III: Aspects of Pre-Carbon Burning	113
Presupernova Evolution, Icko Iben, Jr.	115
The s-Process in Stars, Roger K. Ulrich	139
Nucleosynthesis in Red Giants, R. L. Smith, I-Juliana Sackmann and K. H. Despain	168
Chapter IV: The Carbon Detonation Model	177
Degenerate Carbon Burning, R. G. Couch and W. D. Arnett	179

TABLE OF CONTENTS (continued)

    The Influence of Screening Effects on Carbon
    Ignition, Harold C. Graboske, Jr.   186

    Measurement of the $^{12}C(\alpha,\gamma)^{16}O$ Cross Section,
    P. Dyer   195

    The Carbon Detonation Supernova Model, J. Craig Wheeler   203

    The Carbon Detonation Supernova and Associated
    Remnant Formation, S. Bruenn   213

    Off-Center Detonation Supernovae, Jean-Robert Buchler   229

Chapter V: Massive Stars, Deuterium and Gamma Rays   235

    Some Quantitative Calculations of Final Stages of
    Stellar Evolution, W. David Arnett   236

    Supernova Shock Waves, Stirling A. Colgate   248

    Confirming Explosive Nucleosynthesis with Gamma-Ray
    Telescopes, Donald D. Clayton   264

Chapter VI: Nuclear Reaction Rates: The Basis of the Subject   283

    Problems with Nuclear Reaction Rates, Georges Michaud   285

    Experiments of Relevance to Explosive Nucleosynthesis,
    William A. Fowler*   298

* As summarized by the editors.

Preface

　　While driving to the Las Cruces meeting of the AAS we encountered particularly bad weather in the form of freezing rain. As we slid and spun through West Texas among cacti encrusted with icicles, it occurred to us to have a conference that we could reach by foot rather than by automobile. In this overreaction to the inclement weather our plotting began. Since Willy Fowler had already agreed to give a seminar in Austin on April 2, that seemed an obvious date to expand upon. A few weeks and numerous invitations later we had a full fledged conference scheduled for April 2. In retrospect it seems that the enthusiasm for a conference on explosive nucleosynthesis was rooted in the recent outburst of publication on this topic and in a widespread feeling that such a conference would help clarify and integrate the results of this research.

　　In order to keep the conference to a manageable size in number of participants, facilities required, and duration, we were forced (quite reluctantly) to omit some topics which are of considerable interest. For example, cosmic rays, x-ray sources, stellar and galactic abundance differences, millimeter wave observations and gravitational collapse all figured in various discussions. The absence of review talks on these topics was a price we paid for a conference small enough to be relatively open and informal. The coverage of explosive nucleosynthesis itself was probably more complete than at any previous symposium thus making these conference proceedings a relatively complete picture of the subject.

　　Since much of the material to be discussed was completely new, not even having reached preprint status by conference time, we decided to take on the rather tedious task of publishing the proceedings. In all the areas covered there have been significant (and in some cases revolutionary) developments within the last five years, and no connected discussion of these topics has appeared. We found that the conference clarified our thinking on a number of topics and suggested new and intriguing ways of looking at the subject. We hope the conference proceedings prove equally valuable to the reader.

　　The speakers were requested to give short, intensive talks which emphasized new developments in their areas. We hope that the reader will agree with us that they succeeded admirably. We feel that the collection of papers presented here gives a reasonable picture of current research on this rapidly expanding topic. In order to minimize the time between the conference and publication of these proceedings, this volume has been produced with maximum rapidity. We hope any errors overlooked in this process will be excused by its readers.

DAVID N. SCHRAMM
Assistant Professor of
Astronomy and Physics
University of Texas
Austin, Texas

W. DAVID ARNETT
Associate Professor of
Astronomy and Physics
University of Texas
Austin, Texas

April 1973
Austin, Texas

Acknowledgments

The conference was sponsored by the Department of Astronomy of The University of Texas at Austin with supplementary support from a gift from Lettie Jones. The authors also acknowledge partial support from The National Science Foundation.

The astronomy department staff is to be thanked for their aid in preparing the manuscripts. Particular thanks go to Ms. J. Strong for her aid in bringing the proceedings to publication.

LIST OF CONFERENCE PARTICIPANTS*

W. D. Arnett	Univ. of Texas at Austin
J. Audouze	Calif. Inst. of Technology and Paris-Meudon
C. A. Barnes	Calif. Inst. of Technology
J. B. Blake	Aerospace Corp.
D. Bodansky	Univ. of Washington
A. Boudreaux	Univ. of Texas at Austin
S. W. Bruenn	Florida Atlantic Univ.
J. R. Buchler	Yeshiva University
A. G. W. Cameron	Yeshiva University
G. Chapline	Lawrence Livermore Lab.
D. D. Clayton	Rice University
S. Colgate	New Mexico Inst. of Mining & Tech.
R. Couch	Univ. of Texas at Austin
C. Davids	Univ. of Texas at Austin and Brookhaven Nat'l Lab.
P. Dyer	Calif. Inst. of Technology
S. Falk	Univ. of Texas at Austin
W. A. Fowler	Calif. Inst. of Technology
H. Graboske	Lawrence Livermore Lab.
K. Hainebach	Rice University
C. Hansen	Univ. of Colorado
T. G. Harrison	North Texas State Univ.
W. M. Howard	Los Alamos Scientific Lab.
I. Iben	Univ. of Illinois
O. Johns	Paris, CEN Saclay
R. L. Macklin	Oak Ridge National Lab.
G. Michaud	Univ. of Montreal
F. C. Michel	Rice University
H. Mitler	Smithsonian Astrophysical Obser.
J. Morgan	Ohio State Univ.
M. Newman	Rice University
R. Pardo	Univ. of Texas at Austin
L. Parks	Univ. of Texas at Austin
S. Parsons	Univ. of Texas at Austin
J. Peters	Louisiana State Univ.
N. A. Roughton	Regis College
D. N. Schramm	Univ. of Texas at Austin
P. A. Seeger	Los Alamos Scientific Lab.
M. M. Shapiro	Naval Research Lab.
R. Silberberg	Naval Research Lab.
L. Smarr	Univ. of Texas at Austin
R. Smith	Rensselaer Polytechnic Inst.
I. B. Strong	Los Alamos Scientific Lab.
R. Talbot	Rice University
B. Tinsley	Univ. of Texas at Dallas
J. W. Truran	Yeshiva University

LIST OF CONFERENCE PARTICIPANTS (continued)

B. Ulrich	Univ. of Texas at Austin
R. Ulrich	Univ. of Calif. at Los Angeles
P. Vanden Bout	Univ. of Texas at Austin
G. Wallerstein	Univ. of Washington
T. Weaver	Lawrence Livermore Lab.
J. C. Wheeler	Harvard University
D. Whitmire	Univ. of Texas at Austin
J. Wilson	Lawrence Livermore Lab.
L. Wood	Lawrence Livermore Lab.
S. E. Woosley	Rice University
A. Wootten	Univ. of Texas at Austin
C. Zaidens	Univ. of Colorado

*A participant is defined as one who paid the registration fee.

EXPLOSIVE NUCLEOSYNTHESIS

CHAPTER I. Abundances and Chemical Evolution of Galaxies

   A fundamental question for astrophysics is the understanding of the regularity of chemical abundances in cosmic objects. The last five years have been unusually active ones for research in a set of related areas. The common tie is the idea that most of the observed abundances bear the mark of processing in an exploding star, hence the title "Explosive Nucleosynthesis". The recent developments are based on the continued progress in stellar evolutionary calculations, the wider application of numerical hydrodynamics to stellar problems, the development of numerical techniques for solving complex nuclear networks, and (most important) due to progress in experimental nuclear physics, an increasing understanding of thermonuclear reaction rates.

   Meteoritic abundances are the ultimate test for nucleosynthesis theory. While isotopic ratios can be accurately and reliably determined, chemical separation makes elemental abundances less secure. The advantage of having a laboratory sample must be balanced against the drawback of having only a tiny and perhaps unrepresentative segment of what we want to measure. Cameron has completed a revised table of abundances which represents a recent contribution to this vital area.

   To obtain a wider view of the abundance distribution in the universe we must resort to other sources of information, for example absorption lines in stellar atmospheres, emission lines in nebulae and cosmic rays.

   It is not a trivial problem to relate observed abundances to theoretical models of stars and of a galaxy. The papers by Tinsley and Talbot address this important and as yet poorly understood connection.

Chapter I

References for Chapter I.

Abundances

    Suess, H. E. and Urey, H. C. 1956, Rev. Mod. Phys., 28, 53.
    Pagel, B. E. J. 1972, Symposium on Cosmochemistry, in press, Cambridge, Mass.
    Shapiro, M. M. and Silberberg, R. 1970, Ann. Rev. Nucl. Sci., 20, 323.

Galactic Evolution

    Schmidt, M. 1959, Ap. J., 129, 243.
    Schmidt, M. 1963, Ap. J., 137, 760.
    Salpeter, B. B. 1959, Ap. J., 129, 608.
    Truran, J. W. and Cameron, A. G. W. 1971, Ap. and Sp. Sci., 14, 179.
    Arnett, W. D. 1971, Ap. J., 160, 153.

A CRITICAL DISCUSSION OF THE
ABUNDANCES OF NUCLEI

Summary by the editors of the
talk given by

A. G. W. Cameron
Belfer Graduate School of Science, Yeshiva University
New York, New York

A more detailed discussion of the topics presented in this talk may be found in a new review article by Cameron (1973). Here we will be content to present Cameron's new mass table and to point out some of the significant changes in it.

Table 1 is a compilation of elemental abundances normalized so that $Si = 10^6$. Besides Cameron's new abundances values, the older work by Suess and Urey (1956) and Cameron (1968) is given for comparison. The notes refer to the sources used for the new compilation.

TABLE 1

Compilations of Abundance Normalized to $Si + 10^6$

(Notes refer to sources used for the present compilation)

ELEMENT	Suess Urey	Cameron (1968)	This Work	Notes
1 H	4.00×10^{10}	2.6×10^{10}	3.18×10^{10}	1
2 He	3.08×10^9	2.1×10^9	2.21×10^9	2
3 Li	100	45	49.5	3
4 Be	20	0.69	0.81	25
5 B	24	6.2	350	25
6 C	3.5×10^6	1.35×10^7	1.18×10^7	1, 4
7 N	6.6×10^6	2.44×10^6	3.74×10^6	1, 4
8 O	2.15×10^7	2.36×10^7	2.15×10^7	1, 4
9 F	1600	3630	2450	3
10 Ne	8.6×10^6	2.36×10^6	3.44×10^6	2
11 Na	4.38×10^4	6.32×10^4	6.0×10^4	3

Table 1 (continued)

Element	Suess Urey	Cameron (1968)	This Work	Notes
12 Mg	9.12×10^5	1.050×10^6	1.061×10^6	3
13 Al	9.48×10^4	8.51×10^4	8.5×10^4	3
14 Si	1.00×10^6	1.00×10^6	1.00×10^6	3
15 P	1.00×10^4	1.27×10^4	9600	3
16 S	3.75×10^5	5.06×10^5	5.0×10^5	3, 5
17 Cl	8850	1970	5700	3, 5
18 Ar	1.4×10^5	2.28×10^5	1.172×10^5	5
19 K	3160	3240	4200	3, 5
20 Ca	4.90×10^4	7.36×10^4	7.21×10^4	3
21 Sc	28	33	35	6
22 Ti	2240	2300	2775	7
23 V	220	900	262	7
24 Cr	7800	1.24×10^4	1.27×10^4	3
25 Mn	6850	8800	9300	23
26 Fe	6.00×10^5	8.90×10^5	8.3×10^5	23
27 Co	1800	2300	2210	23
28 Ni	2.74×10^4	4.57×10^4	4.80×10^4	3
29 Cu	212	919	540	3
30 Zn	486	1500	1244	18
31 Ga	11.4	45.5	48	3
32 Ge	50.4	126	115	18
33 As	4.0	7.2	6.6	3
34 Se	67.6	70.1	67.2	18, 24
35 Br	13.4	20.6	13.5	24
36 Kr	51.3	64.4	46.8	8
37 Rb	6.5	5.95	5.88	9, 18
38 Sr	18.9	58.4	26.9	3, 10
39 Y	8.9	4.6	4.8	3
40 Zr	54.5	30	28	3
41 Nb	1.00	1.15	1.4	11
42 Mo	2.42	2.52	4.0	3
44 Ru	1.49	1.6	1.9	3

Table 1 (continued)

Element	Suess Urey	Cameron (1968)	This Work	Notes
45 Rh	0.214	0.33	0.4	12
46 Pd	0.675	1.5	1.3	3
47 Ag	0.26	0.5	0.45	18
48 Cd	0.89	2.12	1.48	18
49 In	0.11	0.217	0.189	18
50 Sn	1.33	4.22	3.6	3
51 Sb	0.246	0.381	0.316	18
52 Te	4.67	6.76	6.42	18
53 I	0.80	1.41	1.09	3
54 Xe	4.0	7.10	5.38	13
55 Cs	0.456	0.367	0.387	18
56 Ba	3.66	4.7	4.8	3
57 La	2.00	0.36	0.445	14
58 Ce	2.26	1.17	1.18	14
59 Pr	0.40	0.17	0.149	14
60 Nd	1.44	0.77	0.78	14
62 Sm	0.664	0.23	0.226	14
63 Eu	0.187	0.091	0.085	14
64 Gd	0.684	0.34	0.297	14
65 Tb	0.0956	0.052	0.055	14
66 Dy	0.556	0.36	0.36	14
67 Ho	0.118	0.090	0.079	14
68 Er	0.316	0.22	0.225	14
69 Tm	0.0318	0.035	0.034	14
70 Yb	0.220	0.21	0.216	14
71 Lu	0.050	0.035	0.036	14
72 Hf	0.438	0.16	0.21	22
73 Ta	0.065	0.022	0.021	15
74 W	0.49	0.16	0.16	16
75 Re	0.135	0.055	0.053	17, 18
76 Os	1.00	0.71	0.75	6

Table 1 (continued)

Element	Suess Urey	Cameron (1968)	This Work	Notes
77 Ir	0.821	0.43	0.717	18
78 Pt	1.625	1.13	1.4	6
79 Au	0.145	0.20	0.202	18
80 Mg	0.284	0.75	0.4	19
81 Tl	0.108	0.182	0.192	18
82 Pb	0.47	2.90	4	3
83 Bi	0.144	0.164	0.143	18
90 Th	-	0.034	0.058	6, 20
92 U	-	0.0234	0.0262	18, 21

Notes to Table 1

1. Normalized to abundances in Type I carbonaceous chondrites by multiplying the solar abundance scale, $H = 10^{12}$, by 0.03175, which is an average of the ratios of the solar scale and meteoritic scale $(Si = 10^6)$ for the elements Mg,Al,Si,P,S,Ca, Fe and Ni.
2. Based upon solar flare cosmic rays (Bertsch et al. 1972), for which $He/O = 103 \pm 10$ and $Ne/O = 0.16 \pm 0.03$.
3. Type I carbonaceous chondrites (Mason 1971).
4. Solar photospheric composition adopted by Withbroe (1971).
5. Choice of abundance influenced by quasi equilibrium interpolation between ^{28}Si and ^{40}Ca.
6. Weighted mean of carbonaceous chondrites (Mason 1971).
7. Weighted mean of Types I and II carbonaceous chondrites (Mason 1971).
8. Average of the geometric interpolations of ^{84}Kr between ^{80}Se and ^{88}Sr and of ^{83}Kr between ^{81}Br and ^{85}Rb.
9. Present abundance 5.77. Adopted value includes correction for ^{87}Rb decay since solar system formation.
10. Present abundance 27. Adopted value allows for growth of ^{87}Sr since solar system formation.
11. Very little meteorite data: 0.3 in L6, 1.6 in LL6 chondrites (Mason 1971). Value is an interpolation between ^{91}Zr and ^{95}Mo.
12. Very little meteorite data: 0.4 in H5, 0.31 in L6, 0.70 in LL6, and 0.36 in E4 chondrites (Mason 1971).

13. Fitted to Te and I to form a continuous abundance peak with the same odd-even abundance ratios centered at mass numbers 127 and 129. Carbonaceous chondrite isotope ratios used.
14. There is little variation in rare earth abundances among meteorite classes, but Type I carbonaceous chondrite measurements are not as good as those in ordinary chondrites. Hence the latter values were chosen (Mason 1971) but multiplied by 1.20 to normalize to Orgueil and Ivuna Type I carbonaceous chondrites (Urey 1964).
15. No data on Cl meteorites. Value chosen is average of H-group meteorites (Mason 1971) and is possibly low.
16. Average of C2 and C3 meteorites (Mason 1971) omitting higher values of Rieder and Wanke (1968).
17. Present value of 0.050 has been corrected for decay of ^{187}Re since formation of solar system.
18. Measurements in Cl chondrites (Krahenbuhl et al. 1973).
19. Mercury has enormous variations in meteorites and the carbonaceous chondrite values are one to two orders of magnitude too high for nuclear systematics. The value chosen is a representative value for enstatites (Mason 1971).
20. Present value 0.046, corrected for decay since formation of solar system.
21. Present value 0.0098, corrected for decay since formation of solar system.
22. The Cl values of Ehmann and Rebagay (1970) are surprisingly high since there is no indication that there should be fractionation of Zr or Hf among carbonaceous chondrites. The adopted value was obtained by dividing the Cl/Zr abundances of Ehmann and Rebagay by the average chondritic Zr/Hf ratio.
23. Average Cl values of Schmitt, Goles, Smith and Osborn (1972).
24. Average Cl values of Goles, Greenland and Jerome (1967).
25. See Cameron, Colgate and Grossman (1973).

The boron abundance is now thought to be considerably higher, a fact which has important implications for a variety of astrophysical problems; for example, galactic evolution (see the paper by Tinsley in this volume) and supernova shock waves (see Colgate's paper in this volume). The reason for the increase in the estimated boron abundance is a combination of higher observed boron abundances in meteorites and a considerable correction for chemical fractionation of boron in the formation of the meteorites from the protosolar nebula. (See Cameron et al. 1973.)

Cameron emphasized that the abundances of elements from sulphur to calcium was uncertain due to their high volatility in certain common chemical forms. The data are so difficult to interpret that he proposed that the elemental abundances in this region might as well be determined using theory (see Woosley's

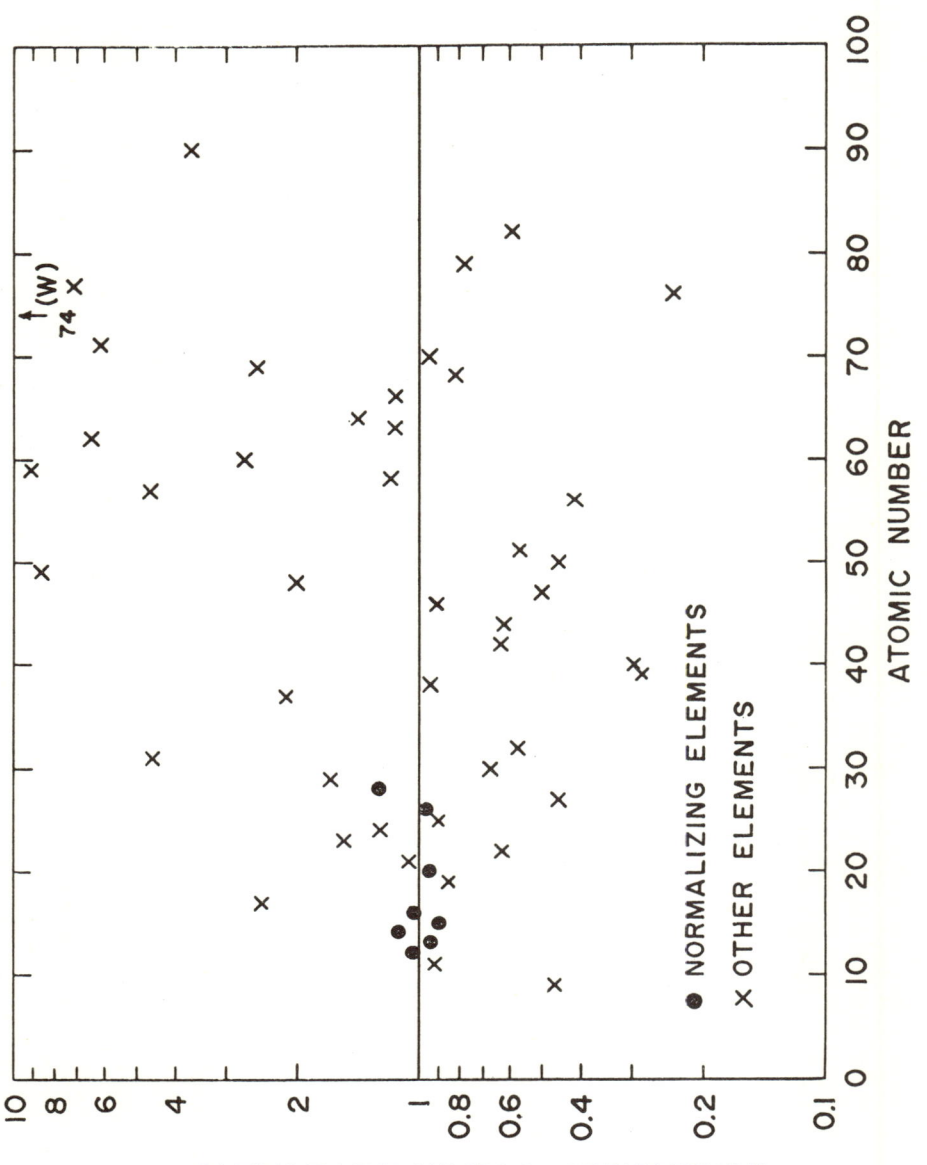

Fig. 1. A comparison of the ratio of solar to meteoritic abundances as a function of atomic number.

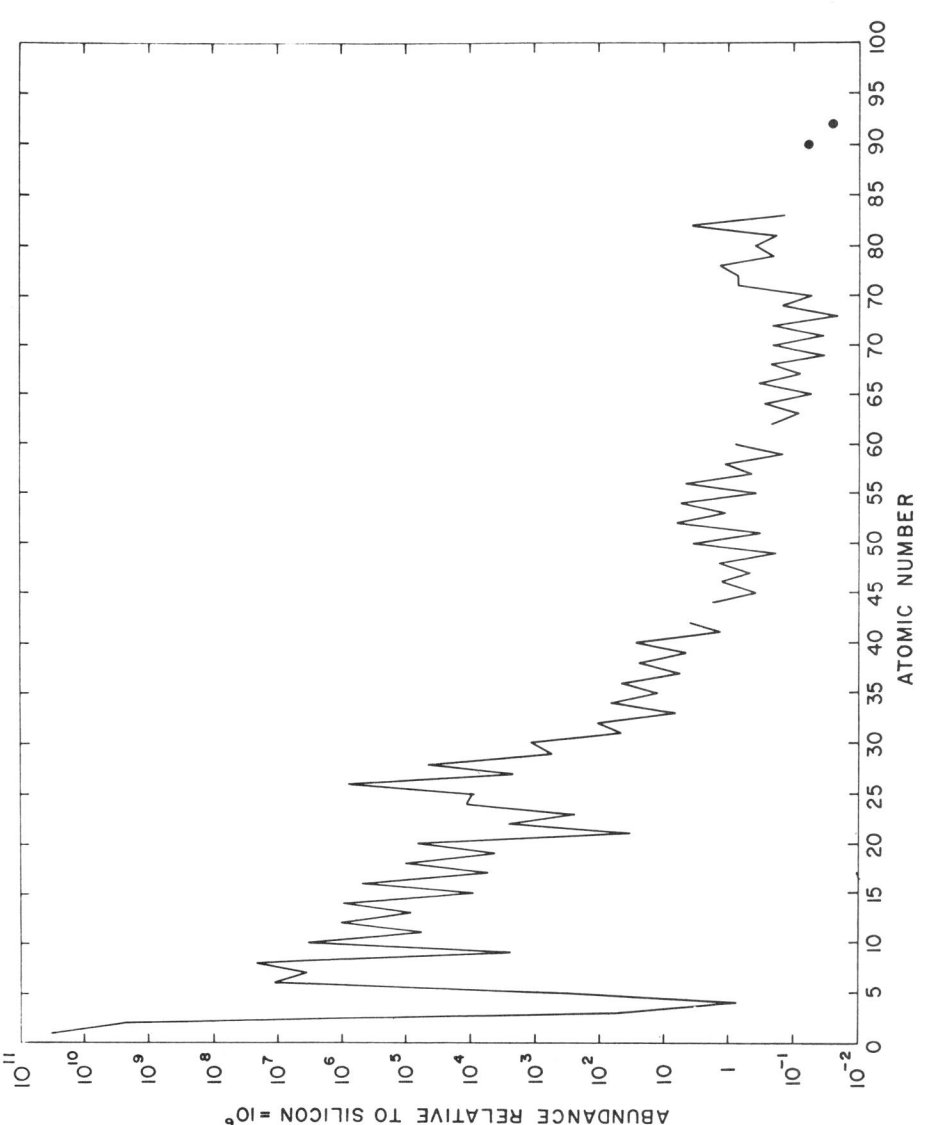

Fig. 2. Elemental abundances (Si=10^6) as a function of atomic number.

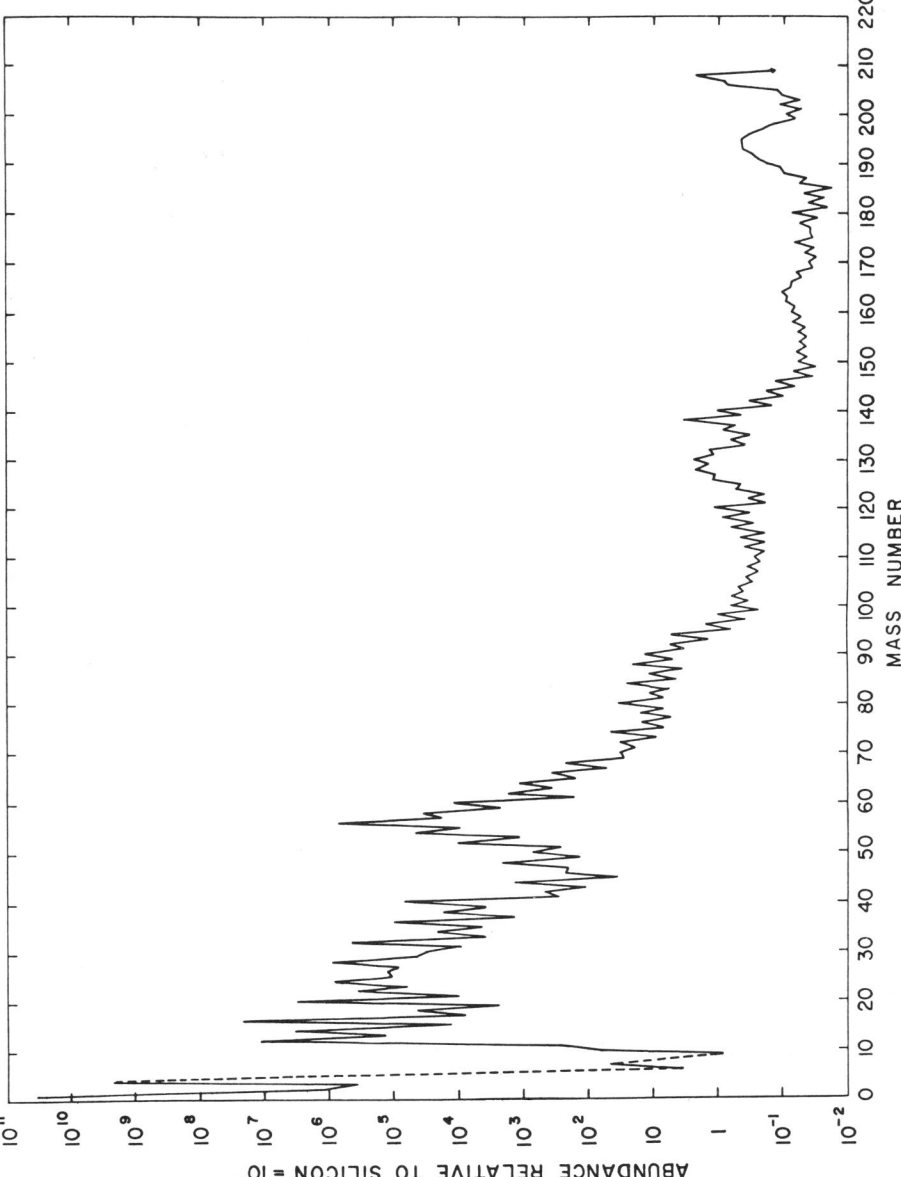

Fig. 3. Nuclear abundances (elemental Si=10⁶) as a function of mass number.

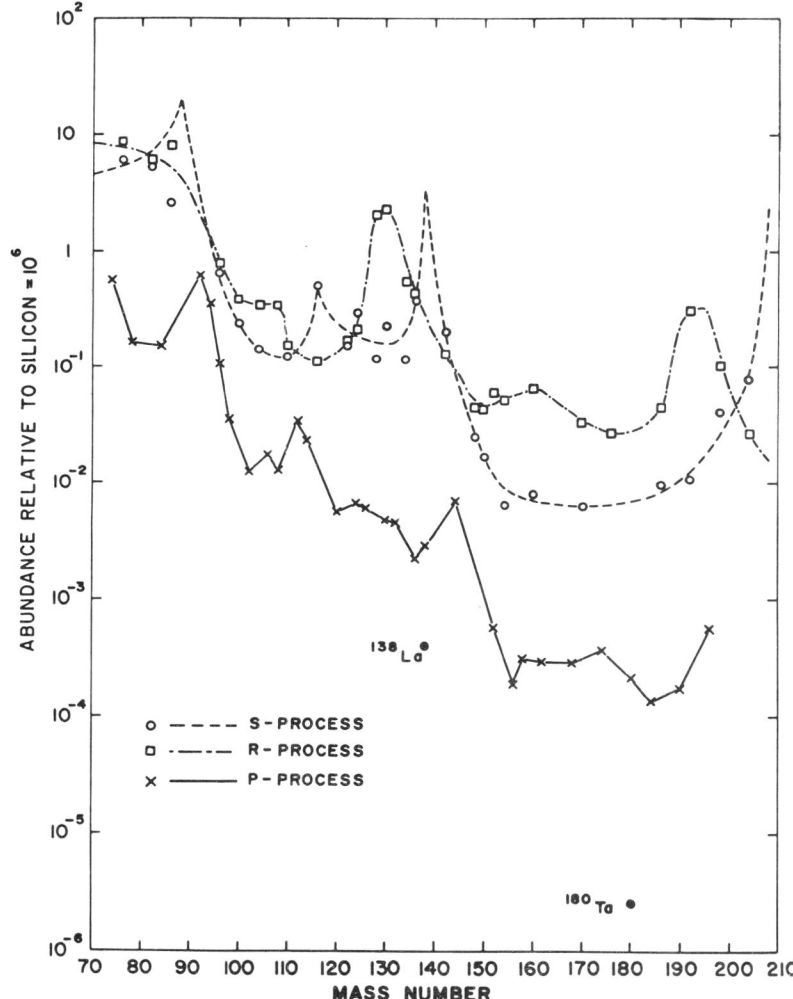

Fig. 4. Abundances of s-, r-, p-process nuclei as a function of mass number, A, $70 \leq A \leq 210$.

discussion of quasiequilibrium in oxygen burning). Such a procedure was adoptoed in Cameron's new table of abundances. While this means that a certain circularity is involved if we compare predicted abundances in this range to the new table, it does emphasize the desperate need for a better understanding of the implications of the experimental results on the abundances of these elements. Fortunately the <u>isotopic</u> ratios for these elements do provide us with significant constraints upon explosive nucleosynthesis.

Table 2 presents the abundance of nuclei (as well as isotopic ratios); the normalization is again that <u>elemental</u> Si = 10^6. The most significant new isotopic abundance estimate is that for deuterium which is based on $HD/H_2 = \frac{1}{61000}$ in Jupiter (Trauger <u>et al</u>. 1973). It is interesting to compare solar and meteoritic elemental abundances. To do this one must normalize using elements which have a reasonably well-determined abundance in both the sun and in the meteorites. Figure 1 gives such a comparison. There is still a considerable scatter; this probably reflects the uncertainty in elemental abundance determinations. Figure 2 gives the <u>elemental</u> abundance curve for Cameron's new abundance compilation while Figure 3 gives the corresponding abundance curve for each mass number. Finally, Figure 4 displays a set of abundance curves for the s-, r-, and p-processes, for $70 \leq A \leq 210$, using the new compilation. These are fundamental data for current explosive nucleosynthesis research. The general character of these abundances, and often their precise numberical values, will be of concern in virtually all subsequent papers in this volume.

TABLE 2

Abundances of Nuclei

ELEMENT	A	% Abundance	Abundance
1 H	1	~ 100	3.18×10^{10}
	2		5.2×10^5
2 He	3		~ 3.7×10^5
	4	~ 100	2.21×10^9
3 Li	6	7.42	3.67
	7	92.58	45.8
4 Be	9	100	0.81
5 B	10	19.64	68.7
	11	80.36	281.3

Table 2 (continued)

Element	A	% Abundance	Abundance
6 C	12	98.89	1.17×10^7
	13	1.11	1.31×10^5
7 N	14	99.634	3.63×10^6
	15	0.366	1.33×10^4
8 O	16	99.759	2.14×10^7
	17	0.0374	8040
	18	0.2039	4.38×10^4
9 F	19	100	2450
10 Ne	20	(88.89)	3.06×10^6
	21	(0.27)	9290
	22	(10.84)	3.73×10^5
11 Na	23	100	6.0×10^4
12 Mg	24	78.70	8.35×10^5
	25	10.13	1.07×10^5
	26	11.17	1.19×10^5
13 Al	27	100	8.5×10^5
14 Si	28	92.21	9.22×10^5
	29	4.70	4.70×10^4
	30	3.09	3.09×10^4
15 P	31	100	9600
16 S	32	95.0	4.75×10^5
	33	0.760	3800
	34	4.22	2.11×10^4
	36	0.0136	68
17 Cl	35	75.529	4310
	37	24.471	1390
18 Ar	36	84.2	9.87×10^4
	38	15.8	1.85×10^4
	40		~ 20 ?
19 K	39	93.10	3910
	40		5.76
	41	6.88	289

Table 2 (continued)

Element	A	% Abundance	Abundance
20 Ca	40	96.97	6.99×10^4
	42	0.64	461
	43	0.145	105
	44	2.06	1490
	46	0.0033	2.38
	48	0.185	133
21 Sc	45	100	35
22 Ti	46	7.93	220
	47	7.28	202
	48	73.94	2050
	49	5.51	153
	50	5.34	148
23 V	50	0.24	0.63
	51	99.76	261
24 Cr	50	4.31	547
	52	83.7	1.06×10^4
	53	9.55	1210
	54	2.38	302
25 Mn	55	100	9300
26 Fe	54	5.82	4.83×10^4
	56	91.66	7.61×10^5
	57	2.19	1.82×10^4
	58	0.33	2740
27 Co	59	100	2210
28 Ni	58	67.88	3.26×10^4
	60	26.23	1.26×10^4
	61	1.19	571
	62	3.66	1760
	64	1.08	518
29 Cu	63	69.09	373
	65	30.91	167
30 Zn	64	48.89	608
	66	27.81	346
	67	4.11	51.1
	68	18.57	231
	70	0.62	7.71

Table 2 (continued)

Element	A	% Abundance	Abundance
31 Ga	69	60.4	29.0
	71	39.6	19.0
32 Ge	70	20.52	23.6
	72	27.43	31.5
	73	7.76	8.92
	74	36.54	42.0
	76	7.76	8.92
33 As	75	100	6.6
34 Se	74	0.87	0.58
	76	9.02	6.06
	77	7.58	5.09
	78	23.52	15.8
	80	49.82	33.5
	82	9.19	6.18
35 Br	79	50.537	6.82
	81	49.463	6.68
36 Kr	78	0.354	0.166
	80	2.27	1.06
	82	11.56	5.41
	83	11.55	5.41
	84	56.90	26.6
	86	17.37	8.13
37 Rb	85	72.15	4.16
	87		1.72
38 Sr	84	0.56	0.151
	86	9.86	2.65
	87		1.77
	88	82.56	22.2
39 Y	89	100	4.8
40 Zr	90	51.46	14.4
	91	11.23	3.14
	92	17.11	4.79
	94	17.40	4.87
	96	2.80	0.784
41 Nb	93	100	1.4

Table 2 (continued)

Element	A	% Abundance	Abundance
42 Mo	92	15.84	0.634
	94	9.04	0.362
	95	15.72	0.629
	96	16.53	0.661
	97	9.46	0.378
	98	23.78	0.951
	100	9.63	0.385
44 Ru	96	5.51	0.105
	98	1.87	0.0355
	99	12.72	0.242
	100	12.62	0.240
	101	17.07	0.324
	102	31.61	0.601
	104	18.58	0.353
45 Rh	103	100	0.4
46 Pd	102	0.96	0.0125
	104	10.97	0.143
	105	22.23	0.289
	106	27.33	0.355
	108	26.71	0.347
	110	11.81	0.154
47 Ag	107	51.35	0.231
	109	48.65	0.219
48 Cd	106	1.215	0.0180
	108	0.875	0.0130
	110	12.39	0.124
	111	12.75	0.189
	112	24.07	0.356
	113	12.26	0.181
	114	28.86	0.427
	116	7.58	0.112
49 In	113	4.28	0.008
	115	95.72	0.181
50 Sn	112	0.96	0.0346
	114	0.66	0.0238
	115	0.35	0.0126
	116	14.30	0.515
	117	7.61	0.274
	118	24.03	0.865

Table 2 (continued)

Element	A	% Abundance	Abundance
	119	8.58	0.309
	120	32.85	1.18
	122	4.72	0.170
	124	5.94	0.214
51 Sb	121	57.25	0.181
	123	42.75	0.135
52 Te	120	0.089	0.0057
	122	2.46	0.158
	123	0.87	0.056
	124	4.61	0.296
	125	6.99	0.449
	126	18.71	1.20
	128	31.79	2.04
	130	34.48	2.21
53 I	127	100	1.09
54 Xe	124	0.126	0.00678
	126	0.115	0.00619
	128	2.17	0.117
	129	27.5	1.48
	130	4.26	0.229
	131	21.4	1.15
	132	26.0	1.40
	134	10.17	0.547
	136	8.39	0.451
55 Cs	133	100	0.387
56 Ba	130	0.101	0.00485
	132	0.097	0.00466
	134	2.42	0.116
	135	6.59	0.316
	136	7.81	0.375
	137	11.32	0.543
	138	71.66	3.44
57 La	138		0.00041
	139	99.911	0.445
58 Ce	136	0.193	0.00228
	138	0.250	0.00295
	140	88.48	1.04
	142	11.07	0.131

Cameron 18

Table 2 (continued)

Element	A	% Abundance	Abundance
59 Pr	141	100	0.149
60 Nd	142	27.11	0.211
	143	12.17	0.0949
	144	23.85	0.186
	145	8.30	0.0647
	146	17.22	0.134
	148	5.73	0.0447
	150	5.62	0.0438
62 Sm	144	3.09	0.00698
	147		0.0349
	148	11.24	0.0254
	149	13.83	0.0313
	150	7.44	0.0168
	152	26.72	0.0604
	154	22.71	0.0513
63 Eu	151	47.82	0.0406
	153	52.18	0.0444
64 Gd	152	0.200	0.000594
	154	2.15	0.00639
	155	14.73	0.0437
	156	20.47	0.0608
	157	15.68	0.0466
	158	24.87	0.0739
	160	21.90	0.0650
65 Tb	159	100	0.055
66 Dy	156	0.0524	0.000189
	158	0.0902	0.000325
	160	2.294	0.00826
	161	18.88	0.0680
	162	25.53	0.0919
	163	24.97	0.08099
	164	28.18	0.101
67 Ho	165	100	0.079
68 Er	162	0.136	0.000306
	164	1.56	0.00351
	166	33.41	0.0752
	167	22.94	0.0516
	168	27.07	0.0609

Table 2 (continued)

Element	A	% Abundance	Abundance
	170	14.88	0.0335
69 Tm	169	100	0.034
70 Yb	168	0.135	0.000292
	170	3.03	0.00654
	171	14.31	0.0309
	172	21.82	0.0471
	173	16.13	0.0348
	174	31.84	0.0688
	176	12.73	0.0275
71 Lu	175	97.41	0.0351
	176		0.00108
72 Hf	174	0.18	0.00038
	176	5.20	0.0109
	177	18.50	0.0389
	178	27.14	0.0570
	179	13.75	0.0289
	180	35.24	0.0740
73 Ta	180	0.0123	0.00000258
	181	99.9877	0.0210
74 W	180	0.135	0.000216
	182	26.41	0.0422
	183	14.40	0.0230
	184	30.64	0.0490
	186	28.41	0.0454
75 Re	185	37.07	0.0185
	187		0.0341
76 Os	184	0.018	0.000135
	186	1.29	0.00968
	187		0.0088
	188	13.3	0.0998
	189	16.1	0.121
	190	26.4	0.198
	192	41.0	0.308
77 Ir	191	37.3	0.267
	193	62.7	0.450
78 Pt	190	0.0127	0.000178

Table 2 (continued)

Element	A	% Abundance	Abundance
	192	0.78	0.0109
	194	32.9	0.461
	195	33.8	0.473
	196	25.3	0.354
	198	7.21	0.101
79 Au	197	100	0.202
80 Hg	196	0.146	0.000584
	198	10.2	0.0408
	199	16.84	0.0674
	200	23.13	0.0925
	201	13.22	0.0529
	202	29.80	0.119
	204	6.85	0.0274
81 Tl	203	29.50	0.0567
	205	70.50	0.135
82 Pb	204	1.97	0.0788
	206	18.83	0.753
	207	20.60	0.824
	208	58.55	2.34
83 Bi	209	100	0.143
90 Th	232	100	0.058
92 U	235		0.0063
	238		0.0199

REFERENCES

Bertsch, D. L., Fichtel, C. E., Reames, D. V. 1972, Ap. J., 171, 169.
Cameron, A. G. W. 1973, Space Sci. Rev., in press.
Cameron, A. G. W. 1968, in Origin and Distribution of the Elements ed. L. H. Ahrens, 125, (New York: Pergamon Press).
Cameron, A. G. W., Colgate, S. and Grossman, L. 1973, Nature, in press.
Ehmann, W. D. and Rebagay, T. V. 1970, Geochim Cosmochim. Acta, 34, 649.
Goles, G. G., Greenland, L. P. and Jerome, D. Y. 1967, Geochim Cosmochim. Acta, 31, 1771.
Krahenbuhl, U., Morgan, J. W., Ganapathy, R. and Anders, E. Preprint, 1972.

Mason, B. 1971, Handbook of Elemental Abundances in Meteorites, (New York: Gordon and Breach).
Rieder, R. and Wanke, H. 1969, Meteorite Research, P. M. Millman ed. (Dordrect, Holland: Riedel Pub.).
Schmitt, R. A., Goles, G. G., Smith, R. H. and Osborn, T. N. 1972, Meteoritics, 7, 131.
Suess, H. and Urey, H. 1956, Rev. Mod. Phys., 28, 53.
Trauger, J. T., Roesler, F. C., Carleton, N. P. and Traub, W. A. 1973, Bull. Am. Astron. Soc., 5.
Urey, H. 1964, Rev. Geophys. 2, 1.
Withbroe, G. L. 1971, The Menzel Symposium on Solar Physics Atomic Spectra and Gaseous Nebulae, National Bureau of Standards Special Publication # 353, 127.

ON THE ORIGIN AND EVOLUTION OF THE LIGHT ELEMENTS

Beatrice M. Tinsley

The University of Texas at Dallas and University of Maryland

Introduction: Our Nucleogenetic Pool

By studying the evolution of abundances of the light elements, deuterium through boron, we hope to understand more about both the history of the Galaxy and the site of the "x-process." Most of this paper will describe recent work on these elements, done in collaboration with Jean Audouze (Audouze and Tinsley 1973).

First, I shall give some introductory remarks on the nature of problems of chemical evolution. To account for solar system abundances, we must study the history of the matter with which the solar nebular material mixed for the $\sim$ six billion years prior to its isolation. This often called "evolution of the solar neighborhood," but it is important in this context not to identify the neighborhood with our present environment, that which Baade called our "local swimming hole" in the Galaxy. Rather, we refer to all parts of the Galaxy with which we have exchanged material, constituting what might be called our "nucleogenetic pool."

The picture I adopt is like that of Schmidt (1959, 1963). Thus I take it as likely, though not certain, that radial mixing has been minor, so not the whole Galaxy contributes. Our pool has vertical extent, since stars formed in the plane long ago and subsequently accelerated out preserve information about the gas from which they formed. And of course, our material orbited many times about the Galaxy in six billion years, implying (by inference from the appearance of other galaxies) that it endured periods of intense star formation as it crossed spiral features, separated by longer quiescent periods. Let us therefore picture our pool as the irregularly populated volume of the cylindrical shell traced by a deep vertical swimming hole at the sun, as it orbits around the Galactic center.

To study chemical evolution in other regions and other galaxies, one must define the relevant pool from similar considerations of motions of the stars and gas.

The Nature of Evolutionary Models

In studies of evolution of the solar neighborhood, one usually takes a bold average in space over the orbital inhomogeneities, and smooths the changes over corresponding time intervals. Searle (1972) has recently shown that some inferences are vitally affected by this unrealistic assumption of homogeneity, in particular those based on the distribution of stellar abundances. Reeves (1972a) discusses effects of the discreteness of supernova events. We consider only homogeneous models in this paper, a simplification which, in spite of problems in other contexts, is unlikely to affect the abundances of light elements significantly.

Models for evolution of any elemental abundances may consider many processes as sources and sinks: synthesis in the big bang; synthesis (and loss) by nuclear reactions in the interiors and envelopes of stars, including "supermassive" objects; gain and loss by spallation in the interstellar medium; loss and gain by radioactivity; motions of gas into and out of the region, either from extragalactic space or through radial Galactic flows; loss by locking into stellar remnants (white dwarfs, neutron stars, black holes) and low-mass stars. Enough for us model-makers, if not for Nature! Most model computations use part or all of a scheme like this:

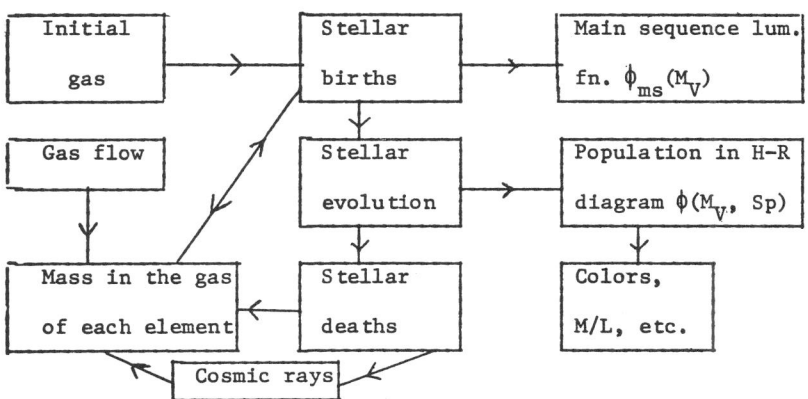

Initially, there is gas from which first-generation stars form. These evolve on time scales determined by mass, composition, and other factors, then die. Meanwhile, other stars form from the interstellar gas. The late evolutionary stages and deaths release astrated matter to the gas, with mass and composition depending on the star and the size of its remnant core. Deaths by supernova apparently produce cosmic rays, which affect the gas composition by spallation reactions. The gas content depends finally on inflow and outflow. Stars spend most of their lives on the main sequence, but the later briefer stages include stellar types of great interest for the synthesis of some elements. The numbers of stars computed to be at each stage can be compared with the distribution $\Phi(M_V, Sp)$ in the local swimming hole (with due account of its departure from an orbital average arising from the uneven distribution of young stars), and determine the ratio of luminosity to mass (relevant to the infamous "missing mass" [Oort 1965]), as well as integrated photometric and spectroscopic quantities for comparison with observations of other galaxies.

It is clearly possible to treat parts of this scheme independently, as has been done in numerous studies. (Space precludes an extensive review, but many earlier references are given in the papers mentioned below.) An outstanding example which uses only a few parameters of abundances, stellar birthrates, and lifetimes, is the derivation of Galactic time scales from the radioactive "aeonglasses" by Fowler and his colleagues (Fowler 1972). Using only the parts of the scheme involving the gas, stellar births and deaths, and the main-sequence luminosity function, comprehensive studies of the abundances of many elements have been made recently by Truran and Cameron (1971) and Talbot and Arnett (1972). My own studies have paid more attention to the lower right-hand column of the scheme, stressing that the evolution of stellar populations and the resulting luminosity, spectra, etc., provide useful constraints and consistency checks on models to account for abundances (Tinsley 1972a). Indeed, the time-dependence of luminosity is closely related to the total energy released by nucleosynthesis, so constraints are set by the integrated light of distant, unresolved galaxies (Peebles 1971, Tinsley 1973).

Models of these types require a startling variety of input data, often with admittedly more unknown parameters than independent observational checks! The following partial list of unknowns makes clear the complexity of our problems:

 Initial abundances? Stellar nucleosynthesis?
 Rates of gas flow? Remnant masses?
 Inflow abundances? Production of cosmic rays?
 Birthrates, b(m,t)? Spallation cross sections?
 Stellar lifetimes? Evolution in H-R diagram?
 Effects of simplifying assumptions?

We have clues to most of these quantities from theory, observation, or experiment, and we hope to gain more during this conference. A particularly important and ill-determined function is the stellar birthrate. Unfortunately, the range of stellar lifetimes ($<10^7$ yr for an early O main-sequence star, to $>10^{10}$ yr for dwarfs later than $\sim$ G2) means that the observed distribution $\Phi(M_V, Sp)$ does not correspond to <u>unique</u>

birthrate functions of time or mass (Schmidt 1963).

Two Contrasted Evolutionary Models

In our study of the light elements (Audouze and Tinsley 1973), we consider two models for evolution of the solar neighborhood. Both are consistent with the data considered by their authors, and should demonstrate the model-dependence of our results since they behave very differently - an example of the non-uniqueness which plagues this field. They are the "consistent" model of Truran and Cameron (1971; TC) and the "infall" model (F) of Quirk and Tinsley (1973; QT). Their main differences are shown by the behavior of key quantities in Figure 1. In TC, the stellar birthrate (proportional to the supernova rate plotted) declines exponentially, as does approximately the gas mass; stars are born with the Limber (1960) initial mass function (IMF). In QT, star formation occurs at rates suggested by considerations of dynamic stability (Quirk 1972): at first it quickly depletes the gas to its present mass, after which it maintains equilibrium with the rate at which gas enters by infall and stellar mass loss. The IMF in QT is adjusted to give rise to the present $\Phi_{ms}(M_V)$, including many faint M dwarfs (Weistrop 1972). In QT, but not in TC, infalling extragalactic gas is included, to represent the observed high velocity clouds (Oort 1970; but see Verschuur [1973] and Hulsbosch and Oort [1973] for a discussion of their interpretation). The gas in TC was enriched at a pregalactic stage by a burst of massive stars, but its "metal" content Z_g changes little thereafter since TC assume that stars above $8M_\odot$ swallow their enriched cores into black holes.* QT start with primordial abundances; the gas is enriched quickly during the period of rapid star formation, but, in spite of the continued output by stars between 9 and $30M_\odot$, is enriched little further because of the dilution by metal-free infalling gas.

Deuterium and Helium-3

First we consider a purely big-bang origin of D. Using Wagoner's (1973) results, we start with roughly the present H and ^{4}He, some D, ^{3}He, and ^{7}Li, but negligible ^{6}Li, Be, or B. To select the big-bang model (entropy per baryon), we find for each evolutionary model the fraction of gas at 6 billion years which has ever been in a star, then assume that D is destroyed in all such gas. Thus the initial D abundance required to

---

*TC show Z_g increasing slightly during evolution. The behavior shown in Fig. 1 results from the slightly different enrichment code used by Tinsley (1972a), based on that of Talbot and Arnett (1972). I apologize to Truran and Cameron for the liberties taken with their model, which will, however, affect our present results negligibly.

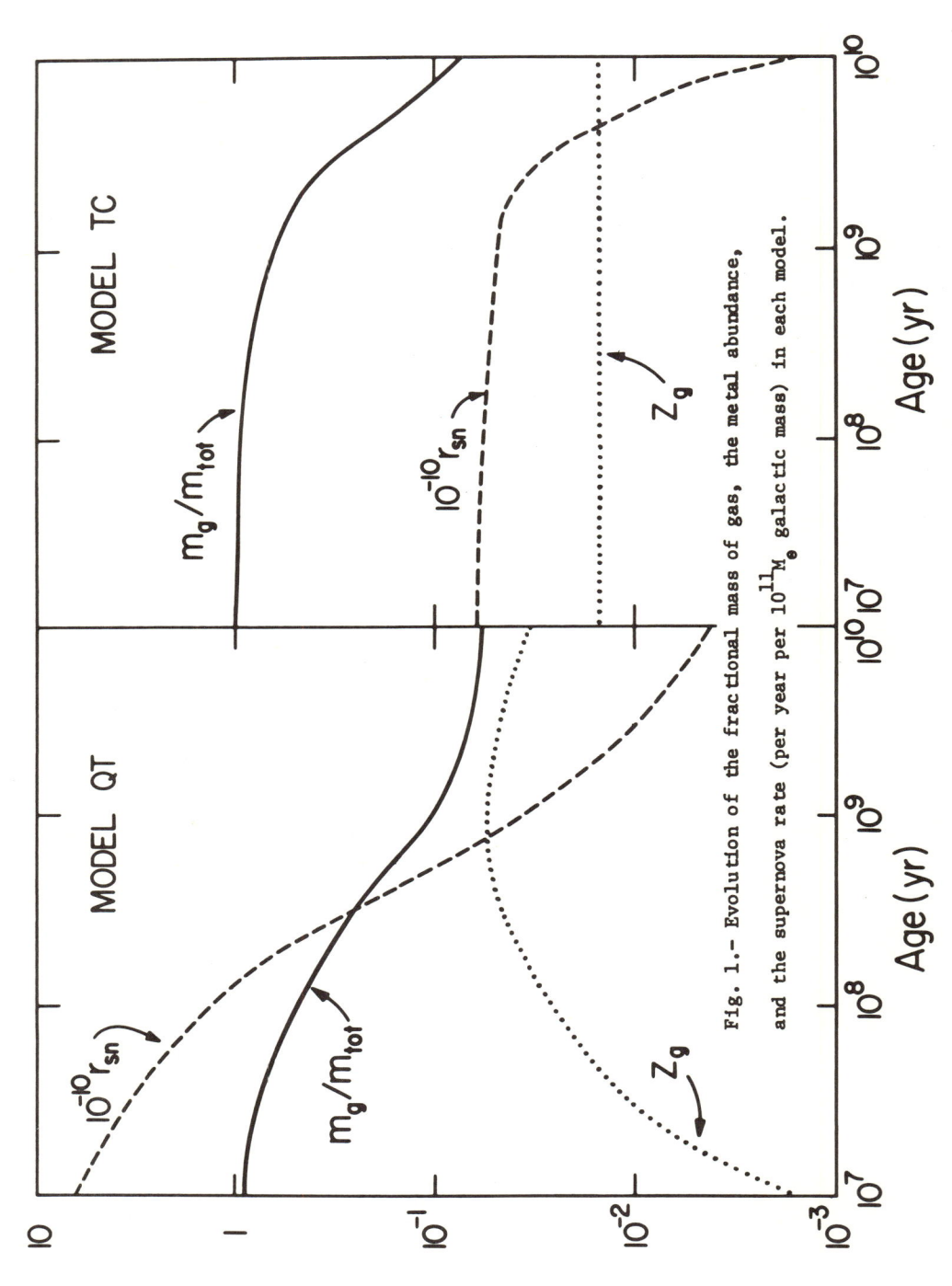

Fig. 1.— Evolution of the fractional mass of gas, the metal abundance, and the supernova rate (per year per $10^{11} M_\odot$ galactic mass) in each model.

give rise to the solar system value* is determined, which in turn determines the big-bang model and other initial abundances. (The corresponding present universal density is $\sim 0.8\%$ (H/100 km s^{-1}Mpc^{-1})$^{-2}$ of the critical density, which is more than the density of galaxies and may be consistent with the magnitude redshift diagram [Sandage 1972] and zero cosmological constant when evolution of galactic magnitudes is considered [Tinsley 1972b].)

Stellar D-burning produces ^{3}He, which is itself destroyed by astration to an extent depending on temperatures in stellar envelopes; we adopt TC's estimates of this destruction. Figure 2 shows the evolutionary abundance curves (EACs). In model TC, D declines slowly while ^{3}He rises to an excessive value. In QT, rapid star formation at first destroys much D, but it is replenished by the infalling gas with primordial composition. ^{3}He also enters in this gas, and is produced and destroyed as in TC. ^{3}He becomes overabundant in model QT too, so the problem appears to be quite model-independent. TC suggested as a solution that extensive mixing in red giant envelopes may expose the ^{3}He to higher temperatures than expected. But the problem may not be too serious, for if we create a D abundance near the lower limit of its uncertainty, the corresponding ^{3}He abundance is near its upper limit of uncertainty.

Alternatively, D may be produced in supermassive stars (Hoyle and Fowler 1973) or supernovae (Colgate 1973), and ^{3}He in red giants (Cameron and Fowler 1971). From the analogous results for ^{7}Li and B, we derive the required production rates per supernova and find these to be quite plausible. We cannot therefore distinguish between big-bang and Galactic production of D or ^{3}He, on the basis of evolutionary calculations.

Lithium, Beryllium, and Boron

Now we turn to the "x-process" elements. We consider several of the current theories for their production (cf. reviews by Reeves et al. [1973] and Audouze and Truran [1973;AT]).

(a) Spallation of the interstellar gas by Galactic cosmic rays has been shown by Meneguzzi et al. (1971; MAR) and Mitler (1972) to explain readily the abundances of ^{6}Li and ^{9}Be. ^{7}Li/^{6}Li is too low by a factor ~ 8, while the agreement for B has been spoiled by the recent 30-fold increase in the solar abundance (Cameron et al. 1973).

---

*We do not discuss the possibly much greater interstellar values of D/H (Cesarsky et al. 1973; Wilson et al. 1973) because of the large ambiguities in interpretation (Solomon and Woolf 1973).

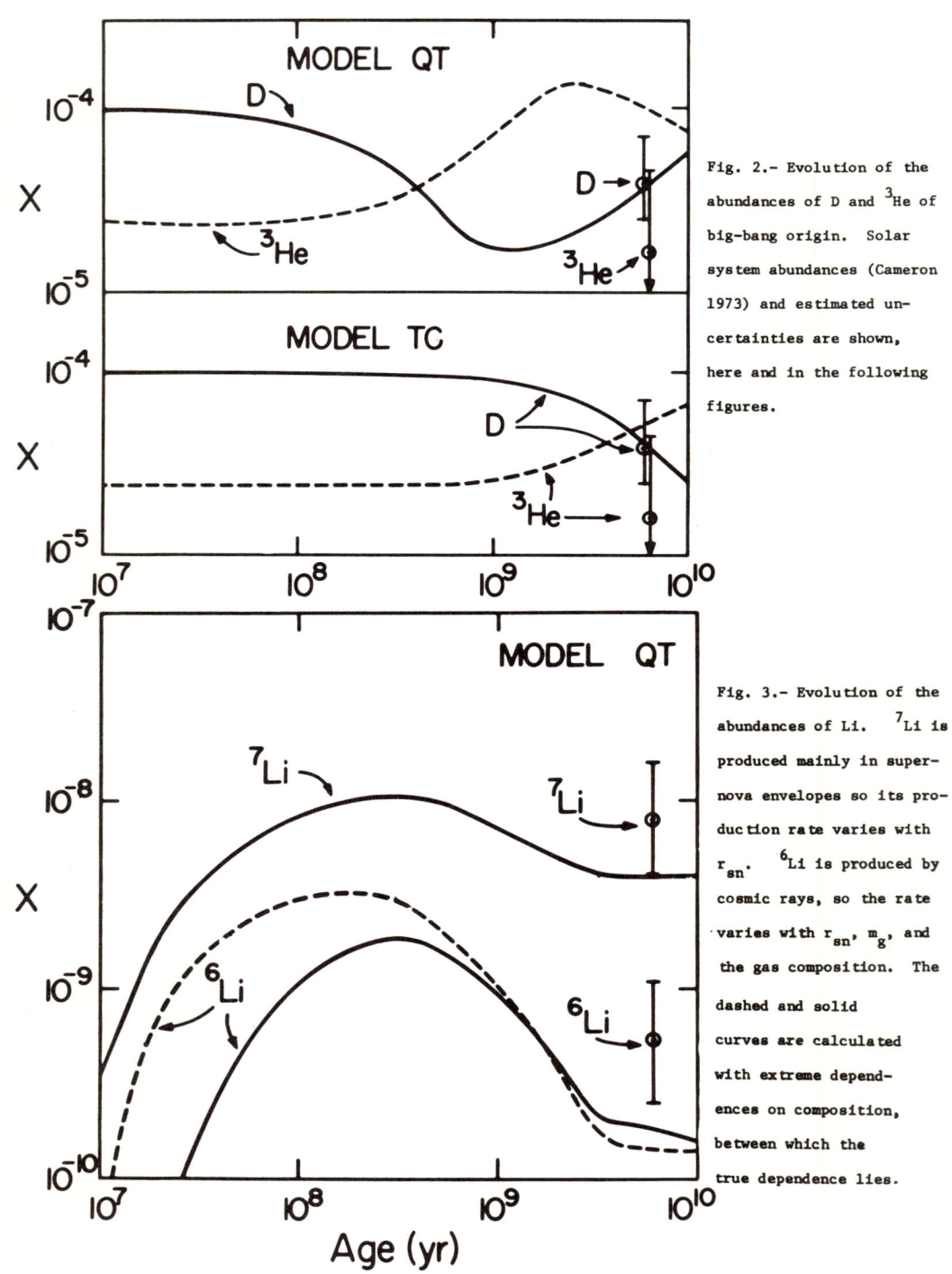

Fig. 2.— Evolution of the abundances of D and ^{3}He of big-bang origin. Solar system abundances (Cameron 1973) and estimated uncertainties are shown, here and in the following figures.

Fig. 3.— Evolution of the abundances of Li. ^{7}Li is produced mainly in supernova envelopes so its production rate varies with r_{sn}. ^{6}Li is produced by cosmic rays, so the rate varies with r_{sn}, m_g, and the gas composition. The dashed and solid curves are calculated with extreme dependences on composition, between which the true dependence lies.

(b) Production by large fluxes of suprathermal particles accelerated by shock waves in supernova envelopes has been shown by AT to give relatively more ^{7}Li, ^{10}B, and ^{11}B, the total requirement per supernova being energetically plausible.

(c) ^{7}Li could originate in the envelopes of red giants (Cameron and Fowler 1971, Scalo and Ulrich 1973, Sackmann et al. 1973), or in a big bang with nonzero leptonic number (Reeves 1972b).

Process (a) alone is not viable because the relative deficiencies of ^{7}Li and B cannot be corrected by adjustment of evolutionary parameters. We therefore consider a combination of processes (a) and (b), using MAR's estimates for the present production rates in (a), and AT's production ratios at $\sim$ 16 MeV/particle for (b); the contribution of (b) is chosen to give the correct B abundances.

The production rates vary in time according to the evolutionary model. Process (a) is greater in the past in proportion to the mass of gas and the supernova rate. It also depends on the composition of the gas (target atoms), which we allow for in a simple way: separate calculations are made with the rate varying as Z_g, then as X_g and Y_g (which are effectively constant); the true composition effect lies between these cases, so, since the resulting abundances at 6 10^9 years are close, we can make an adequate interpolation (cf. **Figures** 3 and 4). Process (b) is simply proportional to the supernova rate.

We assume that all Li is destroyed by astration, while for Be and B we consider cases between survival like ^{3}He and total loss.

The EACs for Li in model QT are shown in Figure 3. ^{6}Li is produced mainly by process (a). ^{7}Li has an unimportant initial and infalling abundance from the big bang. It is produced mainly by process (b), and steadily destroyed by astration. The resulting abundances are somewhat too low, but the discrepancy is within the many uncertainties. Model TC gives similar but not decisive over-abundances, because of the smaller amount of astration in that model. An interesting result is that if process (b) is such as to give enough B, it may also produce enough ^{7}Li, with no need to invoke the alternative sources (c). However, AT's calculations show that ^{7}Li/B depends strongly on the particle energies, so their theory is also consistent with significant other contributions to ^{7}Li.

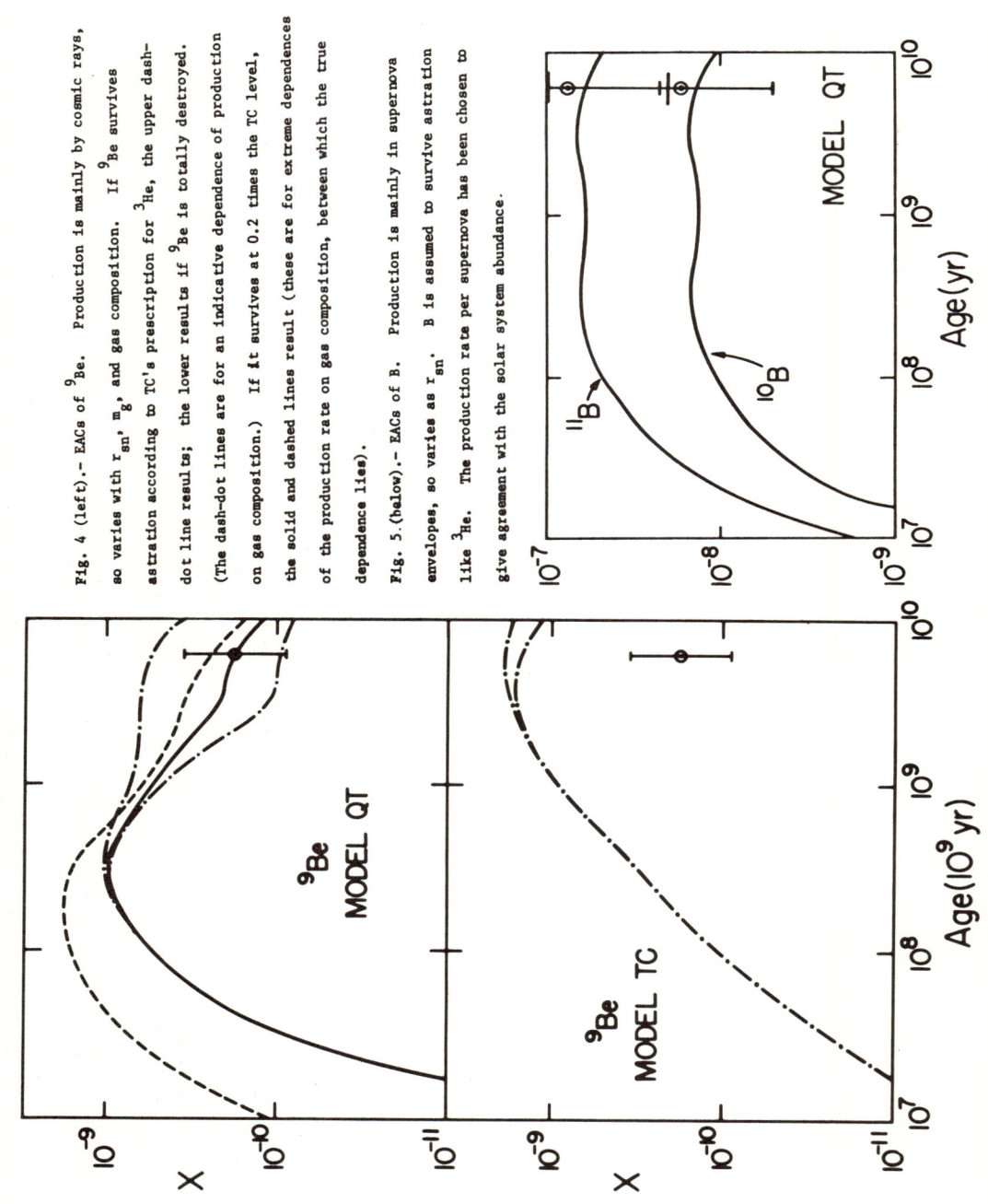

Fig. 4 (left).— EACs of ^{9}Be. Production is mainly by cosmic rays, so varies with r_{sn}, m_g, and gas composition. If ^{9}Be survives astration according to TC's prescription for ^{3}He, the upper dash-dot line results; the lower results if ^{9}Be is totally destroyed. (The dash-dot lines are for an indicative dependence of production on gas composition.) If it survives at 0.2 times the TC level, the solid and dashed lines result (these are for extreme dependences of the production rate on gas composition, between which the true dependence lies).

Fig. 5 (below).— EACs of B. Production is mainly in supernova envelopes, so varies as r_{sn}. B is assumed to survive astration like ^{3}He. The production rate per supernova has been chosen to give agreement with the solar system abundance.

Figure 4 shows results for ^{9}Be. Production is mainly by process (a), and two extreme cases for destruction are illustrated. In model TC, destruction is not great, and we find (as TC did) that their model leads to too much ^{9}Be. But in QT, astration is much more important. Good agreement with the observed abundance is obtained if about 3/5 as much ^{9}Be is destroyed as ^{3}He, a result in accord with their relative cross-sections for destruction. A very pleasing result from the QT model is that the astration and infalling gas (with no ^{9}Be) cause the abundance to decline over the last $\sim 9 \; 10^9$ years, providing an explanation for the mysterious observed increase of stellar Be towards later spectral types (Wallerstein and Conti 1969).

Finally, EACs for B are shown in Figure 5, for the case where B is destroyed by astration to the same extent as ^{3}He (about as expected). As stated above, agreement with solar abundances has been forced. With a small adjustment to the normalization of process (b), to allow for the lesser destruction, equally good results can be obtained in model TC.

Concluding Remarks

As always in this field, we cannot draw definitive conclusions, but can perhaps make headway by showing that certain sets of hypotheses are mutually consistent and others are not. The production of ^{6}Li and ^{9}Be by cosmic rays impinging on interstellar gas is possible, provided that a substantial fraction of the gas entering the solar nebula at ~ 6 billion years had been in a star. This degree of astration, needed to avoid overproduction of ^{9}Be, is consistent with a big-bang origin of the solar system D only if the latter is replenished by an inflow of unastrated gas. However, either a Galactic or a big-bang origin of D and ^{3}He is possible in the evolutionary context. ^{7}Li, ^{10}B, and ^{11}B can originate together by reactions of suprathermal particles in supernova envelopes, but alternative sources of ^{7}Li are not precluded.

Because of their unique nuclear properties, the light elements are sensitive to different aspects of galactic history than most species. For this reason, they remain useful probes of evolution in our solar neighborhood's nucleogenetic pool.

It is a pleasure to thank Jean Audouze for his help in the preparation of this paper. This work was supported in part by The University of Texas at Dallas Research Fund and by NSF grants GP-30455X and GU-4020 at Dallas, and by the Center for Theoretical Physics at the University of Maryland.

REFERENCES

Audouze, J., and Tinsley, B. M. 1973, in preparation.
Audouze, J., and Truran, J. W. 1973, Ap. J., in press.
Cameron, A. G. W. 1973, private communication.
Cameron, A. G. W., Colgate, S. A., and Grossman, L. 1973, Nature, in press.
Cameron, A. G. W., and Fowler, W. A. 1971, Ap. J., 167, 11.
Cesarsky, D. A., Moffet, A. T., and Pasachoff, J. M. 1973, Ap. J. Lett., 180, L1.
Colgate, S. A. 1973, Ap. J., in press.
Fowler, W. A. 1972, in Cosmology, Fusion, and Other Matters, ed. F. Reines (Boulder, Colorado: Associated University Press).
Hoyle, F., and Fowler, W. A. 1973, Nature, 241, 384.
Hulsbosch, A. N. M., and Oort, J. H. 1973, Astr. and Ap., 22, 153.
Limber, D. N. 1960, Ap. J. 131, 168.
Meneguzzi, M., Audouze, J., and Reeves, H. 1971, Astr. and Ap., 15, 337.
Mitler, H. E. 1972, Ap. and Space Sci., 17, 186.
Oort, J. H. 1965, in Stars and Stellar Systems, 5, ed. A. Blaauw and M. Schmidt (Chicago: University of Chicago Press).
_____. 1970, Astr. and Ap., 7, 381.
Peebles, P. J. E. 1971, Comm. Ap. and Space Phys., 3, 20.
Quirk, W. J. 1972, Ap. J. Lett., 176, L9.
Quirk, W. J., and Tinsley, B. M. 1973, Ap. J. 179, 69.
Reeves, H. 1972a, Astr. and Ap., 19, 215.
_____. 1972b, Phys. Rev., D6, 3363.
Reeves, H., Audouze, J., Fowler, W. A., and Schramm, D. N. 1973, Ap. J., 179, 909.
Sackmann, I.-J., Smith, R. L., and Despain, K. H. 1973, preprint.
Sandage, A. 1972, Ap. J., 178, 1.
Scalo, J. M., and Ulrich, R. K. 1973, Ap. J., 183, in press.
Schmidt, M., 1959, Ap. J., 129, 243.
_____. 1963, Ap. J., 137, 758.
Searle, L. 1972, paper presented at I.A.U. Colloquium #17, in press.
Solomon, P. M., and Woolf, N. J. 1973, Ap. J. Lett., 180, L89.
Talbot, R. J., and Arnett, W. D. 1972, preprint.
Tinsley, B. M. 1972a, Astr. and Ap., 20, 383.
_____. 1972b, Ap. J. Lett., 173, L93.
_____. 1973, Astr. and Ap., in press.
Truran, J. W., and Cameron, A. G. W. 1971, Ap. and Space Sci., 14, 179.
Verschuur, G. L. 1973, Astr. and Ap., 22, 139.
Wagoner, R. V. 1973, Ap. J., 179, 343.
Wallerstein, G., and Conti, P. S. 1969, Ann. Rev. Astr. and Ap., 7, 99.

Weistrop, D. 1972, A. J., 77, 849.
Wilson, R. W., Penzias, A. A., Jefferts, K. B., and Solomon, P. M. 1973, Ap. J. Lett., 179, L107.

CHEMICAL EVOLUTION OF THE GALAXY: COEFFICIENTS FROM STELLAR EVOLUTION AND ALTERNATIVE SOLUTIONS TO THE PROBLEM OF FEW METAL-POOR STARS

R. J. Talbot, Jr.

Rice University, Houston, Texas

I. Coefficients for Chemical Evolution

A discussion of the evolution of the chemical composition of galaxies by stellar nucleosynthesis requires specifying the fraction of the mass of a star of mass m which is ejected in the form of chemical species i, R_{mi}. In some recent work (Talbot and Arnett 1971a, 1973a,b; Arnett and Talbot 1973; hereafter called Papers I through IV, respectively) the following form has been employed:

$$R_{mi} = \Sigma_j Q_{mij} X_j \qquad (1)$$

where the summation is over all species. The production matrix Q_{mij} specifies the fraction of primordial mass of species j which is eventually ejected from a star of mass m in the form of species i.

By integrating over the initial mass function ψ_m one obtains the production matrix for a generation of stars

$$q_{ij} = \int_0^\infty \psi_m Q_{mij}\, dm. \qquad (2)$$

The total fraction of mass ejected by a generation of stars, f, and the yield of a species, p_i, are defined

$$f = \Sigma_{ij} q_{ij} X_j \qquad (3)$$

and

$$p_i = (\Sigma_j q_{ij} X_j)/(1-f). \qquad (4)$$

For the initial mass function (IMF), the parametrization of Paper II is adopted:

$$\psi_m = \zeta(\mu-1) m^{-\mu} \qquad (5)$$

for $m \geq 1\ M_\odot$. For models with a constant IMF, $\zeta = 0.25$ matches the solar neighborhood luminosity function and μ is estimated to be 1.55 ± 0.25 (the Salpeter IMF has $\mu = 1.35$). The IMF below $1\ M_\odot$ is thought to play essentially no role in nucleosynthesis other than to fix ζ.

The purpose of this section is to briefly review the specification of the yields of stellar evolution. The <u>relative</u> production of species heavier than ^{4}He is not discussed here. Arnett (1971) discussed the relative production of ^{12}C and ^{16}O. Most of the work on explosive nucleosynthesis (much of which is discussed in other papers at this conference) has been devoted to producing the relative abundances of heavier species as found in the solar system. Here we discuss only the combined production of ^{12}C plus ^{16}O (CO); they comprise about 75% of the heavy elements.

For the production of CO we adopt the recent calculations of massive star evolution by Arnett (1972a,b). Figure 1, adopted from Paper II, shows the mass fractions involved in specifying Q_{mij}. A stellar remnant--either a white dwarf or a neutron star (or a black hole)--is assumed to be left upon the death of a star. Below $4\ M_\odot$ we assume that no primary heavy element nucleosynthesis occurs. The secondary process of converting initial CO to ^{14}N is considered, as are the destruction of initial D and the production and/or destruction of ^{3}He. The picture for $m = 4$ to about 15 $M_\odot$ is quite uncertain at this time. The point at 7 $M_\odot$ is computed from the model of Iben (1972). The region between q_4 and q_c <u>represents</u> the amount of ^{4}He produced and mixed into the envelope of that model. The actual mass fraction between the He-burning shell and the H-burning shell is very narrow.

There are obviously many uncertainties in this picture. The position of the q_c and q_4 lines for $m = 4$ to 15 $M_\odot$ are simply an attempt to make a smooth transition and indicate the reasonable possibility that some primary heavy element production occurs. Models do not yet give much information on this point. The sharp cutoff at 60 $M_\odot$ is imposed because of the main-sequence pulsational instability which sets in at about that mass. This cutoff mass is sensitive to the initial heavy element composition of the star. Some production of ^{4}He and heavy elements may occur in such stars (Talbot and Arnett 1971b). The positions of the other zone boundaries are probably also dependent upon composition in some way yet to be determined.

The amount of CO conversion to ^{14}N in the envelopes of stars is not yet clear. Estimates based upon the late stages of Iben's (1965, 1966, 1967, 1972) models have been used in Paper II. From those models it appears that ^{14}N is underproduced by about a factor of 3 to 6 depending upon μ. If one

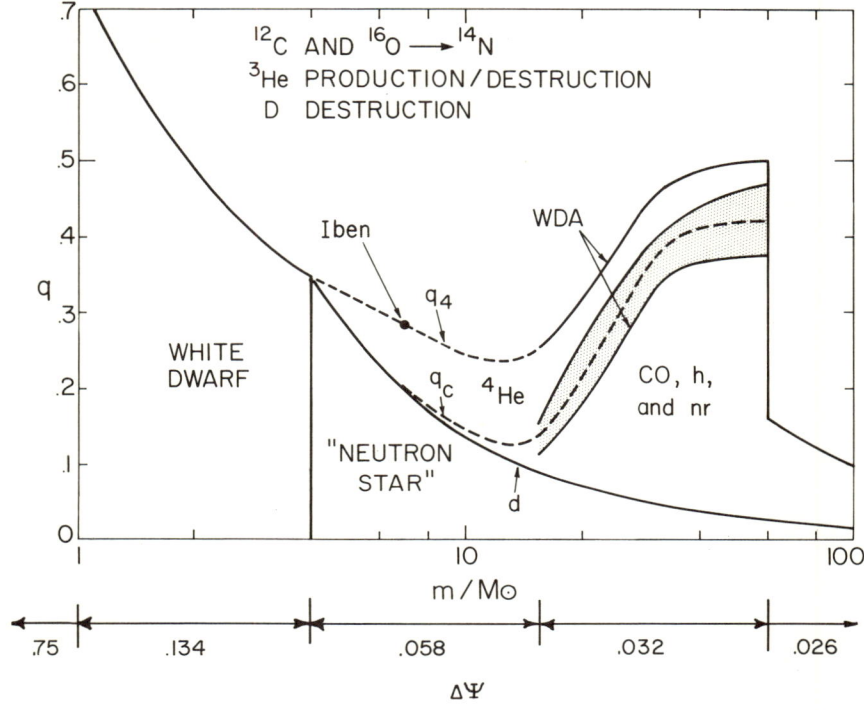

Figure 1--Shown for stars of mass m are the mass fractions (q) which are anticipated to contribute to various final states of stellar matter. The mass fraction within which H to ^{4}He conversion has occurred is q_4; interior to q_c, ^{4}He has been converted to heavier species: CO = ^{12}C + ^{16}O, nr = neutron-rich species heavier than ^{16}O (e.g. ^{18}O, ^{22}Ne, ^{25,26}Mg), and h = all other heavy species. The only primary nucleosynthesis product considered in the envelope is ^{3}He. The secondary process CO → ^{14}N occurs in the region between q_4 and q_c and in the envelope. Also shown is $\Delta\Psi$ which is the fraction of mass in a stellar generation within the indicated mass range; $\zeta = 0.25$ and $\mu = 1.55$. Details of sources of data and other considerations may be found in Talbot and Arnett (1973a), from which this figure was taken.

arbitrarily specifies that all CO in all ejected envelopes is returned as ^{14}N, then ^{14}N is overproduced by a factor of about 1.5 to 2. Since those models had not gone through all of the very late stages with the extended, fully convective envelopes which one anticipates will occur, it is not unreasonable to assume that the true results will fall between the two stated cases. We feel it not unlikely that the abundance of ^{14}N may be fully accounted for as a secondary process; we see no <u>requirement</u> to introduce a primary ^{14}N process, though one cannot be excluded.

We find that Iben's models produce enough ^{3}He early in their lives to explain the cosmic abundance; however, if a big bang source for D is assumed, the accompanying big bang abundance of ^{3}He is sufficiently large that we must conclude that the ^{3}He in Iben's models is destroyed in later stages. This is consistent with the conditions required to produce the ^{14}N. On this basis we find that by solar system formation both D and ^{3}He should be depleted to about 0.6 to 0.8 of their initial values in the Galaxy.

For the yield of CO we find $p_{CO} = (0.85 \pm 0.35) \times 10^{-2}$ where the stated errors are the variation produced by varying μ over the range 1.55 ± 0.25. Uncertainties in ζ and in stellar evolution produce uncertainties in p_{CO} of about the same magnitude.

If one assumes that (1) the IMF is constant, (2) the interstellar gas is well mixed, and (3) there is no mass inflow from outside the system, then one can easily show that for <u>any</u> form of the birthrate function for star formation, the time development of a primary species (e.g. CO) is given by

$$X_{CO}(t) = X_{CO}(0) + p_{CO} y(t) \qquad (6)$$

where $y(t) = \ln[\mathcal{M}_g(0)/\mathcal{M}_g(t)]$, $\mathcal{M}_g(t)$ is the mass in the form of gas at time $\underline{t}$ and $X_{CO}(0)$ is the initial abundance of CO (Searle and Sargent 1972, Paper II). This will be referred to as the <u>standard model</u> (STD). We adopt $\mathcal{M}_g(now)/\mathcal{M}_g(0) = 0.10$ with an uncertainty of a factor of two; $y = 2.3 \pm 0.7$. For the mean metal abundance in young stars $\mathfrak{z}*$ we adopt $1.1\, \mathfrak{z}\odot$ (e.g., Powell 1972) or $X^*_{CO} = 1.3 \times 10^{-2}$ using the Cameron (1973) abundances. This value is uncertain by at least 30%.

Adopting $X_{CO}(0) = 0$ we find that the standard model yields $X_{CO}(STD) = 1.9 (+1.6, -1.1) \times 10^{-2}$ which matches the observed X^*_{CO} to within the uncertainties. Adopting the exponential model of nucleosynthesis [y(t) linear in $\underline{t}$] yields a nucleo-cosmochronology age for the Galaxy of $T = (12.5 \pm 3.3) \times 10^9$ years. This is from the long-lived ^{235,238}U, ^{232}Th, ^{187}Os, and ^{187}Re chronologies. The short-lived ^{244}Pu and ^{127,129}I chronologies are just barely concordant; the addition of a last nucleosynthesis spike makes their concordance more satisfactory. This model of course does not satisfy the observed distribution of metal abundance in G dwarfs (van den Bergh 1962, Schmidt 1963).

II. The Solutions to the Problem of Few Metal-Poor Stars

There are three solutions to this problem which have been discussed; we will review them briefly. Two have been discussed in the literature before. The other is discussed by myself and Arnett in Paper III.

a) Variable Initial Mass Function (VIMF)

This is the now classic solution discussed by Schmidt (1963), subsequently adopted by Truran and Cameron (1971; TC), Fowler (1972), and Quirk and Tinsley (1973; QT). Basically it consists of simply assuming that few, if any, low mass stars formed in the Galaxy until $\mathcal{J}$ rose to about $1/3\, \mathcal{J}_\odot$. We have briefly discussed this solution in Paper III, and find that our CO yield produces X_{CO}(VIMF) = 3.5(+2.9, -2.0) x 10^{-2} and find T = (10.0 ± 2.5) x 10^9 years for the exponential model. Although this model can satisfy the data on G-dwarf metal abundances, it does not explain the observations that stars of the same age (young stars for example) possess an appreciable dispersion in their metal abundances (e.g. Eggen 1964, Dixon 1966). It does not allow the short-lived chronologies to be concordant with the rest unless a last spike model is assumed or very extreme values of the parameters are invoked. Both TC and QT discuss lowering the yield of heavy elements to give the observed abundance. This is accomplished by <u>assuming</u> that most of that material is retained in massive black holes, an assumption which we feel is not substantiated by stellar evolution calculations, but cannot be dismissed.

b) Infall of Low Metal Abundance Matter (INFALL)

Larson (1972), Fowler (1972), QT, and Paper IV discuss models with infall of low metal abundance primordial material. Through dilution, the metal abundance is kept from rising above an asymptotic level which we find to be about X_{CO}(INFALL) $\simeq p_{CO} \simeq (0.85 \pm 0.35)$ x 10^{-2}. T in these models becomes indeterminant, but greater than about 10 x 10^9 years; the short-lived chronologies work out well. Additional (but plausible) assumptions are required to fit the data on the spread of metal abundance in stars of the same age. Appreciable CO consumption by black holes would lead to an unobserved underabundance of heavy elements.

c) Metal-Enhanced Star Formation (MESF)

In Paper III we explicitly incorporate the following--

each of which is supported by observational evidence and/or theoretical arguments:
1) There exists abundance inhomogenieties in the interstellar gas.
2) Stars form preferentially in regions possessing high metal abundance.

It is argued that 2) is to be expected since the thermal instability which produces the interstellar clouds is produced by heavy element cooling agents (e.g. Field, Goldsmith, and Habing 1969). The analyses of observations by Eggen (1964) and Dixon (1966) show 1) to be the case, and in Paper IV we supply theoretical arguments to support it.

We develop a model which has a single coefficient $\underline{a}$ which is theoretically estimated to be $10^{-2.4 \pm 0.5}$. This coefficient describes the distribution of metal abundance in stars of the same age. The value of $\underline{a}$ required to fit the observed distribution of metal abundance in both young and old stars is $10^{-2.10 \pm 0.08}$, a very satisfactory agreement. The model gives $X_{CO}(MESF) = 1.3(+1.5, -0.9) \times 10^{-2}$, also a very satisfactory agreement with observations. The value of T is $(18.6 \pm 5.7) \times 10^9$ years. Concordance of the short-lived chronologies with the long-lived ones is achieved with no additional parameters required.

d) Summary

Table 1 summarizes the basic properties of the four models reviewed in this paper. Figure 2 shows the forms of the theoretical distribution of metal abundance in stars with lifetimes longer than the age of the Galaxy. Observational data are very similar to curve 3, the MESF model.

III. Discussion

The uncertainties in the chemical evolution coefficients stem from three main sources other than the intricacies of explosive nucleosynthesis network theories: hydrostatic evolution defining the initial conditions for explosive (or quiescent) mass loss, details of the hydrodynamic events of mass loss, and the initial mass function (IMF). The specific problems are the following: the amount of mass locked up in stellar remnants, composition of the ejected envelope, whether the IMF has varied, and the fraction of mass locked up in the low mass end of the IMF.

With all uncertainties considered, for the species discussed here, there appears to exist no difficulty between observed abundances and those predicted by Galactic evolution using results of stellar models. Reduction of those uncertainties by about a factor of two would make the model predic-

TABLE 1

	STD	VIMF	INFALL	MESF
Galaxy Age (10^9 yr)	12.5±3.3	10.0±2.5	>10	18.6±5.7
Heavy element yield vs. observed abundances	?	yield too great	?	?
Satisfy with no additional parameters:				
short-lived chronologies?	Just	No	Yes	Yes
distribution of metallicity in				
old stars?	No	Can	No	Yes
young stars?	No	No	No	Yes

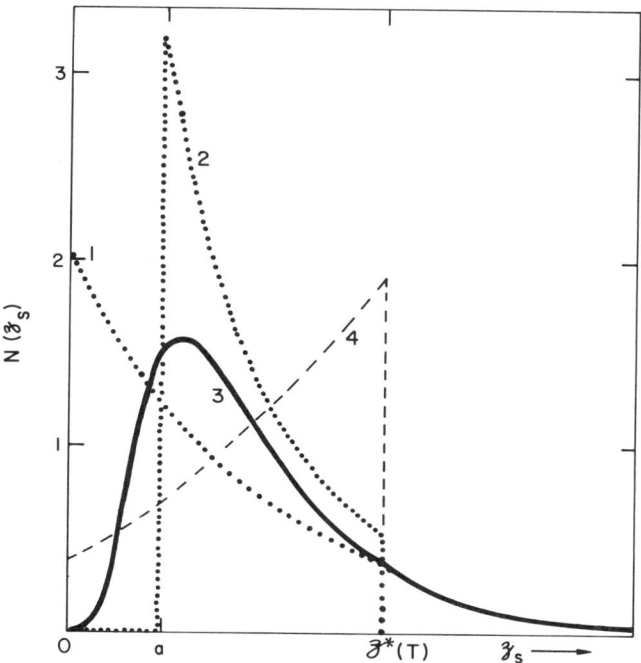

Figure 2--The distribution of the numbers of low-mass low-luminosity stars N with individual metal abundance z_s. Curve 1 is the standard model; curves 2 and 4 respectively are the Truran and Cameron (1971) and Schmidt (1963) models with variable initial mass functions; curve 3 is the metal-enhanced star formation model of Talbot and Arnett (1973b), from which this figure was taken. The INFALL model of Arnett and Talbot (1973) is qualitatively similar to curve 4.

The observational data employed by van den Bergh (1962), Schmidt (1963), Eggen (1964), and Dixon (1966) are represented in a common format in a figure in Talbot and Arnett (1973b); they are of the form of curve 3. The parameters quoted in the text for the MESF model are obtained by fits to those observed $N(z_s)$ distributions.

tions sufficiently precise that observed abundances could be used to distinguish between Galactic models. The parameters in models of the history of Galactic nucleosynthesis are significantly constrained by the observed distribution of metal abundance in long-lived low-mass stars. The three processes discussed to solve this problem -- VIMF, INFALL, or MESF -- are not easily distinguished on this basis alone.

For a single-process solution, we believe that metal-enhanced star formation is attractive because, with no additional hypothesis, it simultaneously resolves several other problems. However, it would not be surprising to discover all three mechanisms operating together. To disentangle these will require consideration of more observational tests than Galactic evolution models have been heretofore subjected. Some of the important ones are:

1) Does the IMF vary in time and/or space? More detailed information is needed than has been yet obtained from mass-to-light ratios. The distribution of metal abundance in old stars does not require that the IMF vary, although it is consistent with that.
2) Is there mass inflow into the Galaxy, and if so what is its composition and distribution over the Galactic disk?
3) What is the intrinsic spread in metal abundance in stars of the same age? Currently, the observational uncertainties are roughly the same size as the observed spread, but there are strong indications that there is an intrinsic spread.
4) Are there abundance inhomogeneities in the interstellar gas? If so, what is the distribution? Is there any evidence that star formation rates depend upon the metal abundance?

REFERENCES

Arnett, W. D. 1971, Ap. J. Lett., 170, L43.
————. 1972a, Ap. J., 176, 681.
————. 1972b, Ap. J., 176, 699.
Arnett, W. D., and Talbot, R. J., Jr. 1973, paper in preparation (IV).
Bergh, S. van den 1962, A. J., 67, 486.
Cameron, A.G.W. 1973, private communication.
Dixon, M. E. 1966, M.N.R.A.S., 131, 325.
Eggen, O. J. 1964, A. J., 69, 570.
Field, G. B., Goldsmith, D. W., and Habing, H. J. 1969, Ap. J. Lett., 155, L149.
Iben, I., Jr. 1965, Ap. J., 142, 1447.
————. 1966, Ap. J., 143, 483; 505; 516.
————. 1967, Ap. J., 147, 624; 650.
————. 1972, Ap. J., 178, 433.
Larson, R. 1972, Nature Phys. Sci., 236, 7.

Powell, A.L.T. 1972, M.N.R.A.S., 155, 483.
Quirk, W. J., and Tinsley, B. M. 1973, Ap. J., 179, 69 (QT).
Schmidt, M. 1963, Ap. J., 137, 758.
Searle, L., and Sargent, W.L.W. 1972, Ap. J., 173, 25.
Talbot, R. J., Jr., and Arnett, W. D. 1971a, Ap. J., 170, 409 (I).
——————————. 1971b, Nature Phys. Sci., 229, 150.
——————————. 1973a, submitted to Ap. J. (II).
——————————. 1973b, submitted to Ap. J. (III).

CHAPTER II. Explosive Processes

Following the lead of Burbidge, Burbidge, Fowler and Hoyle (1957) and Cameron (1958), it has become traditional in nuclear astrophysics to divide nucleosynthesis into "processes". In early stellar evolution these processes correspond to discrete hydrostatic thermonuclear burning stages. In analysis of these processes, one usually made simplifying assumptions about the astronomical situation, and then expended most effort trying to get the thermonuclear aspects right. Recent research in the explosive regime has generally followed the same approach, parameterizing the hypothesized explosive conditions in a "typical" zone, finding what values of these parameters (peak temperature, density and time scale for example) reproduce solar system abundances, and thereby inferring something about the explosive process itself.

It now seems likely that almost all nuclei between helium and zinc were synthesized in processes occurring in these explosive events occurring during the last violent stages of stellar evolution; the same can be said for many nuclei (p- and r-process) which are more massive. The hydrostatic processes which were once thought to explain the synthesis of the nuclei have now largely been superseded by the explosive processes. Only the s-process (for nuclei heavier than zinc) and hydrostatic hydrogen and helium burning remain clearly important* in directly determining abundances ejected.

Audouze begins this chapter with a discussion of how the CNO process can in some cases have an explosive character. Howard then reviews explosive carbon burning. Woosley's paper deals with explosive oxygen and silicon burning, and Schramm shows how he is now able to make the r-process an explosive

---

*See papers by Ulrich and Smith et al. for a discussion of the s-process in a realistic environment and the paper by Dyer for comments on hydrostatic helium burning.

Chapter II   46

process rather than the static process it was once treated as.
The chapter ends with Truran discussing the latest results on
the p-process.

References to Chapter II

Early Work

    Burbidge, E. M., Burbidge, G. R., Fowler, W. A. and
        Hoyle, F. 1957, Rev. Mod. Phys., 29, 547.
    Cameron, A. G. W. 1957, Publ. Astron. Soc. Pac., 169, 201.

Textbooks

    Clayton, D. D. 1968, Principals of Stellar Evolution and
        Nucleosynthesis, (New York: McGraw - Hill).
    Reeves, H. 1968, Stellar Evolution and Nucleosynthesis
        (New York: Gordon & Breach).

Reviews

    Arnett, W. D. and Clayton, D. D. 1970, Nature, 227, 780.
    Fowler, W. A. and Stephens, W. E. 1968, Am. J. Phys., 36, 284.
    Barnes, C. 1971, Adv. Nuc. Phys., 4.
    Allen, B. J., Gibbons, J. H. and Macklin, R. L. 1971
        Adv. Nuc. Phys., 4, 205.
    Arnett, W. D. 1973, Ann. Rev. Astron. & Astrophys., 11.

HOT CNO PROCESS [†]

Jean Audouze [‡]

California Institute of Technology, Pasadena, California

I. Introduction

Since the discovery by Bethe (1939) and von Weizsäcker (1938) of the role of the CNO cycle in hydrogen burning, a great deal of effort has been devoted to the study of nuclear properties and astrophysical implications of this set of reactions. Hydrogen burning in this manner can occur at temperatures $\leq 10^8$ °K in the centers of massive main sequence stars and in the hydrogen burning zones of red giants. The main result of the standard CNO cycle is the transformation of ^{12}C and ^{16}O to ^{14}N. However, the process at these temperatures does not account for the observed concentrations of the relatively rare isotopes such as ^{13}C, ^{15}N, ^{17}O, and ^{18}O.

The purpose of this paper is to report on recent studies of the CNO-Ne cycles at temperatures greater than 10^8 °K. At these temperatures reactions induced by protons and alpha particles become faster than beta-decay reactions. Under these conditions the behavior and result of the nuclear cooking become very different from the CNO cycle at lower temperatures. In particular, one may expect large enhancements of mass 13, 14, 15, and 17 nuclei which have an unstable isobaric parent of lifetime ≥ 50 sec.

First I recall the main characteristics of these nuclear processes in the temperature range of interest here, i.e., $10^8 \leq T \leq 10^9$ °K. Then I mention the role that these processes may play in the chemical evolution of astrophysical objects where hydrogen-rich zones can experience these large temperatures: this is the case of the explosive regions of the novae

---

[†]Supported in part by the National Science Foundation [GP-28027, GP-36687X].

[‡]Present address: Radio Astronomy Department, Observatoire de Meudon, 92190 Meudon, France and SEP CEN Saclay, France.

TABLE 1

DEPLETIONS AND ENHANCEMENTS RELATIVE TO THE SOLAR ABUNDANCES FOR VARIOUS (T_9) CONDITIONS AND SPECIFIC TIME SCALES (5, 50, 500 SEC)

Element	Solar Abundances	Time (sec)	$T_9 = 0.1$		$T_9 = 0.2$		$T_9 = 0.5$		$T_9 = 1$	
			$\rho = 1$ g cm^{-3}	$\rho = 100$ g cm^{-3}	$\rho = 1$ g cm^{-3}	$\rho = 100$ g cm^{-3}	$\rho = 1$ g cm^{-3}	$\rho = 100$ g cm^{-3}	$\rho = 1$ g cm^{-3}	$\rho = 100$ g cm^{-3}
^{13}C	4.6×10^{-5}	5	1.0	1.9	3.0	50	2.0(-2)	2.2(-4)	1.1(-3)	8.1(-8)
		50	1.1	9.3	24	1.8	4.4(-2)	7.8(-4)	5.7(-5)	1.2(-8)
		500	1.8	44	520	5.9	6.5(-2)	1.5(-3)	4.3(-7)	1.2(-9)
^{14}N	9.7×10^{-4}	5	1.0	1.0	1.0	2.4	5.0	5.0	0.90	1.3(-5)
		50	1.0	1.0	1.0	3.8	3.6	3.5	0.01	1.9(-6)
		500	1.0	1.5	1.5	5.6	2.3	4.2	8.0(-5)	1.9(-7)
^{15}N	3.8×10^{-6}	5	0.90	0.23	4.8	270	350	400	300	23
		50	0.20	1.1	64	930	770	1400	170	3.1
		500	0.23	8.7	290	2700	1100	2600	1.5	2.4(-4)
^{17}O	4.3×10^{-6}	5	1.0	1.1	2.3	140	2.0	190	1.0(-3)	0.095
		50	1.0	2.1	15	930	2.0	150	1.1(-3)	0.028
		500	1.1	5.2	110	100	2.0	44	9.0(-4)	2.2(-6)
^{18}O	2.5×10^{-5}	5	1.0	0.3	2.0(-5)	2.4(-5)	4.8(-4)	4.6	6.5(-4)	2.8
		50	0.9	1.0(-5)	5.0(-3)	3.6(-4)	6.0(-4)	4.2	9.4(-4)	0.84
		500	0.44	3.0(-5)	0.16	4.0(-5)	5.7(-4)	1.2	9.0(-4)	6.4(-5)
^{19}F	2.0×10^{-6}	5	0.85	7.0(-4)	5.0(-7)	5.5(-5)	5.0(-3)	0.63	23	220
		50	0.27	3.0(-8)	1.9(-4)	5.0(-3)	4.6(-2)	7.9	75	16
		500	1.1(-3)	4.0(-7)	3.3(-2)	1.0(-3)	0.09	21	0.8	1.0(-3)
^{21}Ne	3.6×10^{-6}	5	1.0	1.0	1.2	70	330	330	110	5.2
		50	1.0	0.83	4.3	140	75	120	48	1.7
		500	1.0	2.3(-2)	4.7	2.5(-3)	2.4	39	31	1.2(-4)
^{22}Ne	1.3×10^{-4}	5	1.0	0.95	1.0	0.22	3.4	1.7	4.9	19
		50	1.0	0.78	1.1	6.5	9.2	5.6	2.2	8.5
		500	1.0	0.22	2.8	11	11	1.8	1.5	5.2(-4)

[as studied by Starrfield et al. (1972, 1973)] or the supermassive stars (Audouze and Fricke 1973). I will conclude this report by giving my present beliefs concerning the origin of the rare nuclear species within the range $12 \leq A \leq 25$.

II. Characteristics of the High-Temperature Hydrogen Burning

A rather extensive study for fixed temperature and density conditions has been performed by Audouze, Truran, and Zimmerman (1973), hereafter referred to as ATZ. Similar calculations are also made by Caughlan and Fowler (1972, 1973). Since the nuclear evolution is governed mainly by the nuclear rates which are extremely temperature dependent, it appears useful to explore first the consequences of this dependence without any further hydrodynamical complications.

ATZ have taken into account the important reactions induced by p, α, n, or γ particles[1] which can influence the abundances of nuclei from ^{12}C to ^{25}Mg. Starting with solar abundances as initial conditions, the time variations of the abundances have been followed up to 10^5 sec. The main results of their study follow:

1 - The transformation of hydrogen into helium takes place on a time scale which varies from 10^7 sec at $T_9 = 0.1$ to $\sim 10^5$ at $T_9 \sim 0.2$; this decrease is mainly due to the increase of the (p,γ) rates with T. For $0.2 < T < 0.5$ the time scale of the hydrogen burning remains $\sim 3 \times 10^4 - 10^5$ sec. It is mainly limited by the slowest beta decays $[^{14}O(\beta^+)^{14}N$ and $^{15}O(\beta^+)^{15}N]$. At the highest temperature studied $T_9 = 1$, the hydrogen exhaustion becomes limited by the increase of the (α,p) and the (γ,p) rates. At this temperature the time scale is $\sim 10^6$ sec. The choice of $T_9 = 1$ as a temperature upper limit is not arbitrary. At temperatures $\geq T_9 = 1$ the elements with $A \leq 19$ are rapidly transformed into species with $A > 20$, thus terminating the hot CNO processes.

2 - Since we are interested in the enhancements of relatively rare isotopes (^{13}C, ^{15}N, ^{17}O, ^{18}O, ^{19}F, ^{21}Ne, and ^{22}Ne) the following discussion will be concentrated on these nuclear species. Table 1 summarizes the depletions and enhancements with respect to the initial (solar) abundances for the various temperature and density conditions and specific time scales 5, 50, and 500 sec [these time scales correspond roughly to the hydrodynamic time scale for temperature given by Fowler and Hoyle (1964) $\tau \propto 446/\sqrt{\rho}$ sec].

---

[1]The nuclear rates have been taken from Fowler et al. (1973), Wagoner (1969) and Wagoner et al. (1967).

a) The production of ^{13}C is strongly temperature dependent. Significant overproduction of ^{13}C occurs only at lower temperature $T_9 \leq 0.2$. At higher temperatures, although ^{13}C is more refractory than ^{12}C ($^{13}C/^{12}C$ remains > 1) the total abundances of both carbon isotopes are strongly reduced primarily due to the rapid chain $^{12}C(p,\gamma)^{13}N(p,\gamma)^{14}O$.

b) The isotopes ^{14}N and ^{15}N are both enhanced particularly at lower temperatures due to the relative stability of their positron-decay progenitors ^{14}O and ^{15}O against proton-induced reactions. But the enhancement of ^{15}N is more important than the ^{14}N increase. ^{15}O has a longer lifetime (~ 178 sec) than ^{14}O (~ 100 sec). ^{15}N is the more easily overproduced isotope in the conditions explored here (the enhancement can be $\sim 10^3$ for $T_9 = 0.2$) but at $T_9 = 1$ this factor is reduced due to the increased importance of $^{15}O(\alpha,\gamma)^{19}F$ forming mass 19.

c) ^{17}O is largely enhanced (factors ~ 100 and more) for $T_9 \leq 0.5$ and relatively large densities ($\rho > 100$ g cm^{-3}). At higher temperatures and lower densities, $^{17}F(\gamma,p)^{16}O$ seriously inhibits production of mass 17. ^{18}O is destroyed at $T_9 \leq 0.2$ by the very rapid $^{18}O(p,\alpha)^{15}N$. However, some enhancement of ^{18}O can occur at intermediate temperature $T_9 \sim 0.5$ and density $\rho > 100$ g cm^{-3} through $^{17}F(p,\gamma)^{18}Ne$. But since ^{18}Ne has a lifetime of 2.1 sec, mass 18 is destroyed if the burning proceeds in time scales long enough for the weak interaction to take place.

d) ^{19}F can be found overabundant if the CNO-Ne cycle proceeds at high T and ρ (enhancements of ≥ 100 for $T_9 \geq 0.5$ and $\rho \geq 100$ g cm^{-3}). This enhancement is due mainly to $^{15}O(\alpha,\gamma)^{19}F$ and in a less extend to $^{18}F(p,\gamma)^{19}Ne$. The enhancement is limited by the short lifetime of ^{19}Ne [^{19}F is easily destroyed by $^{19}F(p,\alpha)^{16}O$].

e) The neon isotopes are produced in various amounts in the hot CNO process. A large overabundance of ^{21}Ne, reaching factors ~ 100-400, was obtained for $T_9 = 0.2$ to 0.5. It results from the reaction $^{20}Ne(p,\gamma)^{21}Na$. Some overabundance of ^{22}Ne results from burning at $T_9 = 1$ and $\rho \geq 100$ g cm^{-3}. This is due to $^{21}Na(p,\gamma)^{22}Mg$. Following the decay $^{22}Mg(\beta^+)^{22}Na$, the long-lived isotope ^{22}Na can survive at somewhat higher temperatures than the progenitors of other rare isotopes.

From the above we see that the rare isotopes of light nuclei ^{13}C, ^{15}N, ^{17}O, ^{18}O, ^{19}F, ^{21}Ne, and ^{22}Ne can all be produced in enhanced concentrations under conditions where hydrogen burning proceeds via CNO-Ne cycles at temperatures higher than 10^8 °K. The precise conditions of maximum enhancement of these various isotopes vary considerably as a function of temperature, density, and time scale. An indication of this was illustrated by J. W. Truran in preliminary calculations for a

TABLE 2

ENHANCEMENTS ACHIEVED IN "EXPLOSIVE HYDROGEN BURNING"

Nucleus	$T_8 = 3$		$T_8 = 5$			$T_8 = 7.5$		$T_8 = 10$	
	$\rho = 10^2$	$\rho = 10^4$	$\rho = 10^2$	$\rho = 10^4$	$\rho = 10^4{}^*$	$\rho = 10^2$	$\rho = 10^4$	$\rho = 10^2$	$\rho = 10^4$
^{13}C	11.	.48	16.	2.6	.30	11.	1.0	4.8	
^{14}N	4.8	5.5	3.8	4.7	4.7	1.4	.33	.48	
^{15}N	920.	550.	1300.	2900.	2900.	920.	1200.	420.	2.9
^{17}O	1400.	2300.	1100.	67.	74.	1900.	130.	1300.	.26
^{18}O				.10	25.				
^{19}F				9.5	17.		550.		.90
^{21}Ne	24.	280.	4.7	210.	310.	2.3	330.	.67	230.
^{22}Ne	11.	3.3	12.	5.2	2.5	4.5	42.	1.2	47.
^{24}Mg	1.1	1.1	1.1	1.1	1.1	3.0	9.5	15.	35.

*Shortened time scale $\tau_{HYD}/10$.

forthcoming paper in collaboration with S. Starrfield and myself. His calculations were made using the theoretical temperature density profile obtained from the Fowler-Hoyle hydrodynamical time scale which is generally considered in the canonical explosive burning calculations. From these preliminary calculations performed for various initial temperatures ($T_9 = 0.3$, 0.5, 0.75, and 1) and densities ($\rho = 10^2$ and 10^4) we see that ^{15}N and ^{17}O reach large enhancements by factors $\propto 10^3$ in almost all the cases studied here while ^{21}Ne can reach, for large ρ, an enhancement of $\propto 300$. ^{19}F is only overabundant by $\propto 500$ for $T_9 = 0.75$ and $\rho = 10^4$. We see that in the preliminary conditions studied ^{13}C, ^{18}O, and ^{22}Ne are generally not well accounted for. In particular, the very short lifetime of ^{18}Ne makes the overproduction of ^{18}O very difficult in the conditions of this study.

All of what I have said clearly illustrates the very complex character of the thermonuclear burning sequences explored in this paper. I will return to this when we try to summarize what can be concluded as far as the nucleosynthesis of these rare isotopes is concerned.

III. Astrophysical Locations of the Hot CNO Processes

The hot CNO processes are not just an exercise in theoretical nucleosynthesis. They are likely to occur in at least two actual astrophysical situations: the novae and the bouncing supermassive stars.

An extensive study of the hydrodynamic conditions of the nova outburst, together with the hot CNO process, is being performed by Starrfield et al. (1972, 1973). Since the novae belong to a close binary system consisting of a large cool star and a small hot white dwarf, they use the model in which the cooler star is losing hydrogen-rich material and where a large part of this matter is accreted by the white dwarf. When the accretion proceeds, a layer of hydrogen-rich material is built at the surface of the white dwarf. This matter, gradually compressed, then heated reaches the ignition temperatures necessary to produce the hot CNO cycle.

Then, in this particular type of object, the thermonuclear runaway is governed by the hot CNO cycle itself: the hot CNO process releases energy at a rate $\geq 10^{18}$ ergs g^{-1} sec^{-1} in time scales governed by the beta-decay rate of ^{14}O and ^{15}O. The net result of this rapid energy release is the ejection of $\sim 10^{28}$ -3 x 10^{29} g (i.e., 10^{-4} - 10^{-3} M$_\odot$) with kinetic energies of 10^{44} - 10^{45} ergs which are values consistent with the observations. The matter ejected has been processed through the hot CNO cycle (at temperatures up to 6-7 x 10^8 °K during a few sec and temperatures of 2-3 x 10^8 °K during $\propto$ 1 hour) resulting in a strong enrichment in ^{13}C, ^{15}N, and ^{17}O. ^{18}O and ^{19}F are formed during the hot temperature peak which lasts a few seconds but are destroyed during the longer period at $2-3 \times 10^8$ °K. In the present stage of the calculations of Starrfield et al., it appears that explosions of novae may create ^{13}C, ^{15}N, and ^{17}O but not, unfortunately, the other species in question. To account for the observed value of ^{15}N, 10^2 novae explosions per year are needed during which $\sim 2 \times 10^{-4}$ $M_\odot$ are expelled enriched in ^{15}N by a factor 10^3. This value is close to the actual observed value ~ 20 novae/yr.

Besides the novae, the supermassive stars of a few 10^5 $M_\odot$ which may be present in the condensed nucleus of many galaxies can also undergo violent implosion-explosion processes. As in the previous case, this explosion is caused by the large energy release during hydrogen burning and proceeds on a time scale of about 10^5 sec. With Klaus Fricke, I have estimated the nucleosynthetic effects which can be predicted during the implosion-explosion of such objects. In these calculations we have used the temperature-density profiles predicted for the 5.2×10^5 $M_\odot$ population I star studied by Appenzeller and Fricke (1972). Four representative zones located at fractional masses $M_r/M = 0$, 0.3, 0.5 and 0.7 have been selected and the nucleosynthesis calculations have been performed for these four shells. Figures 1 and 2 show the abundance variations with respect to time for the inner shell.

The results of our investigation are summarized in Table 3. Since the temperature is always less than 3.4×10^8 °K, only interesting amounts of ^{13}C, ^{15}N, ^{17}O, and also ^{22}Ne can be achieved. In the somewhat more realistic case where the supermassive star is not entirely disrupted (as in case 1) but where only the outer zones are ejected (case 2) similar and interesting enhancements of ^{13}C, ^{15}N, and ^{17}O (by a factor ~ 100) can be achieved. So if 1% of the galactic matter has been processed in the nucleus of the galaxy in supermassive star explosions, supermassive stars are interesting objects to explain the ^{13}C, ^{15}N, and ^{17}O nucleosynthesis.

IV. Conclusion: Origin of the Rare Light Elements $13 < A < 22$

To conclude this presentation of the hot CNO-Ne processes, it is very difficult at present to predict for all the rare nuclei in the $13 < A < 22$ range one single explanation regarding their origin. In Table 4 I have tried to summarize the different possible mechanisms. The hydrogen burning in novae and supermassive stars can easily account for ^{13}C, ^{15}N, and ^{17}O [^{13}C can also be produced in the red giant hydrogen-burning zones - see Truran (1973) for instance]. While the helium explosive burning (Howard, Arnett, and Clayton 1971)

TABLE 3

MASS FRACTIONS AND ENRICHMENT FACTORS FOR CASE 1 (TOTAL MASS EJECTED) AND FOR CASE 2 (OUTER ZONES EJECTED) COMPARED TO THE SOLAR VALUES

	^{13}C	^{15}N	^{17}O	^{22}Ne
Solar abundance*	4.1×10^{-5}	4.8×10^{-6}	3.3×10^{-6}	1.6×10^{-4}
Case 1: abundance	1.9×10^{3}	6.0×10^{-3}	3.5×10^{-4}	1.8×10^{-3}
enhancement factor	45	1200	110	11
Case 2: abundance	1.9×10^{3}	1.0×10^{-3}	3.7×10^{-4}	6.7×10^{-4}
enhancement factor	45	210	100	4

*Cameron (1973)

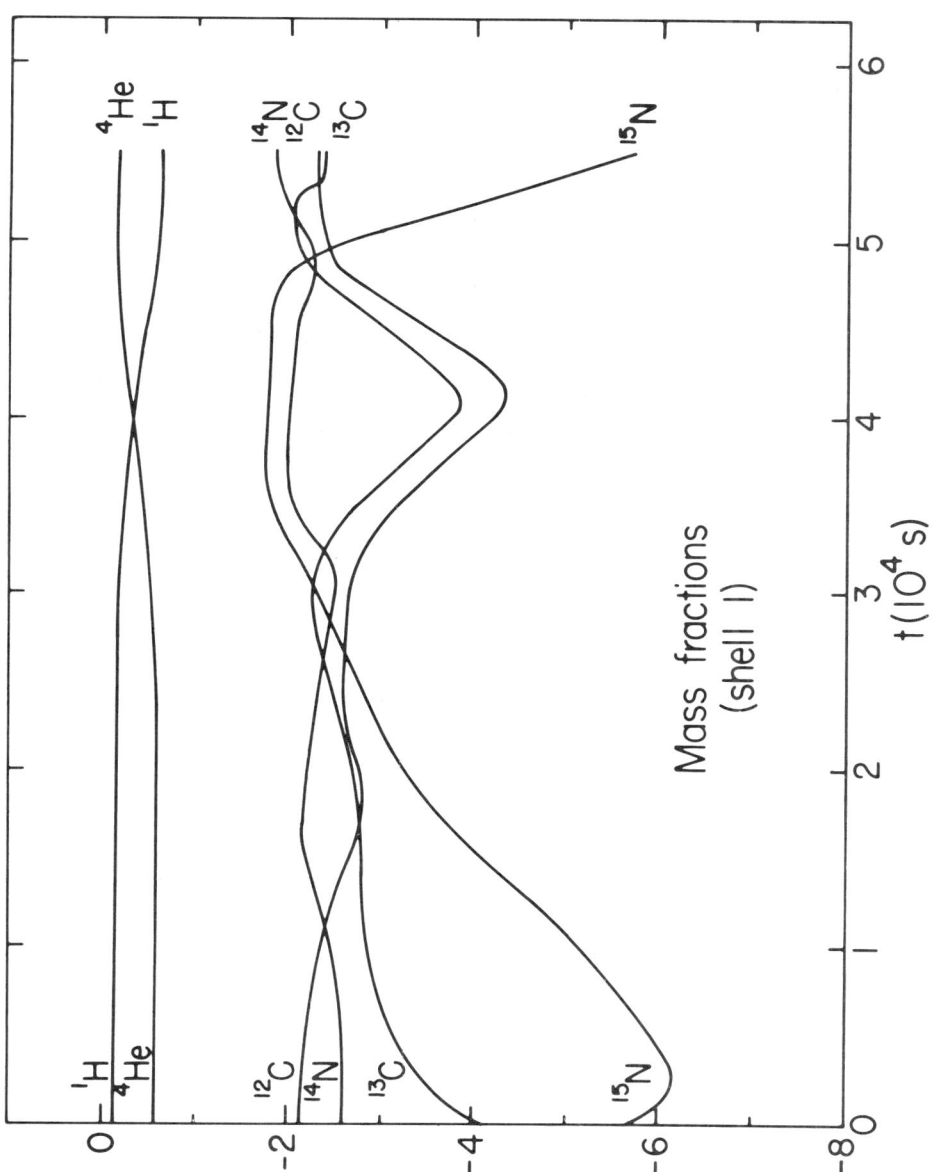

Fig. 1. Chemical evolution at the center of supermassive stars. (Audouze and Fricke 1973)

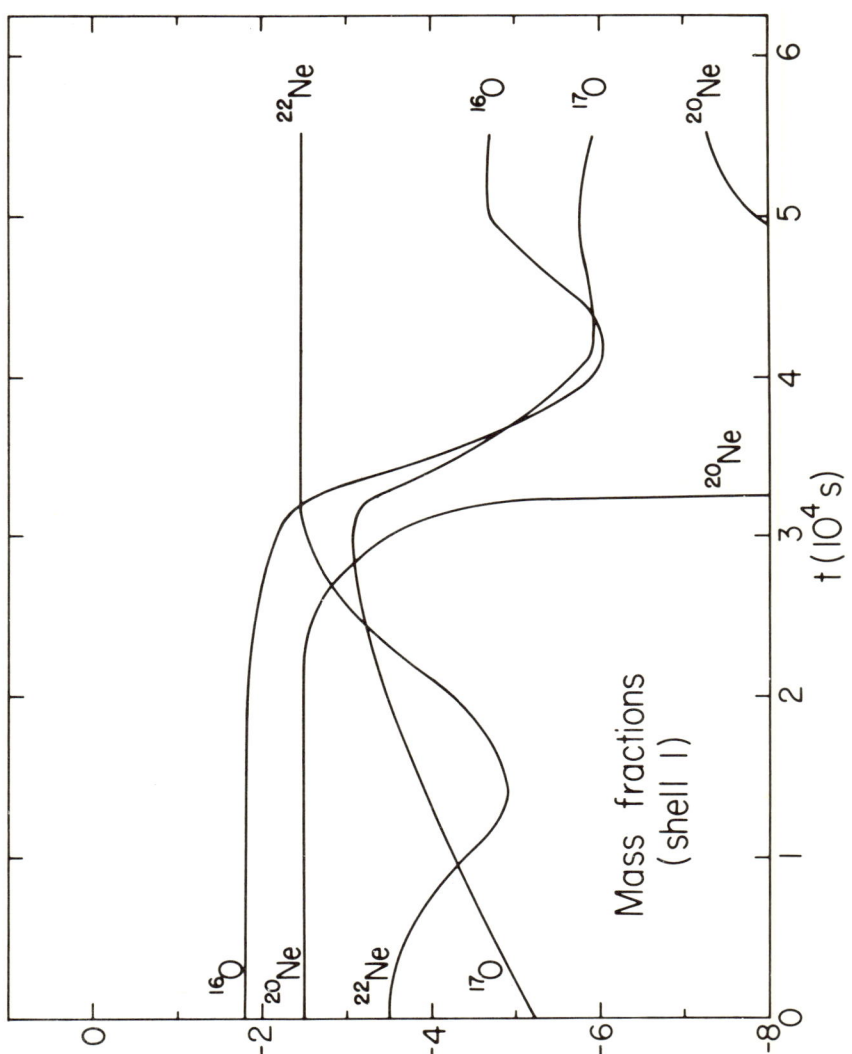

Fig. 2. Chemical evolution at the center of supermassive stars. (Audouze and Fricke 1973)

TABLE 4

CHART OF THE POSSIBLE ORIGIN OF C-Mg ELEMENTS

Element	"Slow" CNO Cycle	"Slow" He Burning	Hot CNO Cycle		Explosive He Burning	Explosive C Burning
			Novae	SMS		
^{12}C		possible				
^{13}C	(possible)		possible	possible		
^{14}N	possible					
^{15}N			possible	possible	possible	
^{16}O		possible				
^{17}O			possible	possible		
^{18}O		possible	?	no	possible	
^{19}F			?	no	possible	
^{20}Ne		possible				possible
^{21}Ne			?	no	possible	
^{22}Ne		possible	?	$\sim 10\%$	unlikely	
^{23}Na						possible
^{24}Mg						possible

may explain the formation of ^{15}N, ^{18}O, ^{19}F, and ^{21}Ne, ^{22}Ne is difficult to account for in these last two theories, and it may be that the more likely hypothesis is the nonexplosive helium burning [as studied recently, for instance, by Couch and Arnett (1972)]. In the present picture three different mechanisms at least seem to be necessary to account for all these rare nuclei. Maybe a good explanation would be the spallation reactions (Audouze 1970) occurring in supernovae (Colgate 1973) or in the supermassive stars (Hoyle and Fowler 1973). It would be exciting to make in the same objects the light elements D and LiBeB (that I particularly like!) together with all the rare and not yet well explained nuclei like those I have spoken of, and possibly the p-process isotopes (Truran, this conference). Perhaps this conference will teach us some simple explanation to understand these complicated but very interesting problems.

ACKNOWLEDGMENTS

This talk is mainly based upon two papers, the first entitled "Hot CNO-Ne Cycle Hydrogen Burning. I. Thermonuclear Evolution at Constant Temperature and Density," written in collaboration with James W. Truran and Barbara A. Zimmerman, and the second "Nucleosynthesis in Supermassive Star Explosions," written in collaboration with Klaus J. Fricke. I have enjoyed fruitful exchanges with my co-workers and also with Sumner Starrfield and William A. Fowler. Part of the research presented here was done while I was supported by a U.S.-French NSF/CNRS exchange fellowship.

REFERENCES

Appenzeller, I., and Fricke, K. 1972, Astr. and Ap., 21, 285.
Audouze, J. 1970, Astr. and Ap., 8, 436.
Audouze, J., and Fricke, K. 1973, in preparation.
Audouze, J., Truran, J. W., and Zimmerman, B. A. 1973, submitted to Ap. J.
Bethe, H. A. 1939, Phys. Rev., 55, 103.
Cameron, A. G. W. 1973, in preparation.
Caughlan, G. R., and Fowler, W. A. 1972, Nature Phys. Sci., 238, 23.
_____. 1973, in preparation.
Colgate, S. A. 1973, preprint.
Couch, R. G., and Arnett, W. D. 1972, Ap. J., 178, 771.
Fowler, W. A., Caughlan, G. R., and Zimmerman, B. A. 1973, in preparation.
Fowler, W. A., and Hoyle, F. 1964, Ap. J. Suppl. No. 91, 9, 201.
Howard, W. M., Arnett, W. D., and Clayton, D. D. 1971, Ap. J., 165, 495.
Hoyle, F., and Fowler, W. A. 1973, Nature, 241, 384.
Starrfield, S., Sparks, W. M., and Truran, J. W. 1973, in preparation.

Starrfield, S., Truran, J. W., Sparks, W. M., and Kutter, G. 1972, Ap. J., 176, 169.
Truran, J. W. 1973, in Proceedings of the Conference on Red Giant Stars, October 4-6, 1972 (Bloomington: Indiana University Press) to be published.
von Weizsäcker, C. F. 1938, Physik Z., 39, 633.
Wagoner, R. V. 1969, Ap. J. Suppl. No. 162, 18, 247.
Wagoner, R. V., Fowler, W. A., and Hoyle, F. 1967, Ap. J., 148, 3.

EXPLOSIVE CARBON BURNING
by

W. Michael Howard

Los Alamos Scientific Laboratory
Los Alamos, New Mexico

When the helium-exhausted zone of a presupernova star burns explosively such that some but not all of the carbon is fused into heavier elements, the nuclei ^{20}Ne, ^{23}Na, ^{24,25,26}Mg, ^{27}Al, ^{29,30}Si, ^{31}P can be produced in amounts close to their solar system abundance ratios. This remarkable result by Arnett (1969) was the beginning of calculations that have given quantitative support to the idea that the bulk of the elements between carbon and germanium have originated in exploding supernovae. Many neutron-rich nuclei between sulfur and germanium are coproduced during the explosive carbon burning from the alteration of the heavy elements present at the formation of the star (Howard et al. 1972).

Unlike the higher temperature explosive processes, the results depend almost explicitly upon the rates of the nuclear reactions that take place. Table 1 lists many of the important reactions. The principal reaction is that of ^{12}C + ^{12}C --it is the fusion of the ^{12}C nuclei that liberates the free protons, neutrons, and alpha particles that, in turn, initiates the complex series of reactions that take place. The effective energy range of interaction for ^{12}C + ^{12}C during explosive carbon burning is $2.9 \leqslant E(MeV) \leqslant 4.8$ and the rate, with appropriate branching ratios, has been experimentally determined over the energies of interest. Most of the rates controlling the production of the primary products are experimentally determined, but most of the rates and nuclear binding energies controlling the production of the neutron-rich species are not.

Table 1. Important Nuclear Reactions

Primary:

^{12}C + ^{12}C	^{22}Ne(p,γ)^{23}Na	^{21}Ne(p,γ)^{22}Na
^{18}O(p,γ)^{19}F	^{15}N(p,γ)^{16}O	^{23}Na(p,γ)^{24}Mg
^{25}Mg(p,γ)^{26}Al	^{26}Mg(p,γ)^{27}Al	^{13}C(p,γ)^{14}N

(Table 1, Primary, cont'd.)

$^{20}Ne(p,\gamma)^{21}Na$	$^{12}C(p,\gamma)^{13}N$	$^{16}O(p,\gamma)^{17}F$
$^{15}N(p,\alpha)^{12}C$	$^{23}Na(p,\alpha)^{20}Ne$	$^{23}Na(\alpha,p)^{25}Mg$
$^{24}Mg(\alpha,p)^{27}Al$	$^{17}O(p,\alpha)^{14}N$	$^{18}O(p,\alpha)^{15}N$
$^{25}Mg(\alpha,n)^{28}Si$	$^{26}Mg(\alpha,n)^{29}Si$	$^{13}C(\alpha,n)^{16}O$
$^{17}O(\alpha,n)^{20}Ne$	$^{24}Na(\alpha,n)^{27}Al$	$^{17}O(n,\alpha)^{14}C$
$^{18}O(\alpha,n)^{21}Ne$	$^{21}Ne(\alpha,n)^{24}Mg$	$^{22}Ne(\alpha,n)^{25}Mg$
$^{16}O(\alpha,\gamma)^{20}Ne$	$^{20}Ne(\alpha,\gamma)^{24}Mg$	$^{14}C(\alpha,\gamma)^{18}O$
$^{18}O(\alpha,\gamma)^{22}Ne$	$^{22}Ne(\alpha,\gamma)^{26}Mg$	$^{12}C(n,\gamma)^{13}C$
$^{13}C(n,\gamma)^{14}C$	$^{13}N(n,p)^{13}C$	$^{14}N(n,\gamma)^{15}N$
$^{16}O(n,\gamma)^{17}O$	$^{18}O(n,\gamma)^{19}O$	$^{18}F(n,p)^{18}O$
$^{20}Ne(n,\gamma)^{21}Ne$	$^{21}Ne(n,\gamma)^{22}Ne$	$^{22}Ne(n,\gamma)^{23}Ne$
$^{21}Na(n,p)^{21}Ne$	$^{22}Na(n,p)^{22}Ne$	$^{24}Na(p,n)^{24}Mg$
$^{24}Mg(n,\gamma)^{25}Mg$	$^{25}Mg(n,\gamma)^{26}Mg$	$^{26}Mg(n,\gamma)^{27}Mg$
$^{28}Si(n,\gamma)^{29}Si$	$^{29}Si(n,\gamma)^{30}Si$	$^{30}Si(n,\gamma)^{31}Si$

Secondary:

$^{32}S(n,\gamma)^{33}S$	$^{33}S(n,\gamma)^{34}S$	$^{33}S(n,\alpha)^{30}Si$
$^{34}S(n,\gamma)^{35}S$	$^{36}S(p,\gamma)^{37}Cl$	$^{36}Cl(n,p)^{36}S$
$^{37}Cl(n,\gamma)^{38}Cl$	$^{37}Cl(p,\gamma)^{38}Ar$	$^{37}Cl(p,\alpha)^{34}S$
$^{44}Cl(\gamma,n)^{43}Cl$	$^{46}Cl(\gamma,n)^{45}Cl$	$^{36}Ar(n,\gamma)^{37}Ar$
$^{36}Ar(n,p)^{36}Cl$	$^{37}Ar(n,p)^{37}Cl$	$^{37}Ar(n,\alpha)^{34}S$
$^{38}Ar(p,\gamma)^{39}K$	$^{40}Ar(p,\gamma)^{41}K$	$^{43}Ar(\gamma,n)^{42}Ar$
$^{45}Ar(\gamma,n)^{44}Ar$	$^{47}Ar(\gamma,n)^{46}Ar$	$^{39}K(n,\gamma)^{40}K$
$^{48}K(\gamma,n)^{47}K$	$^{50}K(\gamma,n)^{49}K$	$^{40}Ca(n,\gamma)^{41}Ca$

(Table 1, Secondary, cont'd.)

$^{41}Ca(n,\gamma)^{42}Ca$	$^{41}Ca(n,p)^{41}K$	$^{41}Ca(n,\alpha)^{38}Ar$
$^{46}Ca(n,\gamma)^{47}Ca$	$^{47}Ca(p,n)^{48}Sc$	$^{48}Ca(n,\gamma)^{49}Ca$
$^{48}Ca(p,\gamma)^{49}Sc$	$^{49}Sc(p,n)^{49}Ti$	$^{49}Ti(p,n)^{49}V$
$^{49}Ti(p,\gamma)^{50}V$	$^{50}V(n,p)^{50}Ti$	$^{62}Fe(p,n)^{62}Co$
$^{64}Fe(p,n)^{64}Co$	$^{65}Co(p,n)^{65}Ni$	$^{67}Co(p,n)^{67}Ni$
$^{68}Ni(p,n)^{69}Cu$	$^{70}Ni(p,n)^{70}Cu$	$^{63}Fe(\gamma,n)^{62}Fe$
$^{65}Fe(\gamma,n)^{64}Fe$		

In addition, the results are suggestive of a solar system distribution of products only if the explosions occur from a very limited range of peak temperatures. Table 2 presents the overproduction factors for product nuclei for various combinations of peak temperature, density, expansion time scale, and whether the neutron excess resides in ^{18}O or ^{22}Ne. In addition, the yield of ^{36}S resulting from heavy elements being present in the carbon-rich zone is also calculated. The overproduction factor is defined as the ratio of the yield by mass X of each nuclear species produced to the mass fraction $X_\odot$ of that species in solar material. For example, the fact that $X/X_\odot = 179$ for ^{24}Mg under condition d means that the ^{24}Mg concentration is 179 times greater after the explosion than it was before; furthermore, it allows (but certainly does not demand) the simple interpretation that all of the natural ^{24}Mg abundance was synthesized in this way when 1/179 of all galactic matter passed through these carbon explosions. That $X/X_\odot = 139$ for ^{20}Ne implies that if that picture were correct the natural ^{20}Ne would also be synthesized in such events. A detailed inspection of the table will reveal that the choice of location of the initial neutron excess makes little difference in the final results. Couch and Arnett (1972) have recently analyzed the $^{14}N(\alpha,\gamma)^{18}F$ and $^{18}O(\alpha,\gamma)^{22}Ne$ reaction rates during hydrostatic helium burning and have determined that ^{22}Ne, rather than ^{18}O, is the proper choice for this neutron excess in the carbon-rich zone. ^{18}O and ^{22}Ne release their excess neutrons rapidly enough that the choice makes little difference to the results. However, variation of the peak temperature, density, and time scale can have sizable effects upon the results. In particular, notice that if the peak temperature is much greater than

Table 2. Overproduction Factors

	^{20}Ne	^{23}Na	^{24}Mg	^{25}Mg	^{26}Mg	^{27}Al	^{36}S
a	83.2	87.9	40.1	49.8	126	79.8	1420
b	107	53.8	78.2	41.9	84.8	110	1030
c	126	46.3	127	28.4	46.4	139	690
d	139	34.8	179	14.1	23.5	170	319
e	138	54.5	144	32.6	56.3	143	665
f	153	44.6	201	18.0	30.7	172	352
g	156	41.7	256	8.26	17.0	202	137
h	144	34.1	289	3.69	9.85	233	37.0
i	34.1	128	5.47	96.6	126	21.0	1010
j	52.8	129	12.8	59.0	141	47.5	1230
l	74.1	89.6	31.2	46.0	123	75.8	1300
m	94.8	17.7	61.8	38.9	81.2	103	1170
n	29.3	127	6.47	75.3	136	30.6	1250
o	145	56.3	142	37.3	63.7	142	677
p	165	49.3	190	22.9	37.3	170	322
q	179	44.1	247	12.6	22.8	195	189
r	186	39.5	276	6.90	15.3	217	80.2
s	77.7	153	30.4	64.5	163	82.9	1320
t	103	82.5	64.0	56.1	129	119	1290
u	125	56.5	110	43.4	75.8	151	1040
v	138	46.1	161	25.4	38.5	181	629
w	146	69.3	124	55.2	100	153	954
x	166	58.2	178	39.0	61.2	185	667
y	181	50.9	224	24.0	37.1	211	396
z	188	44.9	258	14.0	23.8	234	208
aa	137	65.2	128	48.5	87.8	155	958
bb	153	54.4	183	30.9	49.1	186	623
cc	156	45.8	233	15.6	26.2	218	309
dd	141	36.7	273	6.84	14.0	252	110
ee	190	58.9	242	29.6	47.7	211	442
ff	198	51.5	277	18.7	31.5	233	266
gg	198	43.8	301	11.7	21.6	254	149
hh	188	34.4	319	7.4	15.0	275	76.1
ii	189	51.5	261	16.4	29.9	193	240
jj	196	45.7	295	9.6	20.0	214	121
kk	196	39.6	318	5.8	14.2	234	57.0
ll	186	31.3	335	3.7	10.2	254	24.2

(Table 2, cont.)

	$T_9(T/10^9)$	$\rho(g/cm^3)$	$\tau_{exp} \cdot (24\ G\rho)^{\frac{1}{2}}$	Seed
a	2.00	10^5	1	^{22}Ne
b	2.05			
c	2.10			
d	2.15			
e	2.00		$\sqrt{10}$	
f	2.05			
g	2.10			
h	2.15			
i	2.00		$1/\sqrt{10}$	
j	2.05			
l	2.10			
m	2.15			
n	2.00	10^4	1	
o	2.00	10^6		
p	2.05			
q	2.10			
r	2.15			
s	2.00	10^5		^{18}O
t	2.05			
u	2.10			
v	2.15			
w	2.00	10^6		
x	2.05			
y	2.10			
z	2.15			
aa	2.00	10^5	$\sqrt{10}$	
bb	2.05			
cc	2.10			
dd	2.15			
ee	2.00	10^6		
ff	2.05			
gg	2.10			
hh	2.15			
ii	2.00			^{22}Ne
kk	2.05			
jj	2.10			
ll	2.15			

$T_9(T/10^9) \geq 2.10$, the yields of the Mg isotopes becomes distorted from their solar system values. Since the isotopic ratios are determined from studies of meteorites to within an accuracy of a few percent, production of these isotopic ratios provide a severe constraint for the theory of explosive carbon burning.

The (n,γ) reactions on Mg play an important role in the final distribution of the Mg isotopes. The ^{24}Mg is produced directly from the ^{12}C + ^{12}C gamma-ray channel and the reaction sequence ^{12}C(^{12}C,α)^{20}Ne(α,γ)^{24}Mg, while the reaction ^{24}Mg(n,γ)^{25}Mg is primarily responsible for its destruction. The major source of ^{25}Mg is the reaction ^{22}Ne(α,n)^{25}Mg, while ^{22}Ne(α,n)^{25}Mg(n,γ)^{26}Mg is the major source of ^{26}Mg. The ^{18}O(α,γ)^{22}Ne reaction proceeds rapidly enough that it makes little difference whether the initial neutron excess resides in ^{18}O or ^{22}Ne. It is the ^{25}Mg(α,n)^{28}Si and ^{26}Mg(α,n)^{29}Si reactions that destroy the ^{25}Mg and ^{26}Mg during the higher temperature explosive carbon burning and distort the Mg isotopes away from their solar system abundance ratios. Both of these important destructive reactions have estimated rates.

In addition to the primary reactions that take place, if the initial star has formed from a gas nebula with a composition similar to the sun's, it contains approximately 0.27% by mass of nuclei heavier than Mg. These nuclei are assumed to remain unaltered during the hydrostatic evolution of the star and act as "seed" for the free protons, neutrons, and alpha particles released during the explosive burning.

We believe that ^{36}S is formed by the conversion of ^{32}S and ^{36}Ar initially present in the zone by the following reaction mechanisms:

(1) ^{36}Ar(n,p)^{36}Cl(n,p)^{36}S

(2) ^{32}S(n,γ)^{33}S(n,γ)^{34}S(n,γ)^{35}S(n,γ)^{36}S
with branches (p,α) and (p,γ)^{37}Cl from ^{36}S, and (n,α)^{30}Si from ^{33}S.

From ^{36}Ar, the conversion takes place by the slightly exothermic (n,p) reactions. The ^{32}S is converted via (n,γ) reactions to ^{36}S. However, if all of the ^{32}S and ^{36}Ar initially present in the carbon-rich zone is converted into ^{36}S, the ^{36}S is overproduced relative to the other products of the explosive carbon burning by a factor of 100. One would expect from theory then, 100 times more ^{36}S in the solar system than is observed. Thus, the reaction sequences leading from ^{32}S and ^{36}Ar must finely regulate the amount of ^{36}S produced. All the rates used in these sequences are based on optical model calculations by Truran (1972) and, as can be seen from Table 2, current estimates of these rates lead to an overproduction of ^{36}S by as much as a factor of 10.

A study of Table 2 will reveal that there seems to be no consistent way, by varying the temperature, time scale, or density, to reduce this overabundance and also produce the solar system abundance ratios of Mg. Of course, the observed abundances will be some integration over various combinations of peak temperatures and time scales. However, using optical model estimates for these reaction rates, there appears to be no way to produce the solar system abundance ratios of the Mg isotopes without vastly overproducing ^{36}S.

Of particular importance in the reaction sequences leading to ^{36}S is the branching ratio in the compound nuclear states of ^{34}S, ^{37}Ar, and ^{37}Ar formed by ^{33}S + n, ^{36}Ar + n, and ^{37}Cl + p, as well as the rates of ^{36}S(p,γ)^{37}Cl(p,α)^{34}S(n,γ)^{35}S. Experimental investigation of the branching ratios of ^{33}S + n is presently underway at LASL. A resonance at the appropriate energy in the exothermic reaction ^{33}S(n,α)^{30}Si would deplete significant seed nuclei out of this sulfur cycle and could significantly reduce the yield of ^{36}S.

We find in some burning the yield of ^{36}S is inversely proportional to the value taken for the ^{36}S(p,γ)^{37}Cl cross section. The effective energy range for this reaction during explosive carbon burning is $2.0 \leq E(MeV) \leq 5.0$. This reaction has been studied up to proton energies of 2 MeV by Hyder et al. 1968. The ^{36}S has a very small natural abundance so that enriched targets are difficult to make. However, the Oak Ridge National Laboratory has samples that are approximately three percent ^{36}S.

We have concentrated on the yield of ^{36}S from such explosions. Table 3 gives the overproduction factors for other neutron-rich isotopes that are also produced along with the primary seed for each specie. The calculation for Table 3 was carried out a density $\rho = 10^5$ g/cm^3 and for expansion on a hydrodynamic time scale. Notice that the value for the yield of ^{36}S differs from its yield in calculation r of Table 2 by a factor of 3. The calculation of Table 3 does

Table 3. Overabundances

Product	Primary Seed	$X/X_\odot$		
		$T_9 = 2.15$	2.05	2.00
^{20}Ne	C-burn	134	117	77
^{23}Na	C-burn	56	78	102
^{24}Mg	C-burn	186	82	44
^{36}S	^{32}S, ^{36}Ar	180	1000	1100
^{40}Ar	^{36}Ar, ^{32}S	300	220	110
^{40}K	^{36}Ar, ^{32}S	290	28	1.6
^{43}Ca	^{36}Ar	39	46	50
^{46}Ca	^{36}Ar, ^{40}Ca	730	3200	6700
^{48}Ca	^{40}Ca	110	170	140
^{45}Sc	^{36}Ar	95	350	570
^{47}Ti	^{40}Ca, ^{36}Ar	86	230	130
^{49}Ti	^{40}Ca	290	45	23
^{50}Ti	^{40}Ca	68	58	58
^{50}V	^{40}Ca	72	4.2	small
^{62}Ni	^{56}Fe	190	300	320
^{64}Ni	^{56}Fe	225	520	560
^{65}Cu	Fe(p,n)	930	205	57
^{67}Zn	Fe(p,n)	690	510	170
^{68}Zn	^{58,60}Ni	450	140	130
^{70}Zn	^{58,60}Ni	1060	800	750
^{69}Ga	Ni(p,n)	170	small	small
^{71}Ga	Ni(p,n)	1490	110	44
^{73}Ge	Ni(p,n)	770	84	34
^{75}As	Ni(p,n)	127	68	34
^{76}Ge	Zn	200	110	110

not take into account the buildup of ^{36}S from the ^{12}C + ^{12}C reaction which is significant at T_9 = 2.15.

Notice that the ^{46}Ca is also well overproduced at all three temperatures. The ^{46}Ca has the primary sources ^{36}Ar and ^{40}Ca through the sequences ^{36}Ar(n,γ)^{37}Ar(n,p)^{37}Cl(n,γ)... ^{45}Cl(p,n)^{45}Ar(γ,n)^{44}Ar(p,n)^{44}K(n,γ)^{45}K(p,n)^{45}Ca(n,γ)^{46}Ca and also ^{40}Ca(n,γ)^{41}Ca(n,α)^{38}Ar(n,γ)...^{44}Ar(p,n)^{44}K(n,γ)^{45}K(p,n) ^{45}Ca(n,γ)^{46}Ca. The branching ratios of the three decay modes of ^{42}Ca are becoming very important

$$^{41}\text{Ca} + n \begin{cases} ^{42}\text{Ca} + \gamma + 11.47 \text{ MeV} \\ ^{41}\text{K} + p + 1.20 \text{ MeV} \\ ^{38}\text{Ar} + \alpha + 5.23 \text{ MeV} \end{cases}$$

The ^{46}Ca yield also depends on unknown neutron separation energies in neutron-rich matter--especially that of ^{45}Ar. If we reduce this energy by 400 keV from the value of Q = 5.21 given by Garvey et al. (1969) the much smaller concentration of ^{45}Ar, in equilibrium with ^{44}Ar and neutrons, allowed only a much reduced flow into ^{46}Ar. This had no effect on the yield at T_9 = 2.15 which is controlled by (p,n) flows, but it reduced the ^{46}Ca overabundance at T_9 = 2.00 by a factor of 5.

The overabundances of the nuclei ^{65}Cu, ^{67}Zn, ^{68}Zn, ^{70}Zn, ^{69}Ga, ^{71}Ga, and ^{73}Ge show a strong temperature dependence. If we are to have these nuclei coproduced in this way, our results suggest that explosive carbon burning occurs primarily at temperatures less than T_9 = 2.15, but at least as great at T_9 = 2.0. It is interesting that the final yields of the primary products suggest the same conclusion.

For the final yields of these nuclei we shall need good estimates of the (p,n) cross section of ^{62,64}Fe, ^{68,70}Ni, ^{65,67}Co, and ^{69,71,73,75}Cu, as well as good estimates of the relevant neutron separation energies and nuclear statistical weights.

We might also note that the conditions of explosive carbon burning have little effect on seed nuclei with A > 70. In particular, the p-process nuclei cannot be produced under the same conditions that produce the Mg isotopes.

In conclusion, the effects of explosive carbon burning are far reaching in producing many of the observed nuclei in the solar system; however, the results depend upon many nuclear and hydrodynamical details. A consistent theory of explosive carbon burning that produces the Mg isotopes as well as the rare neutron-rich nuclei in their proper solar system abundance ratios will place severe constraints on allowable supernova explosions.

REFERENCES

Arnett, W.D. 1969, Ap.J., 157, 1369.

Couch, R.G., and Arnett, W.D. 1972, Ap.J., 178, 771.

Garvey, G.T., Gerace, W.J., Jaffe, R.L., Talmi, I., and Kelson, I. 1969, Rev. Mod. Phys., 41, S1.

Howard, W.M., Arnett, W.D., Clayton, D.D., and Woosley, S.E. 1972, Ap.J., 175, 201.

Hyder, A.K., Harris, G.I., Perrizo, J.J., and Kendziorski, F.R. 1968, Phys. Rev., 169, 899.

Truran, J.W. 1972, preprint.

NUCLEOSYNTHESIS DURING EXPLOSIVE OXYGEN AND SILICON BURNING

S. E. Woosley

Rice University, Houston, Texas 77001

This lecture analyzes the nuclear evolution that occurs in zones of supernovae which during their explosive ejection are heated to temperatures in excess of 3.0×10^9 °K at densities greater than 10^5 g cm^{-3}. Such zones are thought to be the site of creation of the abundant elements between silicon and nickel, and the nuclear processes involved are called explosive oxygen and silicon burning because of the key roles played by the species ^{16}O and ^{28}Si during the explosive expansion. We will take the approach that the thermodynamic history of the zone may be specified quite simply by characterizing the explosion in terms of an e - folding time for the reduction of the density, so that $\rho(t) = \rho_i \exp(-t/\tau_{HD})$; $\tau_{HD} = 446\chi/\sqrt{\rho_i}$ sec., where ρ_i is the peak density attained during the explosion in g cm^{-3}; τ_{HD} is the hydrodynamic time scale for expansion; and χ is an arbitrary scaling parameter for either testing the dependence of the results upon τ_{HD} or fitting to a specific model. In addition it is assumed that the expansion is approximately adiabatic so that $\rho \propto T_9^3$ where T_9 is the temperature in billions of degrees. Using this prescription we can map the nature of the nucleosynthesis that occurs during a given explosion in terms of the initial chemical composition and the peak temperature and density.

Although most of the calculations reported here were originally obtained using the technique of network evolution as described by Arnett and Truran (1969) it turns out that the results may best be understood in terms of <u>quasiequilibrium</u>. This concept, explored in detail for silicon burning by Bodansky, Clayton and Fowler (1968), assumes the existence of a large cluster of elements which are in a state of mutual equilibrium with respect to the exchange of free particles and photons. When this is so, the abundances of all species in the cluster are known if any four independent quantities (such as the temperature, density, neutron excess, and abundance of any species in the cluster, ^{28}Si for example) are specified. The values of these four parameters which give good agreement with solar (i.e., Cameron 1968) abundances can yield valuable information about the explosive and pre-explosive

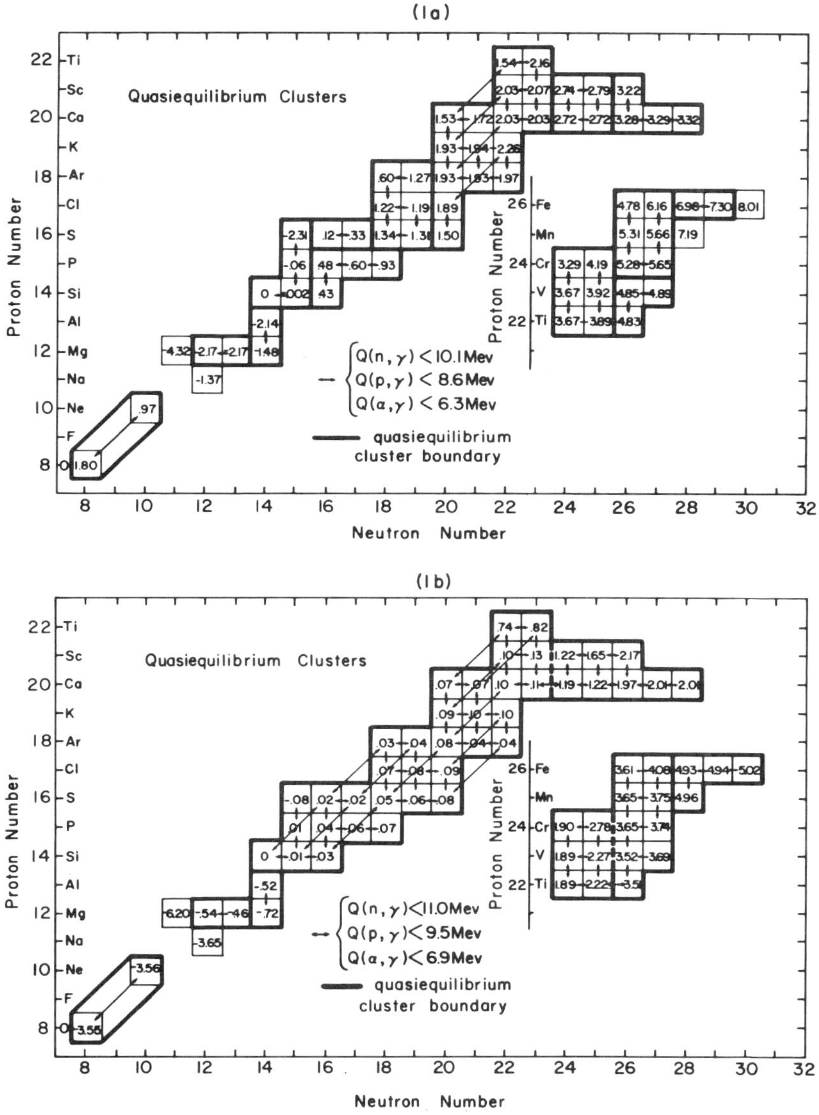

FIG. 1a and 1b — Quasiequilibrium clusters during explosive oxygen burning. The number given for each species is the logarithm of the ratio of the abundance which the species would have if overall ($28 \leq A \leq 62$) quasiequilibrium prevailed to its actual abundance at two different times into the "standard" explosive oxygen burning run. At time $t_a = 2.1 \times 10^{-3}$ sec the temperature is $T_9 = 3.596$ and at $t_b = 7.6 \times 10^{-2}$ sec it is $T_9 = 3.459$. Arrows join species which can be connected by reactions that have Q-values less than the empirical numbers shown. Dark boundries enclose groups of elements which may be interconnected by such links. Species in a given group turn out to have similar values of logarithmic ratios, hence are in approximate equilibrium with one another. (Woosley, et. al. 1973)

history of the material. We will now undertake to show the applicability of quasiequilibrium to explosive oxygen and silicon burning, determine the sets of parameters which give agreement with solar abundances, and relate those parameters to the explosive conditions necessary to produce them.

I. Explosive Oxygen Burning

It was first shown by Truran and Arnett (1970) that expansion of a given initial composition (by mass 54% ^{16}O, 30% ^{24}Mg, 2% ^{26}Mg, and 14% ^{28}Si) from peak conditions of $T_{9i} = 3.6$ and $\rho_i = 5 \times 10^5$ yielded good agreement with certain solar abundances between ^{28}Si and ^{50}Cr. Specifically, the species $\{X_f\}_{EOB}$ = ^{28}Si, ^{32,33,34}S, ^{35,37}Cl, ^{36,38}Ar, ^{39,51}K, ^{40,42}Ca, ^{46}Ti, and ^{50}Cr were all produced within a factor of 2 of their Cameron (1968) values. They also pointed out that explosive oxygen burning has a quasiequilibrium character. Woosley, Arnett, and Clayton (1973) have confirmed these results using a somewhat larger network, although the best-fit value of the density is somewhat shifted (to 2.0×10^5 g cm^{-3} at $T_{9i} = 3.6$). In what follows this is referred to as the "standard" run.

The quasiequilibrium character of the standard run is shown in Figure 1a a short time after the beginning of the explosive expansion. The fact that groups of species which can be interconnected by reactions with low Q-values correspond to quasiequilibrium clusters demonstrates that equilibration occurs slowest for links joining species with large differences in binding energy. In Figure 1b the quasiequilibrium nature of the same run is depicted a short time later. This is the maximum consolidation achieved during this run. Further consolidation would require a longer time scale or higher temperature than is attained in explosive oxygen burning. Examination of a variety of oxygen burning runs shows that the kind of quasiequilibrium shown in Figure 1b will be attained in a time $\tau_{qe} = 0.04 \ (T_9/3.6)^{-33.3}$ sec. fairly independent of initial composition and density. This time scale is much shorter than typical time scales for ^{16}O destruction at the same temperature, hence any reasonable explosive oxygen burning run should achieve approximate silicon-to-calcium quasiequilibrium early in the explosion.

Though this type of quasiequilibrium, once attained, endures until quite late in the evolution ($T_9 \sim 3.0$) the abundances of important species change very little once the ^{16}O abundance freezes out. If we define T_{9f} as that temperature below which the integrated change in the ^{16}O abundance amounts to less than 1%, then the quasiequilibrium distribution calculated at T_{9f} should be a good approximation to the final abundances of species in the silicon-to-calcium region. We may then seek the values of the other 3 parameters (evaluated at T_{9f}) which result in the correct nucleosynthesis. For densities $\rho \geq 10^5$ the distribution is roughly independent of ρ. It turns out that if the neutron and alpha densities are correctly chosen, consistent production of all the species $\{X_f\}_{EOB}$ occurs for $3.1 \leq T_{9f} \leq 3.9$. Outside of this range key species are over or underproduced by a factor of 2. Figure 2 shows the required values of neutron and alpha densities.

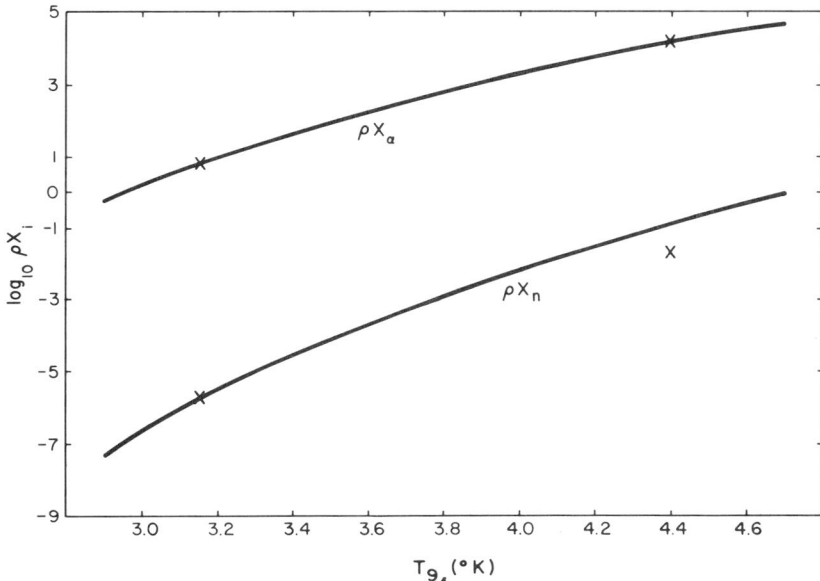

Figure 2. Values of the alpha and neutron mass density required by explosive oxygen burning. The logarithms of the alpha and neutron densities that lead to the production of all species within the set $X_{f\ EOB}$ to within a factor of two of their Cameron (1968) abundances are given as a function of the freeze-out temperature. The "X"'s at $T_9 = 3.17$ are the values obtained by the "standard" explosive oxygen burning run. The "X"'s at $T_9 = 4.4$ are the values used by Bodansky et al. (1968) to fit the solar distribution of elements in the mass range $28 \leq A \leq 62$ (Woosley et al. 1973).

The low value obtained for ρX_n by Bodansky et al. is due to the fact that in silicon burning the quasiequilibrium that is established extends all the way to the iron peak. Larger values of ρX_n would then overproduce species like ^{54}Fe. This problem has been discussed in detail by Michaud and Fowler (1972) and Woosley et al. (1973). It turns out that the restriction that the neutrons remain bound in the $28 \leq A \leq 45$ quasiequilibrium cluster and not be lost to species with high neutron separation energies in the iron peak (like ^{54}Fe) implies an upper bound on the peak temperature during explosive oxygen burning. For peak temperatures in excess of $T_{9i} = 3.9$ so much material penetrates into the iron peak that ^{38}Ar is always underproduced. Decreasing the time scale for expansion by a factor of 10 only raises this limit to $T_{9i} = 4.0$. Increasing ρX_n in these cases only results in the overproduction of ^{54}Fe without noticeably increasing ^{38}Ar; hence we conclude $T_{9i}(EOB) \leq 4.0$.

The values of ρX_n and ρX_α required by Figure 2 can yield information regarding the explosion and presupernova star. Let us define a parameter, $\eta = (N-Z)/(N+Z)$, where N and Z are the total number of neutrons and protons, bound and free, in the initial composition. It turns out that for values of $\eta \approx 0.002$ the neutron density at freeze out will be near what is shown in Figure 2. For values of η a factor of 2 different from this, the species ^{34}S and ^{38}Ar will always be under- or overproduced by a factor of 2. Since the value of η can only be changed by weak interactions which do not occur appreciably in explosive oxygen burning, we conclude that the initial composition should be characterized by $\eta \approx 0.002$.

The value of ρX_α contains much information about the explosion, but it too is in a cumbersome form. It is simpler to consider the results in terms of a parameter

$$\delta_\alpha = \frac{\frac{1}{4} \Sigma' (A(i)-28) n(i)}{\Sigma' n(i)} \quad (1)$$

where n(i) is the number density of the species i having atomic weight A(i) and the sums extend only over species in the set $\{X_f\}_{EOB}$. The parameter δ_α is the total number of alpha particles per ^{28}Si core bound in the set $\{X_f\}_{EOB}$. The value of this parameter at T_{9f} is uniquely, although not simply, related to the value of ρX_α. It turns out that if δ_α falls in the range $0.40 \leq \delta_\alpha \leq 0.70$ at T_{9f} then the set $\{X_f\}_{EOB}$ will be properly produced. For δ_α outside this range key species, in particular ^{40}Ca, will be under- or overproduced by a factor of 2. For $\delta_\alpha = 0.60$ the best fit to Cameron (1968) abundances occurs.

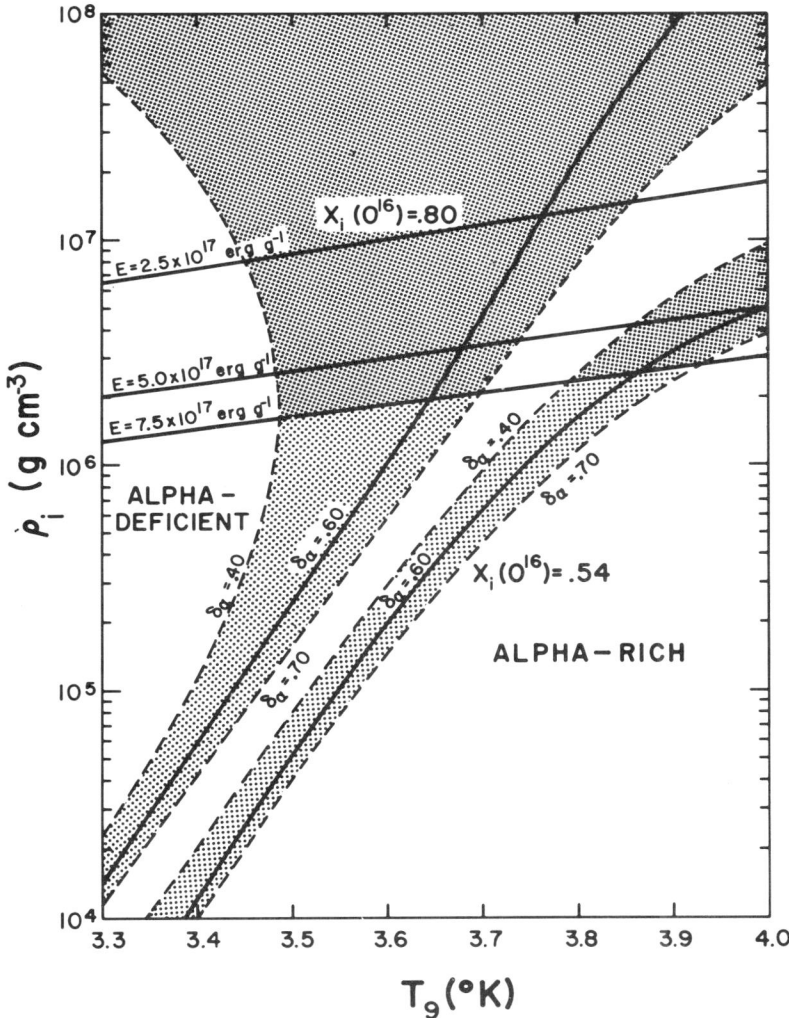

FIG. 3 — Allowed peak conditions for explosive oxygen burning. The values of peak density and temperature which result in the correct nucleosynthesis in explosive oxygen burning zones are shown for two different initial compositions. One contained 54% ^{16}O, 30% ^{24}Mg, 14% ^{28}Si, and 2% ^{26}Mg by mass; the other contained 80% ^{16}O, 12% ^{24}Mg, 6% ^{28}Si, and 2% ^{26}Mg by mass. The best abundance fit occurs along the lines $\delta_\alpha = 0.60$. For regions to the left of the lines $\delta_\alpha = 0.40$ the ratio $^{40}Ca/^{28}Si$ is more than a factor of two less than the solar value; for regions to the right of the lines $\delta_\alpha = 0.70$ the same ratio is overproduced by more than a factor of two. Also shown are lines of various energy densities obtained from an appropriate equation of state. Points to the low density side of the line $E = 7.5 \times 10^{17}$ erg g^{-1} are energetically very difficult to attain. (Woosley, et. al. 1973).

The reason δ_α is a useful parameter is because of the way ^{16}O burns. There are 4 modes of ^{16}O destruction:

(1) $^{16}O + {}^{16}O \to ({}^{28}Si+\alpha); ({}^{31}P+p); ({}^{30}P+d);$ or $({}^{31}S+n)$

(2) $^{16}O(\gamma,\alpha){}^{12}C;\ {}^{12}C+{}^{16}O \to ({}^{27}Si+n); ({}^{27}Al+p);$ or $({}^{24}Mg+\alpha)$

(3) $^{16}O(\gamma,\alpha){}^{12}C({}^{12}C,\alpha){}^{20}Ne(\gamma,\alpha){}^{16}O; \cdots ({}^{12}C,p){}^{23}Na(p,\alpha){}^{20}Ne\cdots;$ or $\cdots ({}^{12}C,n){}^{23}Mg(n,p){}^{23}Na(p,\alpha){}^{20}Ne\cdots$

(4) $^{16}O(p,\alpha){}^{13}N(\gamma,p){}^{12}C$ replaces $^{16}O(\gamma,\alpha){}^{12}C$ in (2) and (3).

Each of these modes have different density and temperature dependences and each produces different numbers of alphas per ^{28}Si core. Mode (1) produces one alpha for each effective ^{28}Si core and dominates at high density. Mode (2), because the formation of ^{12}C from ^{16}O releases an alpha, also creates one alpha per ^{28}Si core. This mode is also favored by high density although never as strong as $^{16}O + {}^{16}O$. Modes (3) and (4) produce only alpha particles and are favored by low density and high temperature. We see that modes (3) and (4) constitute a loop which, centered on ^{16}O, produces alphas which add on to the quasiequilibrium distribution above ^{28}Si. Modes (1) and (2) produce additional ^{28}Si cores as well as alphas. The actual value of δ_α at T_{9f} will be a complicated function of (1) the initial composition; (2) how much ^{16}O is burned; and (3) the branching among the various modes of ^{16}O destruction. These last two quantities will in turn be functions of the peak temperature, density, time scale, and particular nuclear cross sections employed. By integrating the differential equations for the 4 modes of ^{16}O destruction, we can, for a given set of cross sections, determine the range in ρ_i, T_{9i} space that gives $0.40 \leq \delta_\alpha(T_{9f}) \leq 0.70$. The results are shown for two different initial compositions in Figure 3. A fuller discussion of these results appears in Woosley et al. (1973).

II. Explosive Silicon Burning

Explosive silicon burning has been extensively studied by Truran, Cameron, and Gilbert (1966), Bodansky et al. (1968), Michaud and Fowler (1972), Woosley et al. (1973), and others. Explosive silicon burning occurs when the peak temperature during an explosion is so high or time scale so long that the photodisintegration of ^{28}Si becomes the dominant source of free particles for the nuclear evolution. The initial composition may be carbon, oxygen, silicon, or any combination thereof, but by the time silicon burning per se begins, lighter species have been completely consumed producing a distribution of elements having $A \geq 28$. After a short period of readjustment the quasiequilibrium cluster established in explosive oxygen burning (see Figure 1b) merges with the iron-peak cluster so that overall quasiequilibrium in the mass range $28 \leq A \leq 62$ (at least) prevails. By consideration of the nuclear flows which lead to the merging of the two clusters Woosley et al. have shown that full quasiequilibration will occur on a time scale of τ_{qe} ($28 \leq A \leq 62$) = $1.8 \times 10^{-20} \exp(176.3/T_9)$ roughly independent of density and initial composition. This time scale is short enough that any explosion which burns a sizeable fraction of ^{28}Si should achieve total quasiequilibrium quite early during the evolution.

By the time ^{28}Si equilibrates with the iron peak, species below ^{28}Si down to and including ^{24}Mg have also equilibrated with ^{28}Si. If the temperature is higher than $T_{9i} \geq 4.5$ then species down to ^{16}O also equilibrate. Hence the rate of silicon burning becomes determined by the rate of destruction of ^{24}Mg or ^{16}O. Below ^{24}Mg or ^{16}O there is just a quick series of alpha liberating reactions culminating in ^{12}C$(\gamma,\alpha)2\alpha$. The alphas liberated add on to the quasiequilibrium distribution above ^{28}Si and gradually increase its average atomic weight until substantial amounts of iron-peak species are produced.

The final results of silicon burning will depend upon (1) the amount of silicon which is burned (hence the abundance of ^{28}Si at T_{9f}); (2) the value of the neutron excess, η; (3) the freeze-out temperature T_{9f} at which the abundance of ^{28}Si ceases to change; and (4) the density at T_{9f}. These 4 parameters determine a unique quasiequilibrium distribution between $28 \leq A \leq 62$. The dependence of the results upon density and freeze-out temperature is rather weak. Bodansky et al. have shown that for a wide range of ρ and T_{9f} the abundances of the elements $\{X_f\}'_{ESB}$ = ^{28}Si, ^{32,33}S, ^{35,37}Cl, ^{36}Ar, ^{39,41}K, ^{40,44}Ca, ^{45}Sc, ^{47,48,49}Ti, ^{51}V, ^{50,52,53}Cr, ^{55}Mn, and ^{54,56,57}Fe can all be produced within a factor of 2 of their Cameron (1968) values if η and the final mass fraction of ^{28}Si are correctly chosen. Unfortunately, network analysis by Woosley et al. shows that many lower abundance species, specifically ^{33}S, ^{35,37}Cl, ^{39,41}K, ^{44}Ca, ^{45}Sc, and ^{47}Ti are destroyed during the freeze out from $T_{9f} \sim 4.4$ to zero for any reasonable time scale. But the fact remains that a large number of species ($\{X_f\}_{ESB} = \{X_f\}'_{ESB}$ - {species destroyed}) mainly in the iron peak can be produced during explosive silicon burning if $\eta \approx 0.002$ and ρX_α at T_{9f} is near the value shown in Figure 2. This value of ρX_α corresponds to roughly equal amounts of material in the silicon-to-calcium and iron-peak regions. Thus any explosion which attains quasiequilibrium in the mass range $28 \leq A \leq 62$ and freezes out with

$\eta \approx 0.002$ without burning all of ^{28}Si will probably give a good fit to the species $\{X_f\}_{ESB}$. If we may approximate the rate of silicon burning by either ^{24}Mg$(\gamma,\alpha)^{20}$Ne or ^{16}O$(\gamma,\alpha)^{12}$C then the band of peak conditions which will accomplish this goal is shown in Figure 4.

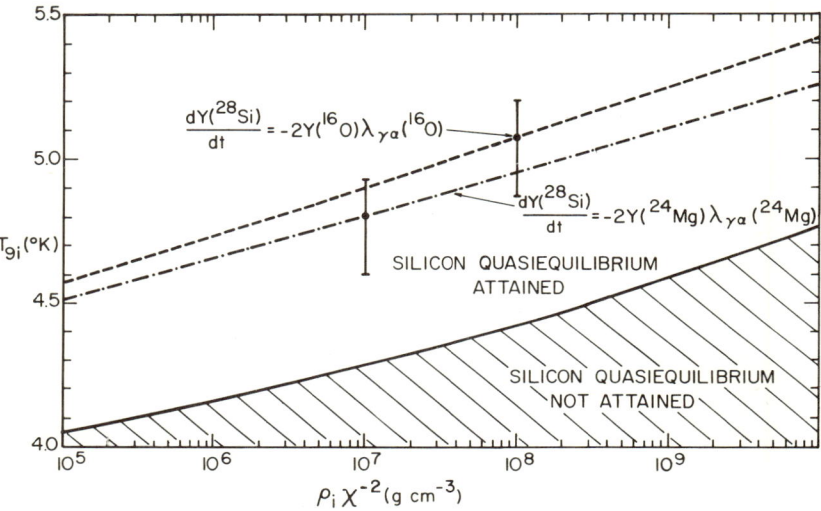

Figure 4. Peak conditions for explosive silicon burning. Those peak thermodynamic conditions which will result in the burning of approximately one-half of the initial ^{28}Si abundance are shown for two approximations. In one case (the dashed line) the rate of silicon burning is approximated by the photodisintegration rate of ^{24}Mg; in the other case (broken line) the photodisintegration of ^{16}O is governing. Error bars indicate an allowed range of the ratio $X_f(^{28}\text{Si})/X_i(^{28}\text{Si})$ of from 0.1 to 0.9. The solid line near the bottom segregates those regions of the ρ_i/X^{-2}, T_{9i} plane where overall (28≤A≤62) silicon quasiequilibrium will be attained during the expansion from those where it will not (Woosley et al. 1973).

III. E-Process

If the peak temperature is so high or time scale so long that ^{28}Si is depleted before the temperature during the explosion has fallen significantly then species below oxygen join on the quasiequilibrium chain all the way down to the alpha particles themselves. This situation, where _all_ strong and electromagnetic interactions are in equilibrium for all species

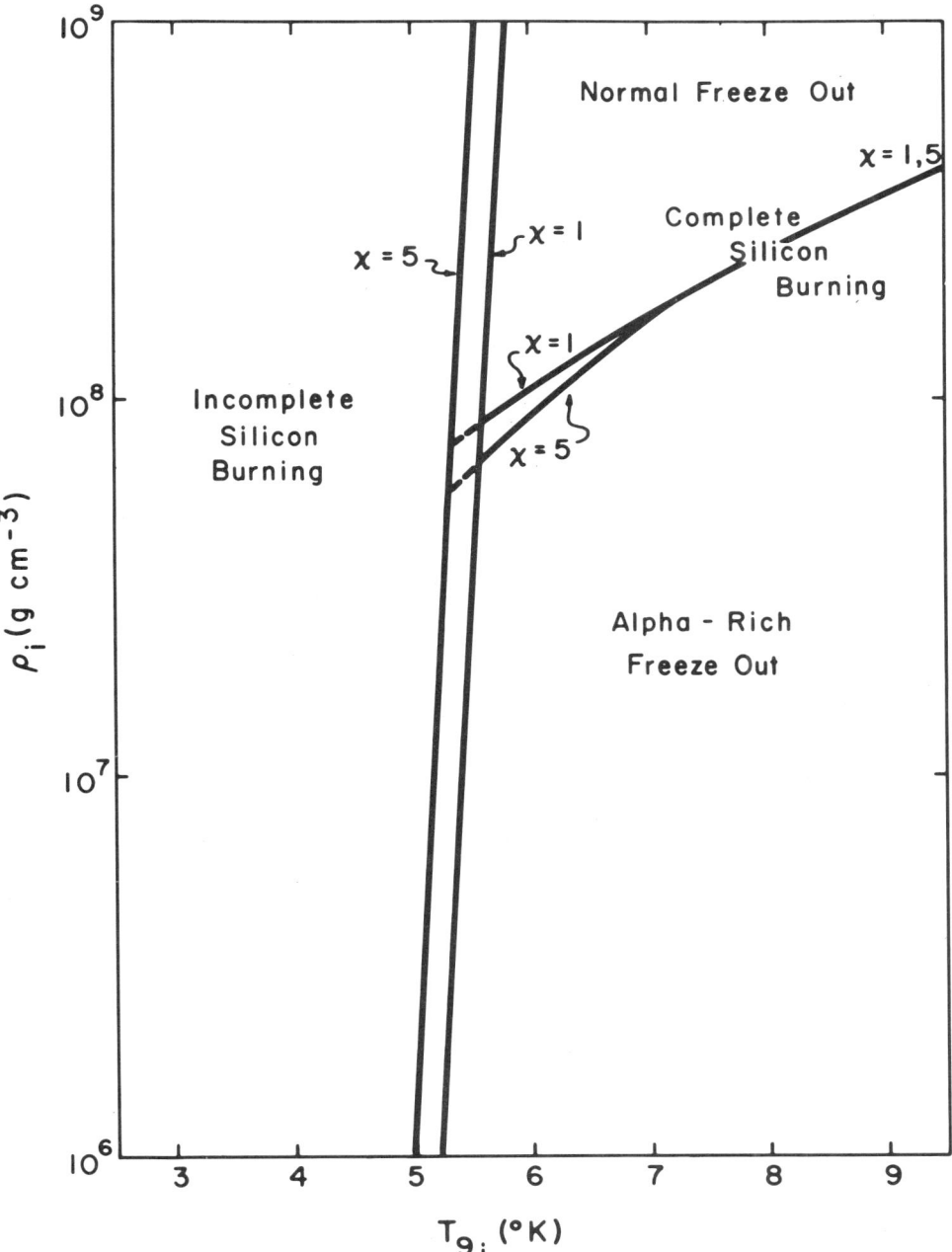

FIG. 5 — Peak thermodynamic conditions for various types of silicon burning. The lines in the vicinity of $T_{9i} = 5.0$ segregate the ρ_i, T_{9i} plane into those regions where silicon exhaustion or normal silicon burning will occur. The delineation varies somewhat with the time scale. Two lines characterized by different time-scale parameters, χ, are shown. The region where silicon exhaustion occurs is further subdivided according to the expected nature of the freeze-out. For densities to the high density side of the lines in the vicinity of $\rho_i = 10^8$ the resultant abundances are similar to those obtained in silicon burning. Lower densities will result in a freeze out characterized by high free-particle abundances. (Woosley, et. al. 1973).

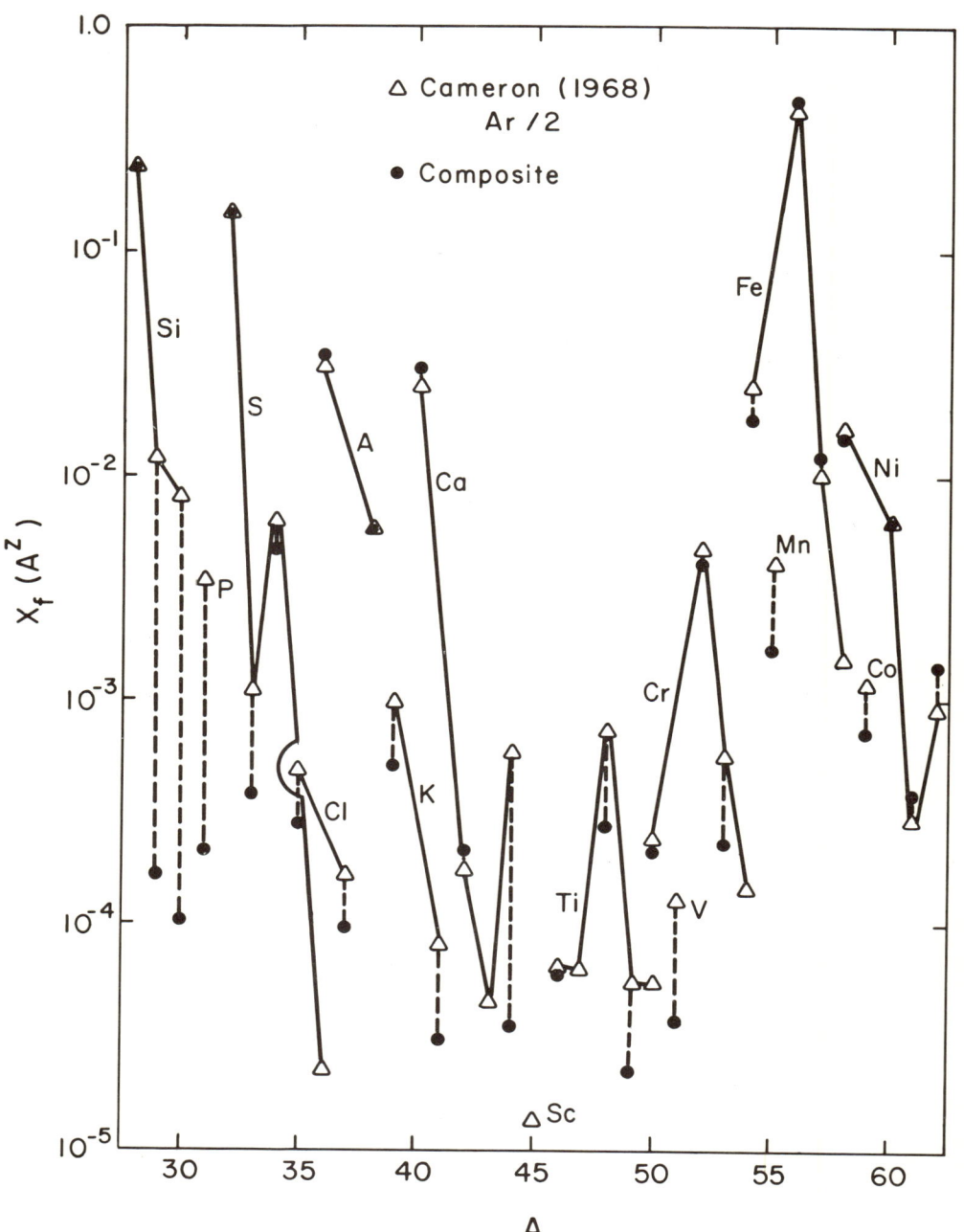

FIG. 6 — Composite of explosive oxygen and silicon burning results. The final abundances by mass fraction of the species $^A Z$ are shown for a composite of equal parts by mass of three different types of nuclear evolution. All three runs employed an initial composition of 54% ^{16}O, 30% ^{24}Mg, 14% ^{28}Si, and 2% ^{26}Mg. One run was the "standard" explosive oxygen burning run ($\rho_i = 2.0 \times 10^5$, $T_{9i} = 3.6$, $\chi = 1.0$); one was silicon burning ($\rho_i = 2.0 \times 10^7$, $T_{9i} = 4.7$, $\chi = 5$); and one was an e-process with an alpha-rich freeze out ($\rho_i = 2.0 \times 10^7$, $T_{9i} = 5.5$, $\chi = 5$). The composite is shown compared to Cameron (1968) abundances. Results have been normalized to ^{28}Si. Abundances less than 10^{-5} by mass fraction are not indicated. A similar result could have been obtained by substituting an e-process with a normal freeze out for the silicon burning portion. (Woosley et. al. 1973).

under consideration, is known as nuclear statistical equilibrium. Instead of 4, we now only require 3 parameters to specify the abundance of every species. These parameters are usually taken to be ρ, T_9, and η.

The peak conditions required for silicon exhaustion are shown in Figure 5. For peak temperatures in excess of $T_9 \sim 5.0$ it appears that nuclear statistical equilibrium will be achieved regardless of initial composition. After that, the nature of the nucleosynthesis depends in a critical fashion on the density. If the density is high (in the region marked "normal freeze out" in Fig 5) then the free-particle abundances are low and the nuclear statistical equilibrium distribution calculated at $3.0 \leq T_{9f} \leq 4.0$ is a fairly good approximation to the final abundances from such an explosion. It turns out that the species $\{X_f\}_{EP}$ = ^{48,49}Ti, ^{50,52,53}Cr, ^{51}V, ^{54,56,57}Fe, ^{55}Mn, and ^{58}Ni can be correctly produced to within a factor of 2 if η is in the usual range, $0.002 \leq \eta \leq 0.004$. This resultant distribution occurs for a wide range of peak temperatures and densities so long as 1) silicon is exhausted and 2) ρ is greater than the value shown in Figure 5. It is interesting to note that the set $\{X_f\}_{EP}$ contains all the species produced in explosive silicon burning that were not already accounted for by explosive oxygen burning, i.e. $\{X_f\}_{EOB}$ + $\{X_f\}_{EP} \geq \{X_f\}_{ESB}$. Hence <u>explosive silicon burning may not even be necessary to the nucleosynthesis of elements in the $28 \leq A \leq 62$ region</u>!

If on the other hand the peak density is so low that we lie in the region marked "alpha-rich" in Figure 5, then there are high free-particle densities present during freeze out. The presence of these free particles greatly alters the nucleosynthesis which results from the <u>e</u>-process. The nuclear details are complicated (see Arnett, Truran, and Woosley, 1971, for discussion) and can only be tracked with a network evolution; however, it appears clear that the species $\{X_f\}_{AP}$ = ^{56,57}Fe, ^{59}Co, and ^{58,60,61,62}Ni will be produced in such zones. Studies are underway to see if heavier species may also be produced. All that is required for the species $\{X_f\}_{AP}$ to all be synthesized in the correct amounts is (1) silicon exhaustion in the zone, (2) η in the range 0.002 to 0.004, and (3) peak densities as shown in Figure 5.

IV. Conclusions

Figure 6 shows a composite of the results of explosive oxygen burning, silicon burning, and an <u>e</u>-process with an alpha-rich freeze out. The overall fit to the Cameron (1968) abundances in the $28 \leq A \leq 62$ range is extremely good. Most of the species not produced here are synthesized in explosive carbon burning (see Howard, Arnett, Clayton, and Woosley 1972) or in types of e-processes not discussed here (see Hainebach 1973). Not only have we found that the species shown can be made by the ascribed processes but it has been pointed out that the ranges of explosive conditions which result in Figure 6 are quite broad. It appears very likely that a typical supernova explosion might reproduce them. To summarize these conditions we have: (1) the best value of η for the nucleosynthesis of elements in the mass range $28 \leq A \leq 62$ is consistently in the range 0.002 to 0.004; (2) of all the material processed to temperatures in excess of $T_{9i} = 3.0$ during explosive ejection roughly one-third by mass has been exposed to peak temperatures in the range $3.0 \leq T_{9i} \leq 4.0$, two-thirds had $T_{9i} \geq 4.0$, and at least 10% had $T_{9i} \geq 5.0$ and $\rho_i \leq 10^8$ g cm^{-3}. More specific values are given in Figures 3, 4, and 5.

Finally, I would like to make a few comments regarding the suggestion by Michaud and Fowler (1972) that <u>all</u> of the abundant species between ^{28}Si and ^{58}Ni might be produced by explosive <u>silicon burning alone</u>. The synthesis ascribed here to explosive <u>oxygen</u> burning they obtain from zones which have burned only a small fraction of ^{28}Si and established quasi-equilibrium only in the cluster $28 \leq A \leq 45$. Obviously if this is true and the free parameters (abundance of ^{28}Si and n in this case) are chosen to give ρX_α and ρX_n at T_{9f} as shown in Figure 2 then a good fit to the set $\{X_f\}_{EOB}$ can result from silicon burning. However, although their paper is quite sound as a model dependent calculation it has serious weaknesses: (1) whereas explosive oxygen burning automatically results in restricted quasiequilibrium in the cluster $28 \leq A \leq 45$ for almost any amount of ^{16}O burned, the explosive conditions for which silicon burning does this are very limited, i.e., the allowed band in the ρ_i, T_{9i} plane would be very small; (2) there should exist oxygen-rich zones overlying the silicon zones and these should explode too, this is an astrophysical question rather than a nuclear one; and (3) the results of Michaud and Fowler are not frozen out. As pointed out previously, low abundance species like chlorine, potassium, scandium, etc. may well be destroyed in the freeze out from temperatures in excess of $T_9 = 4.0$.

Probably the correct answer is that both solutions occur to some extent. Although I believe that explosive oxygen burning is the dominant astrophysical solution to the synthesis of the elements between silicon and calcium, this and other questions can only be answered by the accurate calculation of supernova and pre-supernova models. No more restrictions can be drawn upon the explosions until the explosions themselves are better understood.

This research was partially supported by the National Science Foundation under GP-23459.

REFERENCES

Arnett, W. D., and Truran, J. W. 1969, Ap. J., 157, 339.
Arnett, W. D., Truran, J. W., and Woosley, S. E. 1971, Ap.J., 165, 87.
Bodansky, D., Clayton, D. D., and Fowler, W. A. 1968, Ap. J. Suppl., No. 148, 16, 299.

Cameron, A.G.W. 1968, in The Origin and Distribution of the Elements, ed. L. H. Ahrens (New York: Pergamon Press).
Hainebach, K. 1973, Ph.D. dissertation, Rice University (to be published).
Howard, W. M., Arnett, W. D., Clayton, D. D., and Woosley, S. E. 1972, Ap. J., 175, 201.
Michaud, G., and Fowler, W. A. 1972, Ap. J., 173, 157.
Truran, J. W., and Arnett, W. D. 1970, Ap. J., 160, 181.
Truran, J. W., Cameron, A.G.W., and Gilbert, A. 1966, Can. J. Phys., 44, 563.
Woosley, S. E., Arnett, W. D., and Clayton, D. D. 1973, submitted to Ap. J. Suppl.

THE DYNAMIC r-PROCESS

David N. Schramm

The University of Texas at Austin, Texas

Introduction

The r-process is the rapid neutron capture nucleosynthetic process which is believed to be responsible for the production of the more neutron-rich isotopes of the heavy elements as well as all the trans-Bismuth nuclei (Burbidge, Burbidge, Fowler and Hoyle 1957), including possibly even the predicted superheavy elements (Schramm and Fowler 1971; Schramm and Fiset 1973). Until recently, r-process calculations were done at conditions of constant temperature and neutron density (Seeger, Fowler and Clayton 1965). However, it has always been felt that the only place that the high neutron densities needed in the r-process could be reached was in supernovae, or some similar catastrophic event. It is clear that in exploding objects, the temperature and density will be changing rapidly with time. Therefore more realistic r-process calculations should be done, under dynamic rather than static conditions. It is hoped that by finding the proper dynamic conditions, one will also find the true astrophysical site for the r-process. This method for finding an astrophysical nucleosynthetic site has been found to be quite valuable in the explosive synthesis of the elements from carbon to iron (see for example the review by Arnett 1973). For example, by using the dynamics of particular supernova models, it is possible to examine the possibility of an r-process occurring in that supernova model just as Arnett, Truran and Woosley (1971) were able to do for charged particle nucleosynthesis in the carbon detonation supernova model. Thus, a dynamical r-process calculation enables one to apply the well developed techniques of explosive charged particle nucleosynthesis to the rapid-neutron capture nucleosynthesis of the r-process.

Preliminary attempts at doing dynamical r-process calculations have been made by Cameron, Delano and Truran (1970), Sato, Nakazawa and Ikeuchi (1973), and Schramm (1973a). These preliminary calculations varied the temperature and density with time but at each time step they assumed $n\gamma \leftrightarrows \gamma n$ equilibrium. Schramm (1973a) mentioned that for most of the calculation $n\gamma \leftrightarrows \gamma n$ equilibrium is a reasonable assumption, yielding the same results as the full non-equilibrium calculation. However, an $n\gamma \leftrightarrows \gamma n$ equilibrium calculation is no longer valid during "freeze out", when the bulk of the neutrons have already been captured. It will be shown in this paper that effects during freeze out are of great importance in making a more realistic r-process calculation.

In particular, $n\gamma \leftrightarrows \gamma n$ equilibrium r-process calculations are plagued by the large (orders of magnitude) abundance differences between neighboring r-process nuclei. Because the observed r-process abundances do not show such abundance variations, previous workers were forced to arbitrarily "smooth " the abundances (Seeger, Fowler, Clayton 1965). Seeger et al. tried smoothing by adding together results from calculations using slightly different temperatures, but found that to be insufficient, and had to have additional smoothing. This additional smoothing was done by arbitrarily adding to each abundance 15% of the abundances of each of its neighboring nuclei. Such smoothing was clearly non-physical. It has recently been found (Blake and Schramm 1973a) that some smoothing occurs due to delayed neutron emission during the β-decay of the r-process nuclei back to the valley of β-stability. However, this smoothing is by no means sufficient to smooth out the large abundance variations occurring in a pure $n\gamma \leftrightarrows \gamma n$ equilibrium r-process. It will be shown in this paper that sufficient smoothing occurs naturally during the freeze out of a dynamical r-process.

Another problem that previous r-process calculations had was the need for at least 2 separate sets of conditions in order to fit the three r-process abundance peaks. Although certainly not astrophysically impossible, this need for multiple sites seemed aesthetically unpleasing. It will be shown in this paper that certain dynamical r-process conditions are able to fit all three r-process peaks.

Before describing the calculations, it is worth mentioning an alternative r-process proposed by Amiet and Zeh (1968). Instead of having a huge free neutron density shift the reaction path to the neutron-rich side of the valley of β-stability, these authors proposed shifting the valley of β-stability itself. They would do this by having the material at high density so the chemical potential would change the equilibrium. Such a model has the difficulty of still needing to have the elements ejected into the interstellar medium. As is pointed out by Clayton (1968), such ejection would shift the valley of β-stability back to normal, which would probably erase the effects produced by the high density. This problem is not present in dynamic r-process calculations where one explicitly calculates the ejection.

The Calculation

For the r-process, the rate of change of the abundance, $n(N,Z)$, is given by

$$\frac{d\,n(N,Z)}{dt} = -\lambda_{n\gamma}(N,Z)\,n(N,Z) + \lambda_{\gamma n}(N+1,Z)\,n(N+1,Z)$$
$$+ \lambda_{n\gamma}(N-1,Z)\,n(N-1,Z) - \lambda_{\gamma n}(N,Z)\,n(N,Z)$$
$$+ \lambda_{\beta}(N+1,Z-1)\,n(N+1,Z-1) - \lambda_{\beta}(N,Z)\,n(N,Z)$$

(1)

where $\lambda_{n\gamma}$ is the rate of neutron capture, $\lambda_{\gamma n}$ is the rate of photo-neutron emission and λ_β is the beta decay rate. Fortunately, other reactions can be neglected. However, it is necessary to include ~5000 nuclear species since the r-process path may run ~50 neutrons richer than the valley of β-stability, and it can produce nuclei from the iron peak at Z~26 up to and possibly including the superheavy region at Z~114. Although the network is large, it is relatively simple. For most of a dynamic r-process calculation $\lambda_{n\gamma}$ and $\lambda_{\gamma n}$ are much greater than λ_β, then for each given temperature and neutron density $n\gamma \leftrightarrows \gamma n$ equilibrium is rapidly achieved. Previous r-process calculations (Seeger et al. 1965) assumed constant temperature and neutron density. Thus the assumption of $n\gamma \leftrightarrows \gamma n$ equilibrium was quite valid.

In doing $n\gamma \leftrightarrows \gamma n$ equilibrium r-process calculations, one needed only the neutron binding energies of each nucleus and the temperature and neutron density. Although the r-process path is far from the region of known nuclei, it has been felt that semi-empirical nuclear mass formulae can give reasonable predictions of the neutron binding energies. For the generalized non-equilibrium r-process, actual reaction rates are needed. In these calculations the neutron capture cross sections were calculated using the optical model as described in the Appendix to the paper by Blake and Schramm (1973b). The binding energies were estimated using the Myers and Swiateki (1966) mass formula. The cross-section estimates are admittedly poor. However, for most of the calculation the reactions are in $n\gamma \leftrightarrows \gamma n$ equilibrium where the cross section drops out of the abundance determination and only the neutron binding energy enters (see Seeger et al. 1965).

During the freeze out, the cross sections do enter the calculation. However, changing the cross sections an order of magnitude did not change the results in any significant way. The smoothing during freeze out still occurred. For completeness, it should be mentioned that in addition to the neutron binding energy, the nuclear spins also enter in the calculations. However, the binding energy enters exponentially whereas the spin enters linearly. Therefore, assumptions regarding nuclear spins are relatively unimportant. Calculations were done assuming all spins were alike as well as assuming different spins for odd-odd, even-even and odd-even nuclei. No significant differences resulted.

Since the $\lambda_{n\gamma}$ and $\lambda_{\gamma n}$ are functions of temperature and neutron density (see for example Clayton 1968), it is necessary to know how these quantities change during the dynamic calculation. For the present calculations it was assumed that the density, ρ, changed with time according to the standard prescription

$$\rho = \rho_o \exp(-t/\tau_x) \qquad (2)$$

and temperature, T_9, measured in units of 10^9 °K followed in an adiabatic manner,

$$T_9 = (T_9)_o \exp(-t/3\tau_x) \qquad (3)$$

Given the initial temperature $(T_9)_o$ and initial density ρ_o, as well as the expansion time τ_x, completely specifies T_9 and ρ throughout the calculation. Clearly, one could also use the hydrodynamics from a specific supernova model including non-adiabatic effects on the temperature. In particular, it is apparent that neutrino losses from the β-decays will cause the temperature during freeze out to drop somewhat faster than that estimated by the adiabatic approximation. However, for the present calculation the simple assumptions of Equations 2 and 3 will be assumed. It is felt that a more rapid temperature drop during freeze out will merely enhance the conclusions regarding freeze out smoothing to be presented here.

The expansion time, τ_x, will be expressed in terms of the dynamical expansion time

$$\tau_d = 446/\rho_o^{1/2} \qquad (4)$$

Fowler and Hoyle (1964).

In general,

$$\tau_x \equiv f \tau_d \qquad (5)$$

where f is an arbitrary parameter of the calculation, with f=1 corresponding to the gravitational free expansion of Fowler and Hoyle (1964).

The neutron density, n_n, is related to the total density ρ by the relation

$$n_n(t) = N_a f_n(t) \rho(t) \qquad (6)$$

where $f_n(t)$ is the fraction of free neutrons and N_a is Avagadro's number. As the neutrons are captured, $f_n(t)$ will decrease. The rate of change of $f_n(t)$ can be obtained from mass conservation by noting that

$$f_n(t) + f_s(t) + f_\alpha(t) = 1 \qquad (7)$$

where $f_\alpha(t)$ is the mass fraction of free alpha particles and $f_s(t)$ is the mass fraction of seed nuclei. It will be assumed that all nuclei, other than alpha particles, are seeds. The number of seed nuclei can only be increased by fission or by the synthesis of new seed from alphas and neutrons. Both of these rates can be computed. Any other increase in f_s must be due to the addition of neutrons thus resulting in a decrease in f_n.

In principle, $f_n(t)$ might also be calculated from the $\lambda_{n\gamma}$ and the $\lambda_{\gamma n}$, but in practice such a calculation runs into the numerical difficulties of relying on the small differences of large numbers.

The r-process path terminates either when the neutron captures cease and freeze out occurs or when fission begins competing with the reactions given in Equation 1. Schramm and Fiset (1973) have made a detailed examination of this fission cut off. They showed that the exact position of the fission cut off is sensitive to certain nuclear physics assumptions. In particular, Schramm and Fiset showed that for certain relatively reasonable conditions the r-process might produce the predicted superheavy elements and for other reasonable conditions it might not. Although the actual production of superheavy elements is rather sensitively dependent on the detailed nuclear parameters used, the r-process path is generally not terminated by fission until it at least reaches the magic number N = 184.

Schramm and Fiset's detailed calculations showed that the qualitative behavior of the fission cut off could be well described using the semi-empirical approach of Schramm and Fowler (1971). Since the basic behavior of the dynamic calculations is not dependent on the details of the fission cut off, the Schramm-Fowler treatment of the cut off was used in the calculations presented here.

From the initial conditions ρ_o, $(T_9)_o$, $(f_n)_o$, the initial seed abundances $n_o(N,Z)$ and the expansion rate τ_x, it is possible to determine $n(N,Z)$ at any time by using numerical time differencing techniques. For any value of the neutron density, n_n and T_9 the system of Equation (1) can be solved numerically by a technique developed by Eggleton (1971). This technique uses the fact that λ_β is in general smaller than $\lambda_{n\gamma}$ and $\lambda_{\gamma n}$ thus the matrix at each Z is almost tri-diagonal. A variation on the tri-diagonal solution can then be used as a first order approximation in successive iterations of the complete system. The solution to the abundances $n(N,Z)$ determines the mass fraction of seed nuclei and thus changes f_n (see Equation 7). It was found to be necessary to use a prediction for f_n, then solve the system of equations for $n(A,Z)$, then use this solution to get a better prediction for f_n from equation 7, then reiterate on the above procedure. Such a multiple iteration procedure was carried out at each time step. Convergence was usually within two iterations at each time step except during freeze out, when f_n was dropping rapidly.

Freeze out refers to the time when the abundances, $n(N,Z)$, no longer change with time. This usually occurs when the free neutrons are captured and the temperature is low enough that few new neutrons are produced by photo-ejection. However, it is conceivable that neutron-rich freeze outs exist, where $n(N,Z)$ stops changing because, for example, the expansion rate becomes faster than the reaction rates.

Initial Conditions

As was stated earlier, the astrophysical conditions at the site of the r-process are, at present, unknown. However, it is clear that in order for an r-process to occur, a large ratio of free neutrons to heavy seed nuclei is needed. Normal stellar matter has a nuetron/proton ratio near unity thus making it virtually impossible to free sufficient neutrons relative to seed nuclei in order to produce an r-process. Reactions such as the $^{13}C(\alpha,n)^{16}O$ can produce free neutrons in red giants but the number of these free neutrons relative to seed nuclei is small, thus making an s-process rather than an r-process. Such reaction mechanisms are plagued by the fact that each neutron produced is accompanied by a heavy nucleus (such as ^{16}O) which is capable of capturing neutrons. It is possible to circumvent this problem by having the only charged particles accompanying the neutrons be alphas. Single alpha particles do not capture neutrons. This type of mechanism was used by Seeger et al. (1965). They discussed the fact that at the high temperatures which might be associated with the collapsed core in some supernova models, Fe will photo-disassociate into alphas and neutrons. In particular,

$$^{56}Fe \rightarrow 13\,\alpha + 4\,n$$

As the material expands and cools from these photodisassociation conditions, the alphas recombine to again produce heavy iron-peak nuclei. However, this recombination is hampered by the fact that alphas only synthesize heavy elements via three body interactions. Thus there will be a time during which a few iron-peak seed nuclei have been produced in a sea of alphas and neutrons. The ratio of neutrons to seeds will be large, so that an r-process can take place. Seeger et al. estimated such r-process conditions would occur at

$$\rho \sim 4 \times 10^4 \text{g/cm}^3 \text{ and } T_9 \sim 2.4 \,.$$

They felt these conditions would produce the r-process abundance peaks at $A \sim 80$ and $A \sim 130$. In estimating these r-process conditions, Seeger et al. only took account of the synthesis of seeds via the triple-alpha reaction. Delano and Cameron (1971) mentioned the fact that at high neutron densities the reaction

$$^4He + {}^4He + n \leftrightarrows {}^8Be + n \leftrightarrows {}^9Be + \alpha$$

can occur. However, Delano and Cameron did not emphasize the care one must take in using the α-α-n reaction. In particular, the amount of heavy elements produced via the α-α-n reaction depends on the relative rate of ^{9}Be destruction via the reaction $^9Be + {}^4He \rightarrow {}^{12}C + n$ compared with the photodisintegration of ^{9}Be, $^9Be + \gamma \rightarrow {}^4He + {}^4He + n$. Photodisintegration of ^{9}Be occurs at significantly lower temperatures than photodisintegration of the ^{12}C produced in the triple-alpha process. The reaction rates needed can be calculated from the formulae of Fowler, Caughlan and Zimmerman (1973 and 1968) or Wagoner (1973 and 1969). It is found that the α-α-n reaction is important for the conditions proposed by Seeger et al. Dynamic r-process calculations were carried out with conditions similar to those of Seeger et al. taking into account the α-α-n as well as the 3α reactions. The results will be discussed later.

Another set of conditions where large numbers of free neutrons exist is when the temperature and density get sufficiently high that the reaction $p+\bar{e} \rightarrow n+\nu_e$ dominates over $n+e^+ \rightarrow p+\bar{\nu}_e$. Schramm and Barkat (1972) looked at the evolution of the neutron/proton ratio in the ejection of matter from such conditions. They found that conditions can be found where matter with large n/p ratios is ejected into the interstellar medium. In particular, in the Leblanc and Wilson (1970) model for the formation of a rapidly rotating neutron star with large magnetic fields, some matter is ejected from the poles of the collapsing object. Schramm and Barkat found this matter had an n/p ratio of ~11. It may also be possible that some supernovae eject material from such high densities and temperatures. Delano and Cameron (1971) calculated the subsequent nucleosynthesis in one such supernova model (Arnett 1967, Colgate and White 1966) where matter is ejected with n/p significantly greater than 1. Although this particular supernova model is no longer considered valid (Wilson 1971), the results of the calculation are still quite interesting. It was found that at these high temperature, high density conditions charged particle nucleosynthesis rapidly built up heavy nuclei. If the n/p ratio was ≥ 2, there were still large numbers of free neutrons around when the charge particle reactions froze-out. Thus an r-process could occur. For an n/p ratio of ~4, the dominant nucleus prior to the r-process was ^{78}Ni and the conditions at the start of the r-process were $\rho \sim 6 \times 10^7$ g/cm^3 and $T_9 \sim 4$.

Dynamic calculations using this set of initial conditions will be discussed subsequently. The sequence of nucleosynthesis for these models having ejection from extremely high ρ and T is as follows. When $T_9 \geq 10$, the weak interaction dominates and the n/p ratio varies. As T_9 drops below $T_9 \sim 10$, the n/p ratio becomes fixed and charged particle reactions are important. As the charged particle reactions freeze out, if excess neutrons are still around, an r-process can occur.

Beta-rates

As can be seen from Equation 1, movement to higher Z in the r-process is controlled by β-decay. In static r-process calculations, the resulting abundances are only dependent on the relative β-rates, not on the absolute value of the rates. However, in the dynamical r-process, the absolute rates as well as the relative rates are important. In particular, the value of the β-lifetime compared with the expansion time τ_x is critical. If the β-lifetimes are much greater than the expansion time, the r-process path will not be able to progress to higher Z and will thus be restricted to the production of relatively low atomic masses. If the

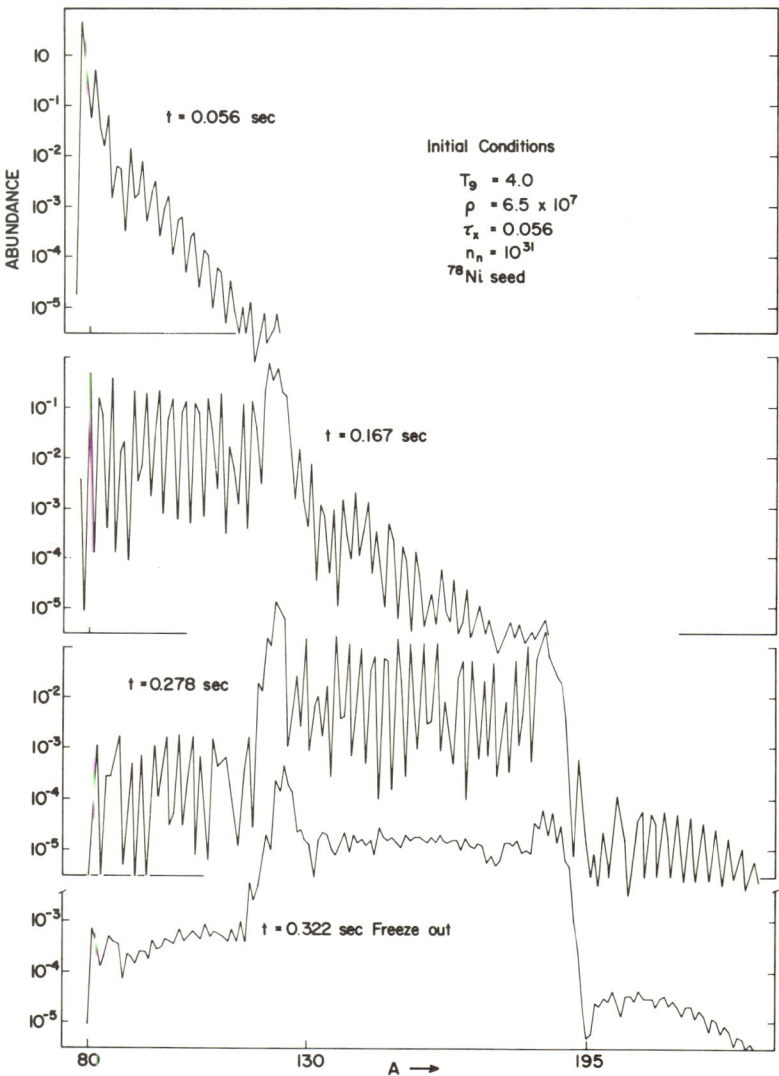

Fig. 1 A time sequence, showing the calculated abundances (arbitrary units) in a dynamic r-process calculation (see text). The initial conditions are based on the work of Delano and Cameron (1971). The Seeger et al. (SFC) beta rates are used and the initial n/p ratio was assumed to be 4. Notice that when freeze-out occurs, there is a smoothing of the abundances.

β-lifetime is much shorter than the expansion time, no such restrictions take place.

Unfortunately, β-rates far from the valley of β-stability are not known. Seeger, et al. (1965) used a β-rate calculation based on the Fermi formula but neglecting the Z dependence. They also assumed the parameter ft=10^5. Senbetu (1973) has revised the Seeger et al. formula, taking into account the Z dependence. This Z dependence yields relatively faster β-rates for higher Z, which means a relatively lower r-process abundance of the higher Z nuclei (Brueckner et al. 1973). The absolute value of the rates developed by Senbetu is somewhat faster than the Seeger et al. rates. They are also still dependent on the assumed ft value. The ft value probably also has some additional Z dependence (Blake and Schramm 1973c). Current average ft values range from the 10^5 used by Seeger et al. to $10^{6.5}$ used by Fiset and Nix (1973). In this work, calculations with both the Seeger et al. rates and the Senbetu rates with ft=10^5 and ft=$10^{6.5}$ have been carried out.

Results

Fig. 1 shows a time sequence of r-process abundances. The initial conditions for this sequence were T_9=4, ρ=6.5x10^7 g/cm^3 which were taken from the calculation of Delano and Cameron (1971). The n/p ratio is assumed to be 4 and all the protons are in seed nuclei, with ^{78}Ni being dominant as was indicated by Delano and Cameron. This means that the fraction of free neutrons, f_n, is initially 0.443. The expansion time τ_x for this particular calculation is just equal to τ_d=0.056 sec. The Seeger et al. (1965) beta rates are used in this calculation. Initially, the temperature is so high that photodisintegration prevents any significant number of neutron captures on the ^{78}Ni seed. As the expansion continues, the neutrons begin to be captured and move the abundance distribution to higher mass. In the second plot of the sequence the abundance peak due to the r-process flow crossing the magic neutron number N=82 can be seen. The initial abundance peak at A∼78 decreases as the r-process flow moves toward higher mass. This is because there is only a fixed number of seed nuclei in this calculation, and they are being moved to higher mass by neutron capture. Also, notice the first 3 plots in the sequence show large differences in the abundances of neighboring nuclei. This is because at those conditions, nγ ⇆ γn equilibrium holds. Therefore the abundance distributions show the same sort of roughness that the unsmoothed Seeger et al. calculations showed. The last plot in the sequence is during the freeze out. It can be seen that these large abundance differences are smoothed out as the free neutron densities drop. This smoothing naturally

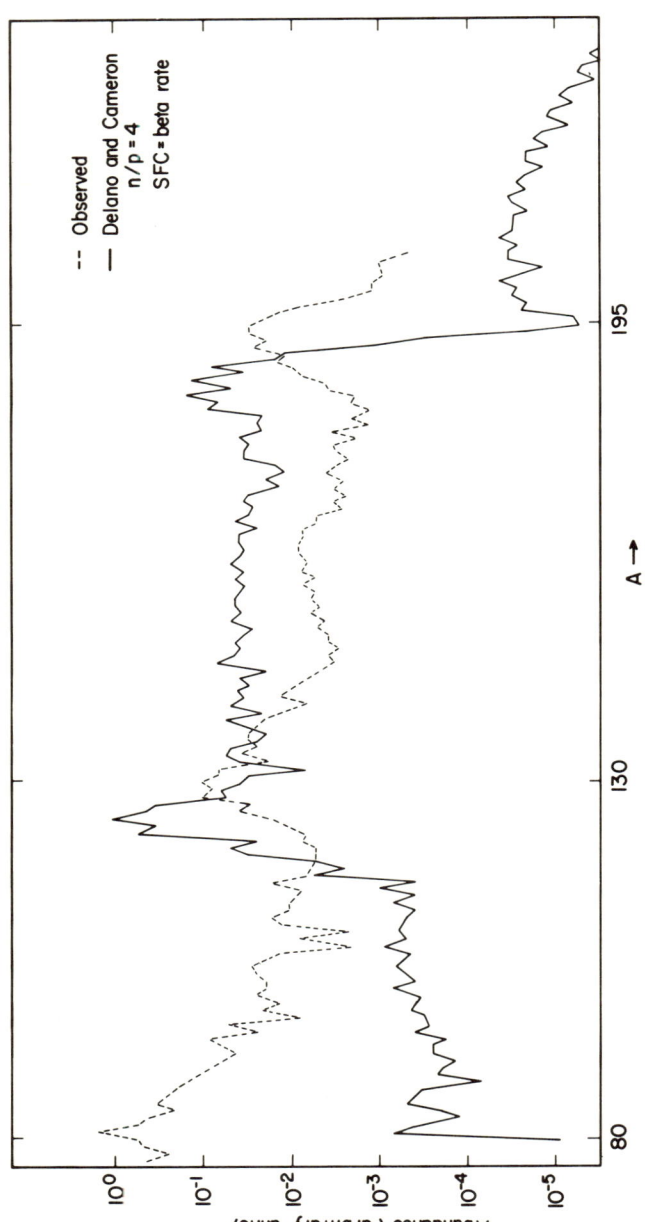

Fig. 2 Shows a comparison between the freeze-out abundances (arbitrary units) of Fig. 1 and the observed (Seeger et al. 1965) r-process abundances. Notice that the calculated peaks miss the observed peaks at A ~130 and A ~195 and the lower mass r-process region is depleted.

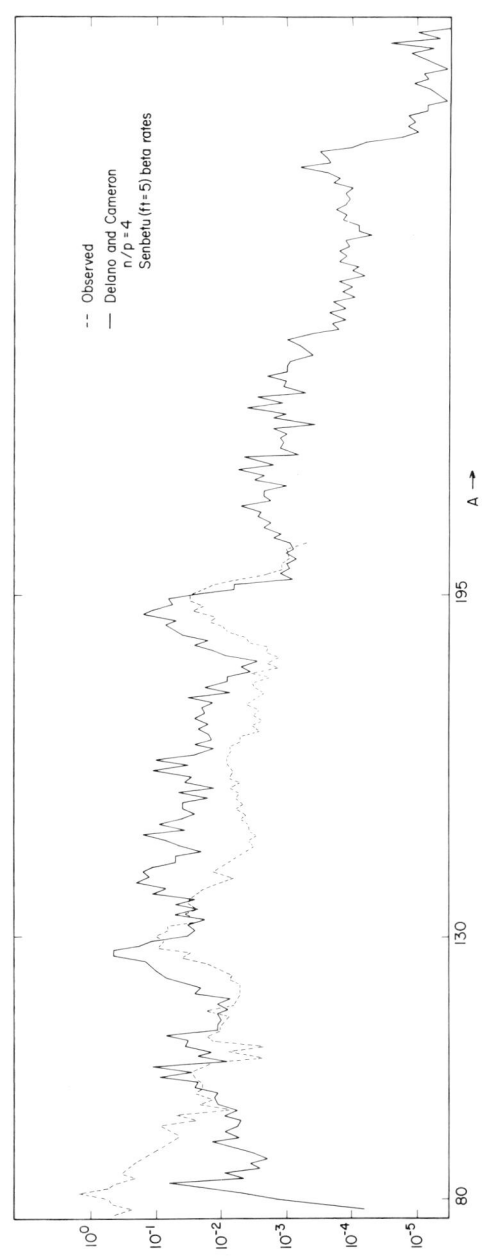

Fig. 3   This is similar to Fig. 2, except the dynamic calculation used the Senbetu (1973) β-rates which decrease the high Z abundances relative to the low Z.

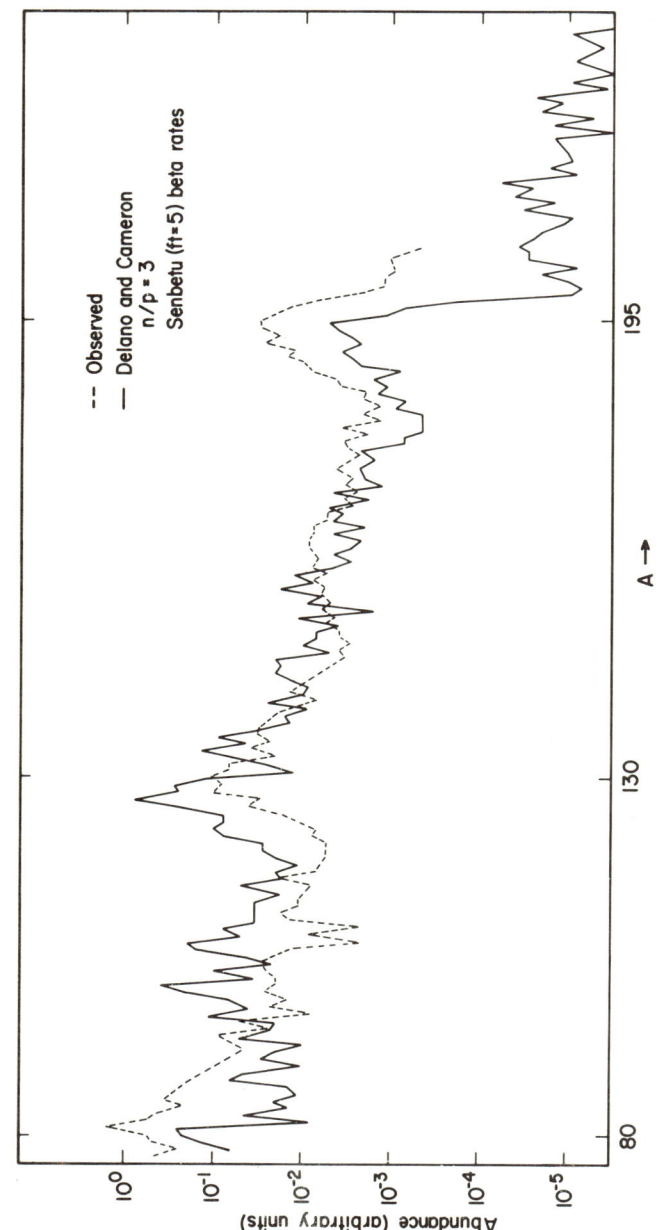

Fig. 4 This is similar to Fig. 3, except the dynamic calculation used an initial n/p ratio of 3.

occurs in the dynamic calculations, whereas such smoothing had to be arbitrarily included in the Seeger <u>et al.</u> calculations. Blake and Schramm (1973a) have also found that additional smoothing occurs due to delayed-neutron emission during the β-decay back to the valley of β-stability. It is clear that there is no longer a need for any arbitrary smoothing.

Fig. 2 shows a comparison between the freeze-out abundances of Fig. 1 and the observed r-process abundances (Seeger <u>et al.</u> 1965). Notice that this calculation fails to fit the observed abundance peaks. The N=82 magic number peak occurs below the observed peak at A∼130 and the N=126 peak occurs below the observed peak at A∼195. This indicates that the assumed initial conditions are not those which produced the r-process nuclei in the solar system. However, it is clear that these initial conditions do produce an r-process just not the correct one.

By varying the initial conditions it is possible to fit the observed peaks at A∼130 and A∼195.

Fig. 3 shows a comparison between the observed abundances and the freeze-out abundances, calculated as above with the Delano and Cameron r-process conditions, but this time using the Senbetu (1973) β-rates with ft=10^5. It can be seen that such a change in the β-rates moves the peaks. This is because the beta times which determine the r-process flow have changed relative to the expansion time, τ_x. Also the beta times have a Z dependence, causing a decrease in the production of the higher mass nuclei relative to the lower mass nuclei. The calculation in Fig. 3 agrees slightly better with observation; however, it is still somewhat off.

Fig. 4 shows the results of a calculation similar to that of Fig. 3 but with an initial n/p=3 rather than 4. With a ^{78}Ni seed this yields a fraction of free neutrons, f_n of 0.31. It can be seen that this lowering of the excess neutrons enables a better agreement with the positions of the abundance peaks at A∼130 and A∼195. In fact, for these conditions the first peak at A∼80 has not been totally destroyed. The non-destruction of the A∼80 peaks is not only due to the lower neutron excess but also to the Z dependence of the Senbetu β-rates. It is also possible to shift the calculated peaks by varying the initial temperature and density. However, orders of magnitude changes in the density are required to do the same job as a small change in the number of excess neutrons. Lower neutron excesses move the r-process peaks toward the right as do higher entropy conditions (higher temperature and/or lower densities).

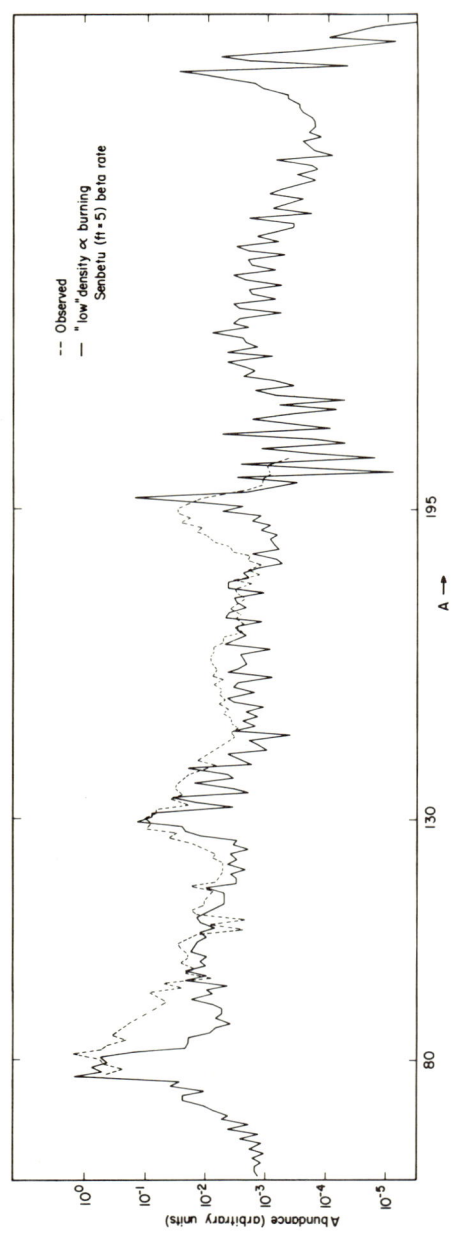

Fig. 5 The calculation shown here used initial conditions similar to those proposed by Seeger et al. (1965), where the bulk of the initial mass is in alpha particles. The alphas make seed nuclei via 3α and α-α-n reactions. The A ~80 peak is not depleted as the higher mass peaks are built up, because new seed nuclei are being produced. The initial conditions used here were $\rho=10^4$ g/cm^3 and $T_9=1.2$ with the Senbetu (1973) Beta-rates (ft=10^5).

Calculations were also carried out using the Leblanc and Wilson (1970) jet temperature and density expansion profile with the n/p ratio of 11 as calculated by Schramm and Barkat (1972). At these densities and temperatures (somewhat similar to the Delano and Cameron conditions) the charged particles rapidly build up to seed nuclei such as ^{78}Ni. The fraction of free neutrons, f_n, for such conditions is ~ 70. These neutron-rich conditions yield an r-process which cycles due to fission (Seeger et al. 1965) and does not produce the low mass peaks at A$\sim$80 and A$\sim$130. By decreasing the n/p ratio the Leblanc and Wilson case yields an r-process similar to that of the Delano and Cameron case.

One interesting point to mention here is that for many n/p ratios (for example n/p=3 in Fig. 3) the neutrons are exhausted before fission cycling takes place. If such conditions were representative of the actual r-process, the actinide region may not be uniformly populated. This could effect the assumptions of the production ratios used in the nucleochronometer calculations (Seeger and Schramm 1970 and Schramm 1973b).

Fig. 5 shows the results using conditions similar to those proposed by Seeger et al. (1965). This r-process started following ^{56}Fe photodisintegration to 13 α's and 4 neutrons. Thus the initial fraction of free neutrons, f_n, is 0.0714 and the initial fraction of free α's, f_α is 0.9286. At "low" densities the alphas recombine slowly via the triple-alpha process and the alpha-alpha-neutron reactions. Once reaching ^{12}C it is assumed that the seed rapidly captures alphas and neutrons to rebuild back to the vicinity of ^{56}Fe. The iron then serves as a seed for the r-process. The bulk of the charged particles are still tied up as alphas, thus the ratio of free neutrons to seed is high. For the calculation of Fig. 5 the initial density, ρ_0, was $10^{4.6}$g/cm^3 and the initial temperature $(T9)_0$ was 1.2. The peak at A$\sim$80 still remains, because new seed nuclei are continuously produced during the expansion. Seeger et al. used a fixed number of seed nuclei thus their A$\sim$80 peak is depleted as the higher masses are populated. From Fig. 5 it can be seen that for the stated conditions all three r-process peaks are populated. The peak at A$\sim$195 is rather sharp and high. However Blake and Schramm (1973a) showed that delayed neutron emission during the β-decay following the freeze out spreads the calculated peaks out. Such spreading toward the lower A is just what is needed to help the calculation in Fig. 5 fit the observed abundances.

In all of the calculations presented there are considerable uncertainties with regard to the nuclear mass formula, the β-decay rates, the fraction of free neutrons and of course the temperature, density and expansion rate. However, the resultant rough agreement with observation is quite suggestive.

Conclusion

Dynamic r-process calculations naturally smooth the calculated r-process abundances during freeze out. It is also possible to find dynamic conditions which produce all three r-process peaks in the same event. A thorough investigation of all possible r-process conditions needs to be carried out before definitive statements can be made with regard to the actual r-process site. However, the present results are certainly a major step forward in r-process nucleosynthetic calculations. It is important to remember that the basic character of the r-process is still dominated by the $n\gamma \leftrightarrows \gamma n$ equilibrium as discussed more than 16 years ago by Burbidge, Burbidge, Fowler and Hoyle (1957). The new dynamic calculations merely remove some of the uncertainties surrounding this basic process.

ACKNOWLEDGEMENTS

I greatfully acknowledge discussions with J. B. Blake and W. D. Arnett. I also thank J. W. Truran for discussing at various times some of his unpublished work with me. This research was supported in part by NSF grant GP-32051 at The University of Texas.

REFERENCES

Amiet, J.P. and Zeh, H.D. 1968, Z. Physik, 217, 485.
Arnett, W.D. 1967, Can.J.Phys., 45, 1621.
Arnett, W.D. 1973, Ann. Rev. Astron. Astrophys., 11,
Arnett, W.D., Truran, J.W. and Woosley, S.E. 1971, Ap.J., 165, 87.
Blake, J.B. and Schramm, D.N. 1973a, submitted to Astrophys. Lett.
Blake, J.B. and Schramm, D.N. 1973b, Ap.J., 179, 569
Blake, J.B. and Schramm, D.N. 1973c, in preparation.
Brueckner, K.A., Cherico, J.H., Jorna, S., Meldner, H.W., Schramm, D.N., and Seeger, P.A. 1973, submitted to Phys. Rev. C.
Burbidge, E.M., Burbidge, G.R., Fowler, N.A., and Hoyle, F. 1957, Rev. Mod. Phys. 29, 547.
Cameron, A.G.W., Delano, M.D., Truran, J.W. 1970, Conf. on Properties of Nuclei far from the Region of β-stability, Leysin, Switzerland.

Clayton, D.D. 1968, Principles of Stellar Evolution and
    Nucleosynthesis (McGraw Hill, New York)
Colgate, S.A., and White, R.H. 1966, Ap.J. 143, 626
Delano, M.D. and Cameron, A.G.W. 1971, Astrophys. and Space
    Sci. 10, 203.
Eggleton, P. 1971, unpublished
Fiset, E.O. and Nix, J.R. 1972, Nuc. Phys. A193, 647.
Fowler, W.A. and Hoyle, F. 1964, Ap.J. Suppl. 9, 201
Fowler, W.A., Caughlan, G.R. and Zimmerman, B.A. 1973,
    to be published
Fowler, W.A., Caughlan, G.R. and Zimmerman, B.A. 1968,
    Ann. Rev. Astron. Astrophys. 5, 525
Leblanc, J.M. and Wilson, J.R. 1970, Ap.J. 161, 541
Myers, W. and Swiatecki, W.J., 1966, Nuc. Phys. 81, 1
Sato, K., Nakazawa, K. and Ikeuchi, S. 1973, Kyoto University preprint
Schramm, D.N. and Fowler, W.A. 1971, Nature, 231, 103
Schramm, D.N. and Fizet, E.O. 1973, Ap.J. 180, 551
Schramm, D.N., 1973a, Bull. Am.Astron. Soc. 5, 27
Schramm, D.N., 1973b, Space Sci. Rev., in press
Schramm, D.N. and Barkat, Z. 1972, Ap.J. 173, 195
Senbetu, L. 1973, Phys. Rev. C. 7, 1254
Seeger, P.A., Fowler, W.A., Clayton, D.D. 1965, Ap.J. Suppl.
    11, 121
Seeger, P.A. and Schramm, D.N. 1970, Ap.J. 160, L157
Wagoner, R.V. 1973, Ap.J. 179, 343
Wagoner, R.V. 1969, Ap.J. Suppl. 18, 247
Wilson, J.R. 1971, 163, 209

p-PROCESS NUCLEOSYNTHESIS*

J. W. Truran

Belfer Graduate School of Science
Yeshiva University
New York, New York

Introduction

The dominant role played by neutron-capture processes in the synthesis of nuclei heavier than iron has been recognized since early abundance compilations (Suess and Urey 1956) revealed major features which correlated with the positions of neutron shell closures. It is also clear that a number of isotopes of heavy nuclei cannot be formed under the typical conditions predicted for s-process or r-process nucleosynthesis. These "bypassed" nuclei lie on the neutron-deficient side of the valley of beta stability, shielded by stable isotopes on their respective isobars from any contributions from neutron-capture mechanisms. The formation of these nuclei is generally attributed to some combination of (p,γ), (p,n), (γ,n) and perhaps positron decay and photobeta reactions proceeding on the products of earlier neutron-capture synthesis (Burbidge, Burbidge, Fowler and Hoyle 1957; Cameron 1957).

A number of investigations of the p-process mechanism have previously been conducted. The role of (p,n), (p,γ) and (γ,n) thermonuclear reactions in the synthesis of the bypassed nuclei was examined first by Frank-Kamenetskii (1961) and Ito (1961) and more recently by Macklin (1970) and Truran and Cameron (1972). The conclusions which may be drawn from these various studies include: (1) the general level of abundance of the bypassed nuclei relative to s-process and r-process products may be understood as a consequence of such thermonucelar burning; (2) thermodynamic equilibrium considerations are probably not relevant; (3) temperatures $T \geq 2 \times 10^9$ °K are demanded for significant thermonuclear burning on a hydrodynamic expansion time scale; and (4) at these temperatures, proton initiated reactions are primarily responsible for the production of the lighter bypassed nuclei while (γ,n) reactions play an increasingly important role as one proceeds to

---

*Supported in part by the National Science Foundation (GP-30289).

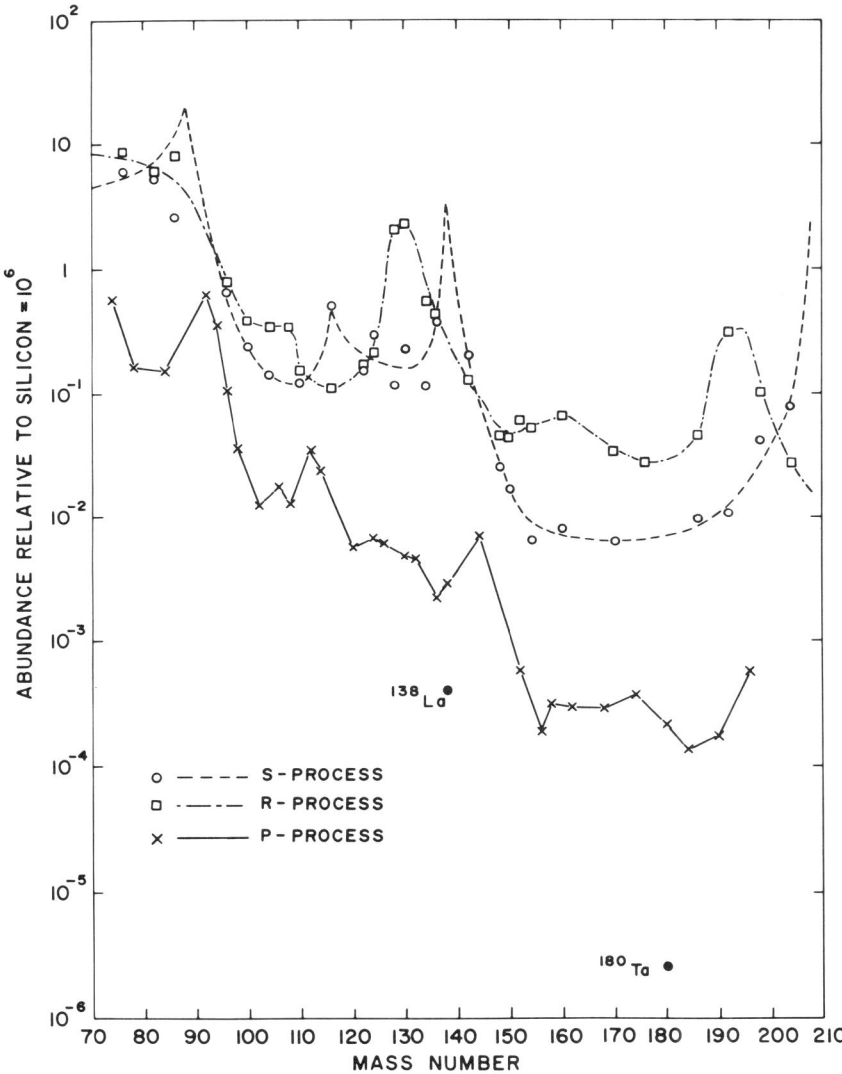

Fig. 1

The abundances of heavy nuclei formed by specific nucleosynthesis processes.

heavier masses. Estimates of the contributions from photobeta reactions (Arnould 1967) and positron capture reactions (Reeves and Steward 1965; Arnould and Brihaye 1969; Agnese, LaCamera and Watoughin 1969) have indicated that, while significant enhancements may occur in a few specific cases, these weak interaction processes cannot account generally for the production of the bypassed nuclei.

In the following sections, a brief summary of current studies of p-process nucleosynthesis by this writer and his collaborators (Drs. J. Audouze and A.G.W. Cameron) is presented.

General Considerations

The observed abundances of the bypassed nuclei provide important clues concerning their formation mechanism. The abundances of the three major classes of heavy nuclei--s-process, r-process and p-process products--are shown as a function of mass number in Fig. 1 (Cameron 1973). Note the striking similarity between the abundance variations of the products of neutron capture and those of the bypassed nuclei. Several interesting features characterize the abundance curve for the bypassed nuclei:

(1) The abundances of the bypassed nuclei are significantly lower than those of heavy elements produced by either of the two neutron-capture processes. (The approximate masses of s-, r- and p-process products in the range $A \geq 70$ are in the ratio $1:0.5:0.02$.)

(2) The abundance ratio of bypassed nuclei to neutron-capture products decreases somewhat with increasing mass number.

(3) The abundance level of the bypassed nuclei remains high right up to the positions of closed neutron shells, beyond which it falls abruptly. The peaks in the bypassed curve are displaced forward in mass number relative to those of neutron capture products by $\Delta A \sim 2-4$. Note particularly the peaks at ^{92}Mo and ^{144}Sm (neutron numbers $N = 50$ and 82) which may be formed by 2-4 proton-capture reactions occurring on closed-shell s-process nuclei (^{90}Zr and ^{88}Sr, and ^{142}Nd and ^{140}Ce, respectively). Furthermore, the p-process peak at ^{112}Sn may be associated with the r-process feature at $A \sim 103-110$.

These abundance features are all consistent with the view that the bypassed nuclei are formed by proton-initiated reactions involving the products of neutron capture as seed nuclei. This might be expected to occur if these neutron-capture products were immersed in a proton-rich medium and heated briefly to high temperatures. The passage of a supernova shock wave through the hydrogen-rich outer envelope of the presupernova star thus provides a promising environment for p-process synthesis. Investigations of two distant mechanisms which can occur in such an environment are in progress: (1) nucleosynthesis by thermonuclear reactions ($T \sim 2 \times 10^9$ °K) proceeding in hydrogen-rich stellar envelopes following the passage of the shock and (2) nucleosynthesis by high energy ($kT \sim 10$ MeV) proton-initiated reactions occurring in the shock front itself prior to the achievement of radiative equilibrium.

Galactic nucleosynthesis requirements impose additional restrictions on the nature of the p-process mechanism. Assuming that stars in the mass range 4 - 8 $M_\odot$ have given rise to supernovae throughout galactic history (with a current rate of occurrence of approximately one per 50 years), Truran and Cameron (1971) determined that the conversion of all <u>primordial</u> neutron-capture products to bypassed nuclei is required in $\sim$ 10-15 percent of the mass of these stars. Two factors can significantly influence this estimate. First, it is unlikely that proton-induced reactions could be 100 percent efficient in converting neutron-capture products to bypassed nuclei without introducing distortions into the distribution in mass number. A reasonable upper limit of $\sim$ 25-50 percent efficiency would imply that $\sim$20-60 percent of the mass of the star must be processed in this manner - in other words, a substantial fraction of the hydrogen envelope. In the opposite sense, the fractional mass in which p-process synthesis must have occurred would be reduced if the abundances of seed s-process nuclei have previously been enhanced in the envelopes of these stars (S-stars typically shown enhancements of s-process elements of factors $\sim$ 10-100). As will become clear from our subsequent discussion, these considerations may prove critical in deterimining which of the p-process mechanisms currently under investigation is the more promising.

Post-Shock Synthesis

Recent calculations have demonstrated that nuclei from carbon to nickel can be formed in solar system proportions by means of charged-particle reactions proceeding under explosive burning conditions (Truran, Arnett and Cameron 1967; Arnett 1969; Truran and Arnett 1970; Arnett, Truran and Woosley 1971). In these studies one assumes an initial post helium-burning composition in the stellar core and then explores the range of temperature and density consistent with the expulsion of the matter in a supernova event for those conditions which provide the best fits to solar system abundances. An equivalent approach has been adopted in recent studies of p-process nucleosynthesis by Dr. J. Audouze and this writer. Lacking detailed numerical predictions of the post-shock conditions, our studies proceed as follows: the peak temperature and density are

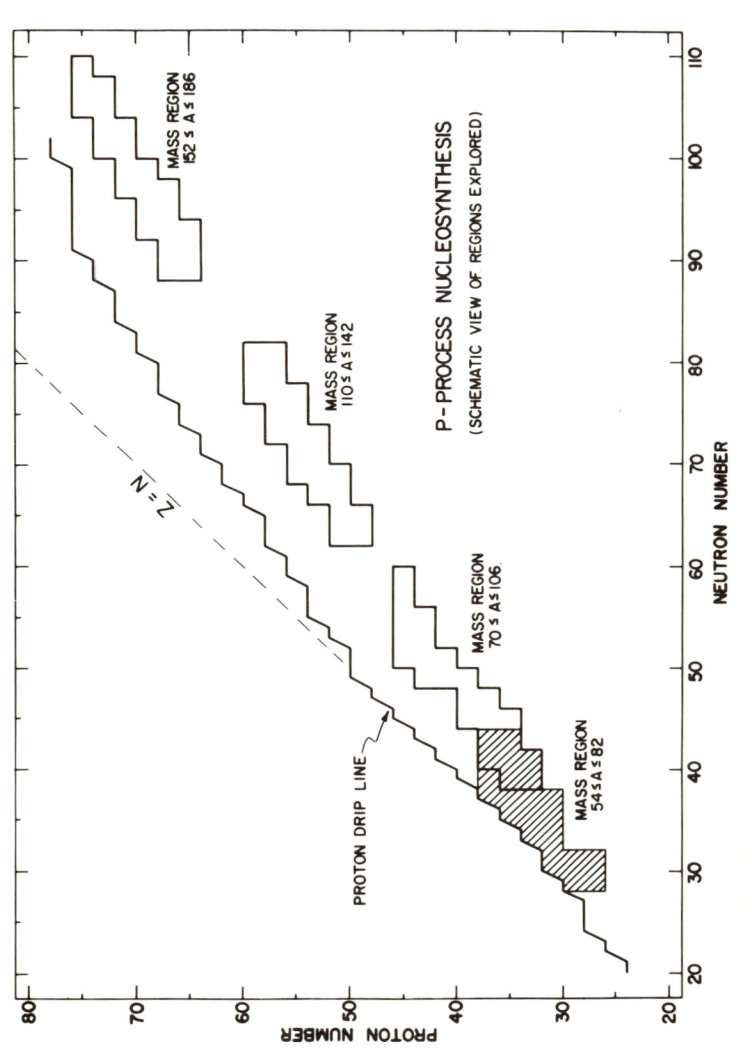

Fig. 2

Survey of networks defined for p-process studies. Three additional overlapping networks (not shown) cover the mass regions $90 \leq A \leq 120$, $130 \leq A \leq 160$ and $170 \leq A \leq 200$.

specified, the expansion time scale is equated to the hydrodynamic time scale $\tau_0 \sim 446/\rho^{1/2}$ seconds, solar system (Cameron 1968) proportions of hydrogen and heavy (r-process and s-process) nuclei are assumed and the thermonuclear transformations are followed in time through expansion and cooling of the ejected matter.

As the bypassed nuclei range broadly in mass ($74 \leq A \leq 196$) and the thermonuclear buildup must be followed from the valley of beta stability to the vicinity of the proton drip line, an enormous number of nuclei must be included in a detailed study of p-process synthesis. In our exploratory studies, seven overlapping reaction networks have been defined, each involving approximately 125 nuclei (Fig. 2). This procedure should not severely distort the resulting abundances so long as we consider only solar-like compositions for which the number of protons greatly exceeds the number of seed nuclei (some distortion of the abundance pattern may result if a large free neutron density is achieved). One such network, constructed to trace the nuclear transformations in the mass range $110 \leq A \leq 142$, is illustrated in Fig. 3.

The temperatures and densities appropriate for our studies are dictated by nuclear and hydrodynamic considerations. Calculations of the rates of charged-particle reactions and neutron-photodisintegration reactions involving heavy nuclei have recently been performed (Truran 1972). In Table 1, the lifetimes of the immediate progenitors of many of the p-process nuclei against proton capture and neutron photodisintegration are presented at several temperatures (T_9 is the temperature in units 10^9 °K) for an assumed proton mass density $\rho_p = 10^4$ g cm^{-3}. The ratios of the abundances of the p-process nuclei to their proton-capture progenitors [n(Z,A)/n(Z-2, A-2)] are also shown. The limiting conditions consistent with p-process synthesis in supernova envelopes are apparent from an examination of these lifetimes. Temperatures $T_9 \gtrsim 1.5$-2 are clearly required; at lower temperatures neither (p,γ) nor (γ,n) reactions proceed rapidly enough to form significant abundances of proton-rich nuclei on a hydrodynamic expansion time scale ($\tau \sim 10$ seconds for a density $\rho \sim 10^4$ g cm^{-3}). Furthermore, post-shock temperatures in this range can only be achieved in matter at higher densities. While our calculations indicate that the bypassed nuclei can be formed at temperatures $T_9 \sim 2$ and densities $\rho \gtrsim 10^2$ g cm^{-3}, our choice of a hydrogen mass density $\rho_p = 10^4$ g cm^{-3} corresponds roughly to the minimum density for which temperatures greater than a billion degrees can be achieved in radiative equilibrium following the passage of the shock.

Calculations of the formation of bypassed nuclei under these post-shock supernova envelope conditions have been performed for temperatures $T_9 = 1.5$, 2.0 and 2.5 at a density

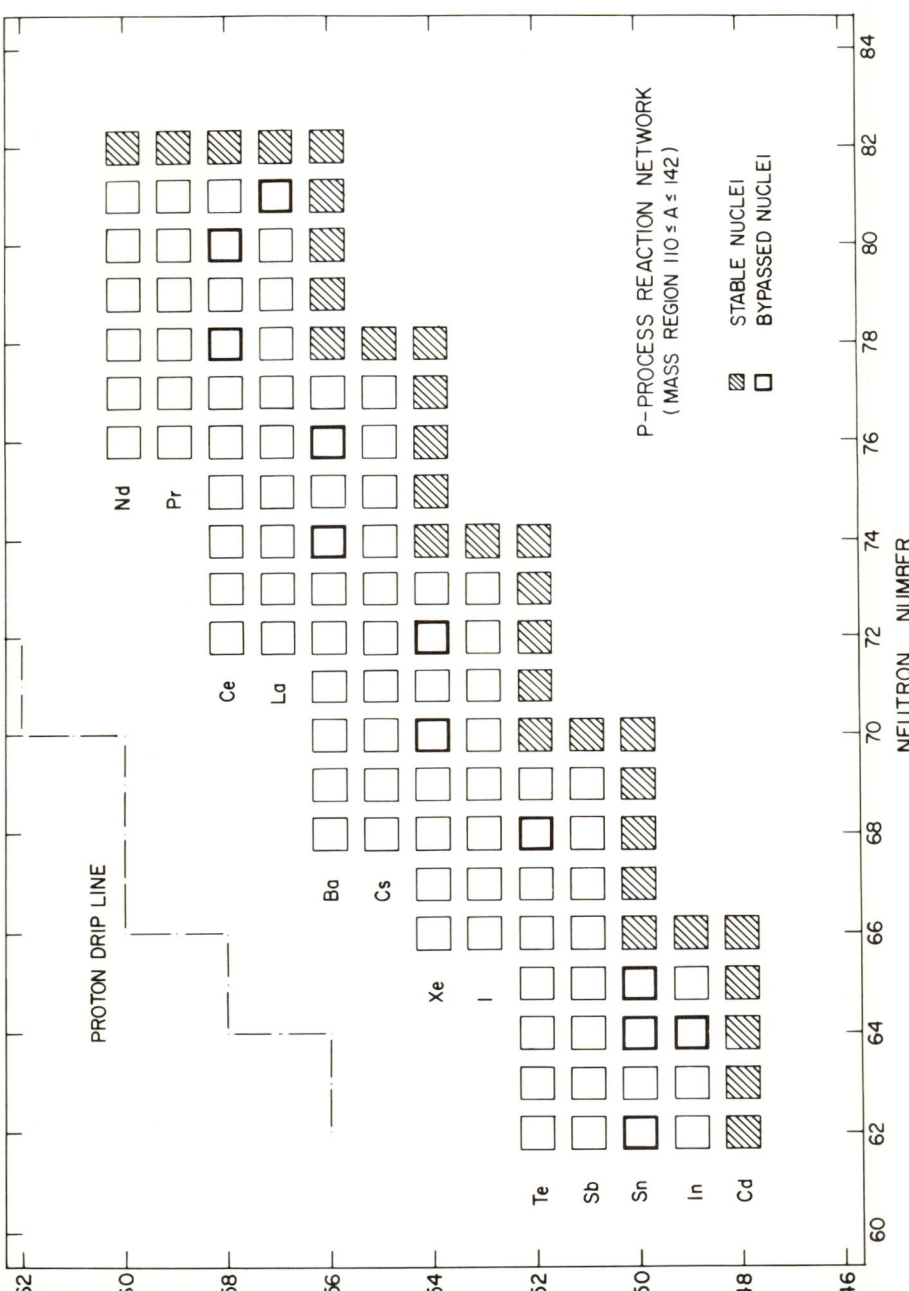

Fig. 3 Reaction network constructed to study the mass region $110 \leq A \leq 142$.

$\rho = 10^4$ g cm^{-3}. At $T_9 = 2.5$, serious distortions of the entire heavy element abundance distribution result from the thermonuclear burning and the p-process abundances are not reproduced. Our preliminary results for the $T_9 = 2.0$ case are presented in Table 2 in the form of ratios of our calculated abundances to the solar system abundances ("calculated enhancement"). For purposes of comparison, characteristic "local" average enhancement, determined as the ratio (for solar system matter) of the average abundance over the five neighboring isobars to that of the bypassed nucleus, are also tabulated.

A particularly encouraging feature of our results is that the predicted enhancements for the bypassed nuclei do not vary greatly (on the average) with mass number. Specifically, our calculations have been successful in predicting comparable relative enhancements of 26 of the 35 candidates for p-process synthesis throughout the mass table. In contrast, the local average enhancements (which do not take coulomb barrier and shell effects into account) differ systematically for the lighter and heavier mass regions by a factor ~5 and tend to be noticeably lower at the closed-shell positions. The failure of our calculations to produce comparable enhancements of odd-A and odd-odd nuclear species (^{113}In, ^{115}Sn, ^{138}La and ^{180}Ta) is understandable since their production is particularly sensitive to temperature, binding energies and relative reaction rates. In subsequent calculations, we hope to use these nuclei as a guide to the appropriate temperature, density and time scale conditions. Some expansion of our reaction networks toward the proton drip line may be required to obtain proper measures of the production of ^{94}Mo, ^{124}Xe and ^{126}Xe (found not to be enhanced in these preliminary surveys). However, uncertainties in our estimates both of the binding energies of the proton-rich progenitors (Truran and Cameron 1973) and of thermonuclear reaction rates can be responsible for the low calculated abundances of these nuclei.

Shock Front Synthesis

The high hydrogen densities ($\gtrsim 10^2$-10^4 g cm^{-3}) required for p-process synthesis following the passage of supernova shocks are considerably greater than the densities typical of the hydrogen envelopes in evolved stars. Only a relatively small mass fraction at the base of the hydrogen envelope might be expected to be exposed to these conditions--far less than the ~0.5 - 1 $M_\odot$ requirement imposed by galactic nucleosynthesis considerations. Significant production of p-process nuclei can occur here only if the abundances of seed nuclei have been systematically enhanced.

An alternative mechanism for the production of proton-rich heavy nuclei involves nuclear transformations occurring in the shock fronts themselves, rather than in the medium behind the shock. Colgate (1973) has recently pointed out that a high ion temperature supernova shock precursor exists for a time scale of the order of a few hundred Thompson collision periods prior to the achievement of radiative equilibrium. This results from the fact that the first event occurring in the shock consists of collisions between the rapidly moving ions in the shock and ions in the medium ahead of the shock; subsequently the system relaxes toward equilibrium on a longer time scale by means of bremsstrahlung and Compton scattering processes. Temperatures $kT_i \sim$ 1-10 MeV characterize the medium in this shock precursor, in contrast to the considerably lower temperatures predicted following shock passage.

A distinguishing and important feature of this mechanism is that <u>the number of nuclear collisions per heavy nucleus is essentially independent of the density</u>. The proton collision lifetime per heavy nucleus is given by:

$$\tau_p = \frac{1}{n_p \sigma_p v} \approx \frac{10^{-9}}{\rho_p \sigma_p \text{ (barns)}} \text{ seconds}$$

The high ion temperature precursor (Colgate 1973) exists for a time of the order of $100\, T_i^{1/2}$ (T_i in MeV) Thompson scattering periods ($1/n_e \sigma_T c$) or

$$\tau_T = \frac{100\, T_i^{1/2}}{n_e \sigma_T c} \approx \frac{10^{-8} T_i^{1/2}}{\rho_p} \text{ seconds}$$

The number of nuclear collisions is thus

$$\frac{\tau_T}{\tau_p} \approx 10\, T_i^{1/2} \text{(MeV)}\, \sigma_p \text{ (barns)}$$

For 10 MeV proton energies, the total proton cross sections on heavy nuclei range ~ 0.1 - 1 barn, hence the number of collisions expected on the lifetime of the shock precursor ranges ~ 0.3-3 per heavy nucleus. The partial buildup of a Maxwellian tail will favor increased collisions since the nuclear cross sections increase with energy. This is particularly important with regard to the relative roles of (p,n) and (p,γ) reactions. Above their thresholds ($\sim$ 4-10 MeV), the (p,n) reactions typical will dominate the (p,γ) reactions. (p,α) reactions contribute significantly only at higher energies in the lighter mass region.

Studies of this p-process mechanism by Dr. A.G.W. Cameron and this writer are now in progress. The insensitivity of the number of collisions to the density suggests that p-process synthesis may proceed in a large fraction of the hydrogen envelopes ejected in supernova events. The low number of collisions per seed nucleus suggests, however, that this mechanism may be relatively inefficient in converting neutron-capture products to bypassed nuclei. In order to satisfy galactic nucleosynthesis requirements, the prior enrichment of s-process nuclei in the envelopes of supernovae will very likely be demanded.

Further calculations pertaining to both of the p-process nucleosynthesis mechanisms described in this paper are clearly required. If either approach ultimately provides a successful theory of the p-process, we will have gained valuable information concerning the pre-explosion envelope structure of supernovae.

ACKNOWLEDGMENTS

I am indebted to my collaborators Drs. J. Audouze and A.G.W. Cameron for discussions of many aspects of this work. I also wish to thank the National Science Foundation for their support and Dr. R. Jastrow for the hospitality of the Goddard Institute for Space Studies, where most of these calculations were performed.

REFERENCES

Agnese, A., LaCamera, M. and Wataghin, A. 1969, Mon. Nct. R. Astro. Soc., 146, 57.
Arnett, W. D. 1969, Ap. J., 157, 1369.
Arnett, W. D., Truran, J. W. and Woosley, S. E. 1971, Ap. J., 165, 87.
Arnould, M. 1967, Nucl. Phys., A100, 657.
Arnould, M. and Brihaye, C. 1969, Astr. and Ap., 1, 193.
Burbidge, E. M., Burbidge, G. R., Fowler, W. A. and Hoyle, F. 1957, Rev. Mod. Phys., 29, 547.
Cameron, A.G.W. 1957, Chalk River Report CRL-41.
Cameron, A.G.W. 1968, in Origin and Distribution of the Elements, ed. L. H. Ahrens (London: Pergamon Press).
Cameron, A.G.W. 1973, in preparation.
Colgate, S. A. 1973, "The Formation of Deuterium and the Light Elements by Spallation in Supernova Shocks", preprint.
Frank-Kamenetskii, D. A. 1961, Soviet Astr.-AJ, 5, 66.
Ito, K. 1961, Progr. Theoret. Phys., 26, 990.
Macklin, R. L. 1970, Ap. J., 162, 353.
Reeves, H. and Stewart, P. 1965, Ap. J., 141, 1432.
Suess, H. E. and Urey, H. C. 1956, Rev. Mod. Phys., 28, 53.
Truran, J. W. 1972, Ap. Sp. Sci., 18, 306.
Truran, J. W. and Arnett, W. D. 1970, Ap. J., 160, 181.
Truran, J. W., Arnett, W. D. and Cameron, A.G.W. 1967, Can. J. Phys., 45, 2315.
Truran, J. W. and Cameron, A.G.W. 1971, Ap. Sp. Sci., 14, 179.
Truran, J. W. and Cameron, A.G.W. 1972, Ap. J., 171, 89.
Truran, J. W. and Cameron, A.G.W. 1973, in preparation.

CHAPTER III. Aspects of Pre-Carbon Burning

Stellar Evolution for $M \leq 8\ M_\odot$

The ultimate theoretical test of explosive nucleosynthesis ideas is whether or not the required conditions are a direct, natural and necessary outcome of stellar evolution. Current research has concentrated on the mass range $4 \geq M/M_\odot \geq 3$. Following Paczynski it is thought that lower mass stars eject their envelopes and form white dwarfs. The s-process is thought to occur during the pre-carbon-burning red giant phase. S-process calculations for realistic stellar models provide both a mode for extensive nucleosynthesis and a probe of what goes on in the late stages of the evolution of these stars.

Does carbon ignition occur first at the center or away from the center of the star? Do the thermal pulses eventually cause the envelope to be lost, prior to carbon ignition? Do the observed properties of Cepheid variables give us any clues as to what goes on in the core? These are some of the questions we would like to answer.

In this chapter Iben reviews pre-supernovae stellar evolution examining whether stars actually evolve to the conditions assumed in supernova calculations. Ulrich and Smith et al. discuss nucleosynthesis during the helium flash phases of stellar evolutions. It is during these flashes that it is thought the s-process takes place. Some ^{7}Li may also be produced during these flashes to supplement the Li produced by the galactic cosmic rays. Ulrich uses a "plume mixing" theory of convection to explore the s-process. Smith et al. look at the possible effects large scale convection can cause on certain nuclear abundances.

References to Chapter III.

Stellar Evolution

Iben, I. 1967, Ann. Rev. Astron. Astrophys., 5, 571.
Paczynski, B. B. and Ziolkowski, J. 1968, Acta Astron., 18, 255.
Rose, W. K. and Smith, R. L. 1970, Ap. J., 159, 903.

s-process

Clayton, D. D. 1968, Principals of Stellar Evolution and Nucleosynthesis, and references therein. (New York: McGraw - Hill).

PRESUPERNOVA EVOLUTION

Icko Iben, Jr.

University of Illinois

I. Criteria for Acceptable Supernova Models

Which stars become supernovae? And how? Theory alone cannot yet answer either question to everyone's satisfaction. Perhaps the best approach is to ask first, what do observations have to say about the properties of supernovae that we might legitimately require our models to mimic? and to then ask, with which aspects of our theoretical models is tampering required in order to meet the constraints imposed by the observations?

What do the observations say?

(1) <u>Frequency</u>. The average frequency with which supernovae occur in our Galaxy is not known. We must rely on statistics accumulated by monitoring external galaxies and argue that an average frequency per galaxy obtained in this way might also be relevant to our own Galaxy. Unfortunately, statistics for external galaxies are rather meager and quoted rates differ significantly from one another, varying from 1 every 25 years to 1 every 700 years per galaxy. It is fair to say that a rate anywhere between $1/30$ years/10^{11} stars and $1/300$ years/10^{11} stars is not inconsistent with the known facts.

(2) <u>Remnants</u>. Some supernovae may leave condensed remnants (pulsars) which are now commonly accepted to be rapidly rotating neutron stars. However, the circumstantial evidence is compelling for only one supernova (the Crab). For all others, the evidence is statistical.

Some supernovae may not leave condensed remnants. For example, there are no signals from pulsars in the direction of two historical supernovae (Kepler and Tycho). Since the liklihood is small that the beam from any pulsar will intersect the earth, this absence of signals is no evidence against the existence of remnant pulsars. Nevertheless, the absence of a detectable signal leaves open the possibility that some supernovae do not leave condensed remnants.

Finally, there is a strong possibility that the formation of a neutron star remnant may not always be associated with the supernova phenomenon. Evidence for this possibility is provided by several compact x-ray sources (e.g., Her X-1) with close companions. The evidence suggests that the compact sources may be neutron stars, and it is difficult to see how a close binary system could survive an explosive event of supernova magnitude.

(3) <u>Energy</u>. The total energy emitted as a consequence of the supernova outburst is an unknown quantity. A lower limit is given by optical observations which suggest that, during the outburst, an energy of $10^{49} - 10^{50}$ erg is emitted in the visual. It is contended by some that the state of excitation of matter surrounding old, extended supernova remnants may be explicable only if an energy of $10^{51} - 10^{52}$ erg is emitted in the ultraviolet (see recent discussions of the Gum nebula and the Vela pulsar). This energy need not, of course, be associated with the supernova event itself, but could be emitted by the remnant pulsar after the event. Finally the velocities associated with matter streaming outward in a direction away from some supernovae imply energies on the order of 10^{51} erg per solar mass of ejected matter or of accelerated interstellar-medium matter, whichever the case may be.

(4) <u>Elements</u>. It is probable that a large fraction of the elements in our Galaxy that are heavier than carbon and oxygen were formed in supernova events. First, Fe-peak and r-process elements cannot be made in sufficient abundance in conventional Big Bang models of the Universe, and may therefore be made in stars in our own Galaxy. Secondly, according to nucleosynthesists, the formation of Fe-peak and r-process elements can be achieved in the proper relative abundances under conditions of dynamic expansion on a time scale that is comparable to the free-fall time (or free-expansion time) appropriate to the condensed core of stars during their last, <u>calculated</u>, quasi-static phases. It should be emphasized, however, that, among elements heavier than helium, the trio carbon, oxygen, and nitrogen together constitute over half of the total. Any model of supernovae that creates and ejects r-process and Fe-peak nuclei in sufficient total abundance and in the proper relative abundances must also preserve a large fraction of the CNO elements that have been built up during the quasi-static phases that precede the supernova explosion.

(5) <u>Types</u>. There are at least two distinct supernova "types". The danger of a classification scheme that admits only two types is well known; most members exhibit characteristics that fall somewhere in between those of the two hypothetical archetypes. Nevertheless, for the sake of analysis, there is some virtue in idealizing the actual situation. Type I super-

novae are presumed to be characterized by: (a) a larger energy (in the optical) per outburst than are type II supernovae; (b) a near uniqueness in the shape of the light curve; (c) high frequency of occurrence in giant elliptical galaxies. In contrast, type II supernovae are presumed to: (a) have lower energy per outburst; (b) exhibit a large variety of spectrum shapes and light curves; (c) do not occur in giant ellipticals, but only in spiral and irregular galaxies.

II. Possible Candidates for Supernovae

(1) <u>Low-mass stars</u>. Theory suggests that single stars of mass less than about $1.4M_\odot$ do not encounter a dynamical instability in their cores at any stage of their careers but evolve quasi-statically during all nuclear burning stages and beyond. After they have developed a substantial carbon-oxygen core and are burning helium in one shell and hydrogen in another, a thermal instability appears in the nuclear burning region of such stars. The ensuing oscillations repeat, with ever growing amplitude, at intervals of a few thousand years. It has been conjectured that ejection of a planetary nebula may be the outcome of an oscillation of sufficiently large amplitude. Once most of the original helium in the star has been completely burned, the carbon-oxygen remnant cools off to become a white dwarf, whether or not an outer shell has been ejected.

There is strong observational support for the theoretical conclusion that single, low-mass stars do not become supernovae. In our Galaxy there are roughly 10^8 old stars that are red giants of less than one solar mass. Since the lifetime of a red giant of solar mass or less is between 10^7 years and 10^8 years, the frequency with which low-mass stars die (exhaust all nuclear fuels burnable under conditions reached in such stars) is between 1/year and 10/year, a value perhaps 30 to 3000 times higher than the inferred supernova rate. Thus, such stars cannot become supernovae.

It must not be forgotten, however, that perhaps half of the stars in the Galaxy are in binary systems. It will be argued in a subsequent section that the frequency of binary systems consisting of a carbon-oxygen white dwarf of mass less than $1.4M_\odot$ and a red giant of mass less than or comparable to $0.8M_\odot$, both at an appropriate separation for mass transfer from the lighter star to the heavier star, is comparable to the inferred frequency of type I supernovae in our Galaxy.

(2) <u>Stars which lose considerable mass</u>. There is observational evidence that some single stars of initial mass greater than $1.4M_\odot$ do not become supernovae. In the Hayades there are

twelve white dwarfs coexisting with active nuclear burning stars whose mass is about $2M_\odot$. We infer that at least some stars of mass equal to or greater than about $2M_\odot$ shed enough mass to become white dwarfs of near solar mass.

Theoretical considerations suggest that all stars with an initial mass in the range 1.4-$3M_\odot$ may lose about half of their initial mass during the phase of shell hydrogen and shell helium burning, leaving a remnant carbon-oxygen white dwarf of mass less than $1.4M_\odot$. The argument relies on the fact that the luminosity-to-mass ratio is such stars reaches a critical value beyond which radiation pressure would drive off matter in the envelope if this matter were in radiative equilibrium (energy flow by radiation). The weak point in the argument is the fact that most of the matter in the envelope is in convective equilibrium (energy flow predominantly by convection) so that radiation is not strongly coupled to the matter. However, the observations suggest that $2M_\odot$ stars lose about $1M_\odot$ at some point after the main sequence phase. Therefore, some mechanism does the trick, and the mechanism just described appears promising.

(3) <u>Intermediate-mass stars</u>. Stars with an initial mass in the range 3-$8M_\odot$ may have a chance to develop a carbon-oxygen core of mass $1.4M_\odot$ in which electrons are relativistically degenerate. What happens next is anybody's guess. Early visionaries suggested that these stars blew to smithereens. Dave Arnett's classic model ignited carbon-burning reactions at the center. A subsequent detonation wave swept through the carbon-oxygen core, leaving iron-peak elements in its wake. At least $1.4M_\odot$ of junk metal flew into space with a kinetic energy of nearly 10^{51} erg. The early Arnett model has a number of features to recommend it but also possibly a number of deficiencies. First of all, the frequency with which stars in the 3-$8M_\odot$ range are now being born in our Galaxy is about the same as the inferred supernova rate. This statement is based on a Paranego estimate of 2×10^4 Cepheids in the Galaxy and a theoretical lifetime of about 10^6 years for Cepheids in the $3.8M_\odot$ group. A major deficiency of the model, however, is the destruction of all of the carbon and oxygen in the core and the overproduction of iron-peak elements. If every star in the 3-$8M_\odot$ group were to behave in this way, the abundance of iron-peak nuclei in the Galaxy would be at least ten times greater than indicated by the observations. If carbon and oxygen were not supplied by another source, we would also be faced with a severe underabundance of these two elements. The failure to form a remnant and therefore the inability to produce altogether more than about 10^{51} erg might also be considered deficiencies. However, since it has not been established that all supernovae produce remnants or that more than 10^{50} erg characterizes the supernova event, these are weak deficiencies.

In any case, developments that have occurred subsequent to the appearance of the Arnett model suggest that a remnant may very well be formed in stars that develop a carbon-oxygen core of about $1.4M_\odot$. Paczinski suggested a version of the Cameron-Tsuruta Urca shell concept that might have the effect of damping out the detonation and forcing the star to burn carbon quietly. Arnett and Couch have shown that one member of the Urca pair, $Na^{23} - Ne^{23}$, is produced in sufficient abundance during the early carbon-burning phase to produce the predicted effect. Urca neutrinos emitted near the edge of the convective region formed around the carbon ignition region carry off sufficient energy to quench the detonation. The edge of the shell occurs as a density such that the electron Fermi energy equals the threshold energy for beta captures on Na^{23}. Arnett and Couch suggest that the increase in electron molecular weight, μ_e, during the phase of quiet carbon burning in the convective core is sufficient to cause the entire carbon-oxygen-magnesium-neon-sodium-etc. core to exceed the critical mass ($M_{crit} \sim 5.76 M_\odot/\mu_e^2$) for stability against dynamic collapse.

The picture one then adopts is that suggested by Bodenheimer, Gunn and Ostriker. As the core adopts a pulsar configuration, $10^{52} - 10^{53}$ erg of gravitational potential energy goes into rotational energy. The spinning beam from the pulsar pumps the cavity between it and the envelope that it leaves behind full of electromagnetic radiation which presses against this envelope until it is expelled. The difficulty with this picture is that no new products of nucleosynthesis are expelled in the envelope to satisfy future generations of spectroscopists. However, we do achieve a neutron-star remnant in conjunction with a supernova event and there is ample energy stored initially in the remnant to ionize a large mass of surrounding matter in the interstellar medium.

(4) <u>Massive stars</u>. To get the metals, Dave Arnett tells us that we must look to stars more massive than $8M_\odot$. This is where the vision of Stirling Colgate, Willy Fowler and Fred Hoyle may pay off. Since Dave Arnett will talk at length about this, I will limit my contribution to the statement that the statistics are favorable. Using the Salpeter birth-rate function, $\frac{d}{dm}(dN/dt) \propto m^{-1.35}$, one finds that the formation rate, $\frac{dN}{dt}$, for stars in the range $8-60 M_\odot$ is greater than $\frac{dN}{dt}$ for stars in the range of $3-8 M_\odot$. If, then, stars more massive than $8M_\odot$ do produce and eject only a tenth or so of a solar mass of newly synthesized elements heavier than helium (in the proper relative proportions) there are enough such stars to account for the heavy element component of our Galaxy. Whether the matter is expelled as a consequence of a supernova type event or in a more quiet manner is a question yet to be answered.

(5) <u>Binaries and type I supernovae</u>. Single stars are probably progenitors of type II supernovae. It has been established theoretically that stars which develop a core of mass $\sim 1.4 M_\odot$ in which electrons are relativistically degenerate suffer some sort of dynamic fate. The mass in the envelope outside of the core of magic mass $1.4 M_\odot$ will vary from star to star, depending on the vagaries of the environment at birth. Since the detailed characteristics of the supernova event might be expected to vary as a function of the initial mass, angular momentum, and magnetic field configuration of the progenitor star, we guess that progenitors of type II supernovae, which do exhibit a wide range of characteristics, may be single stars of initial mass greater than about $3 M_\odot$.

The relative uniqueness of the light curves characterizing supernova outbursts of type I suggests that the mass and angular momentum characteristics of the immediate precursors of supernovae of type I span a much smaller range than in the case of type II supernovae. The fact that type II supernovae do not occur in giant ellipticals suggests that, if single stars of mass greater than $3 M_\odot$ are currently being born in such galaxies, they are being born less frequently (by an order of magnitude) than in spiral galaxies. Further, if most stars in elliptical galaxies are on the order of 10^{10} years old, then most single stars therein that are still burning nuclear fuel are less massive than about $0.8 M_\odot$ and can therefore not be progenitors of supernovae. We are led to conclude that the precursors of type I supernovae must occur in binary systems of total mass greater than $1.4 M_\odot$.

A promising configuration that immediately preceds a possibly explosive final event consists of a carbon-oxygen white dwarf of mass near $1.4 M_\odot$ separated by several hundred A.U. from a red giant of mass near $0.8 M_\odot$ that is either on its first or second ascent up the giant branch. We imagine that the precursor of the white dwarf component had an initial mass in the range $1.8 M_\odot - 3 M_\odot$ and that the precursor of the red giant component had a mass less than $0.8 M_\odot$. Whelan and Iben have shown that it is possible to choose initial orbital parameters in such a way that the proposed final configurations follow from a fairly wide variety of initial conditions and that, in our Galaxy, the frequency of occurrence of binaries with appropriate initial conditions is comparable to the frequency of occurrence of type I systems inferred for our Galaxy.

The type I supernova event is postulated to occur after the red giant companion swells up to fill its Roche surface and transfers hydrogen-rich matter to its white dwarf companion. Hydrogen and then helium burning is ignited at the base of the newly deposited matter. The carbon-oxygen core of the newly

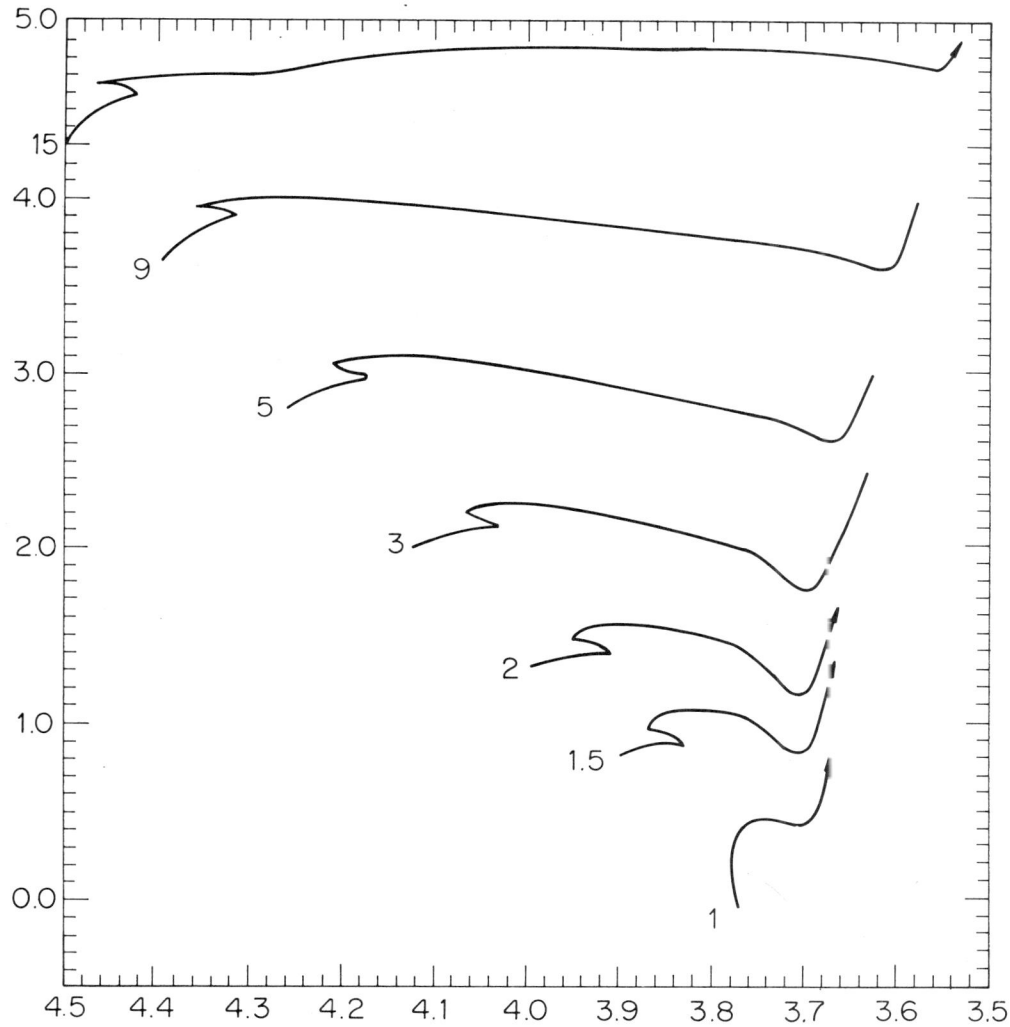

Fig. 1. Paths in the H-R diagram for stars in the mass range $1 M_\odot$ - $15 M_\odot$. Only the $15 M_\odot$ star has burned helium at the center (between main sequence and giant branch); central carbon burning has begun at the center of the $15 M_\odot$ star on the giant branch. Central helium burning has begun in stars of mass $(3,5,9) M_\odot$ that are on the giant branch. Red giants of mass $(2,1.5,1) M_\odot$ are burning hydrogen in a shell. Coordinates: $\log(L/L_\odot)$ versus $\log(T_e)$.

Iben 122

activated companion grows until it reaches the magic mass of
$1.4 M_\odot$, whereupon a dynamic event takes place.

Uniqueness of the ensuing event in the binary model follows
naturally from the lack of an envelope of variable mass. (In
more massive single stars that suffer a similar dynamic event
in the carbon-oxygen core, the mass in the envelope varies from
star to star.) Uniqueness is also favored by tidal coupling
between the binary components. Over the long history preceding
the supernova event, this coupling will tend to reduce the rotation period of each component to a value comparable to the orbital
period of the system. Since this latter period is on the order of
years, one might expect the influence of angular momentum differences to be minimal in affecting the course of the explosive
event in the type I system. Finally, the larger energy associated with the type I outburst follows from the lack of a
massive envelope that can absorb the energy liberated from the
core.

III. Characteristics of Single Stars that Develop Large Carbon-Oxygen Cores.

(1) <u>Early stages in the H-R diagram</u>. Shown in Fig. 1 are
paths in the H-R diagram for stars of mass between $1 M_\odot$ and $15 M_\odot$
during the main sequence and first giant branch stages. All of
the stars on the giant branch, except the $15 M_\odot$ star, have developed
a nearly pure helium core, a thick hydrogen-burning shell, and a
deep convective envelope. The paths of these stars during the
following core helium-burning phase are not shown. For the $15 M_\odot$
star, the core helium-burning phase occurs near the main sequence.
At the last point shown on the giant branch, the $15 M_\odot$ star has
developed a carbon-oxygen core of mass $1.6 M_\odot$ surmounted by a
thick helium-burning shell centered at $2.35 M_\odot$, an **inert** helium
layer of mass $2 M_\odot$, and another layer of mass about $10 M_\odot$ in which
abundances are similar to those in the initial main sequence
star. Convection extends over the outer $9 M_\odot$ of the star. Carbon
burning is about to begin at the stellar center.

Shown in Fig. 2 are paths in the H-R diagram during the
core helium-burning phase for several stars of intermediate mass.
The extent to which an evolutionary track extends to the blue
during this phase depends on the reduced alpha width, θ_α^2, attributed to the reaction $C^{12}(\alpha,\gamma)O^{16}$. The amount of carbon and
oxygen left at the center of the star when helium vanishes there
is also a function of this reduced width. With $\theta_\alpha^2 \sim 0.25$, the
final result in a $7 M_\odot$ star is $X_{12}/X_{16} \sim 1/97$; with $\theta_\alpha^2 \sim 0.08$,
$X_{12}/X_{16} \sim 27.71$. Here, X_{12} and X_{16} are, respectively, the
abundance by mass of C^{12} and of O^{16}.

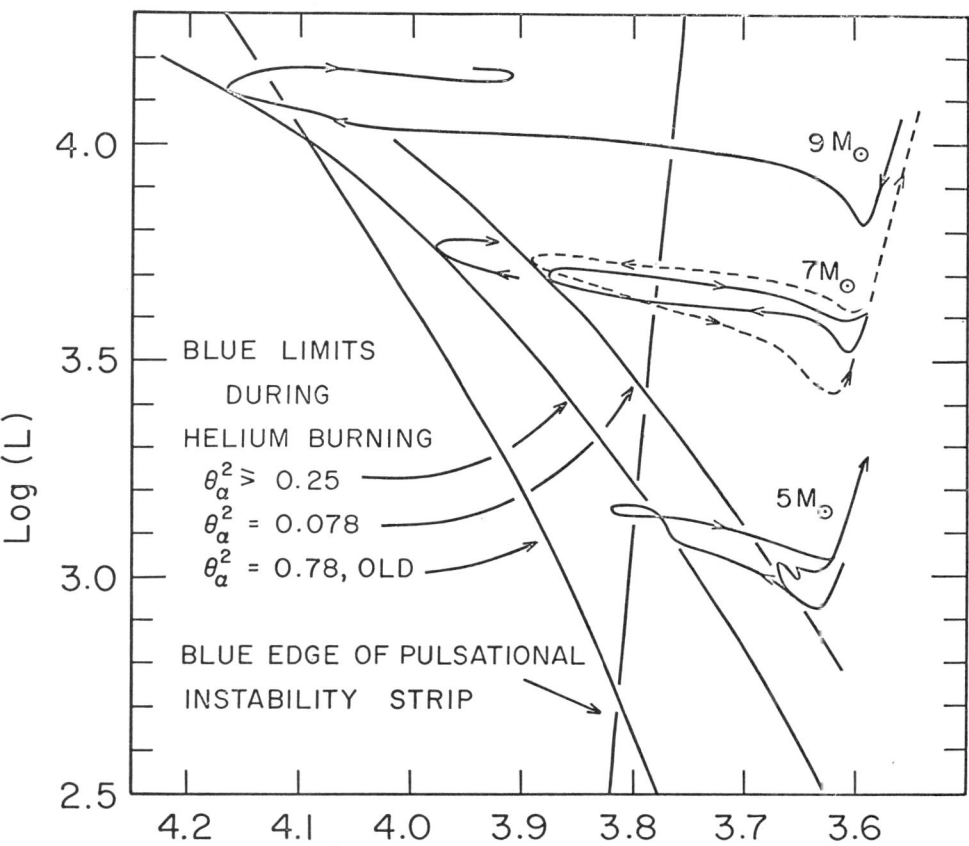

Fig. 2. Evolutionary tracks during core helium-burning, blue limits, and the blue edge of a theoretical instability strip for radial pulsation in the fundamental mode. Initial composition paramaters are $X = 0.7$, $Y = 0.28$, $Z = 0.02$. Three choices of reduced width θ_α^2 are represented. For the $5M_\odot$ models, $\theta_\alpha^2 = 0.078$ and 0.25; for the $7M_\odot$ models, $\theta_\alpha^2 = 0.078$ and 0.78; for the $9M_\odot$ model, $\theta_\alpha^2 = 0.25$. The blue limits connect, for different masses, the bluest points reached during the major core helium-burning phase. A blue limit defined by an earlier set of models (see Iben, Ann. Rev. of Ast. & Ap., 5, 571, 1967) is also shown.

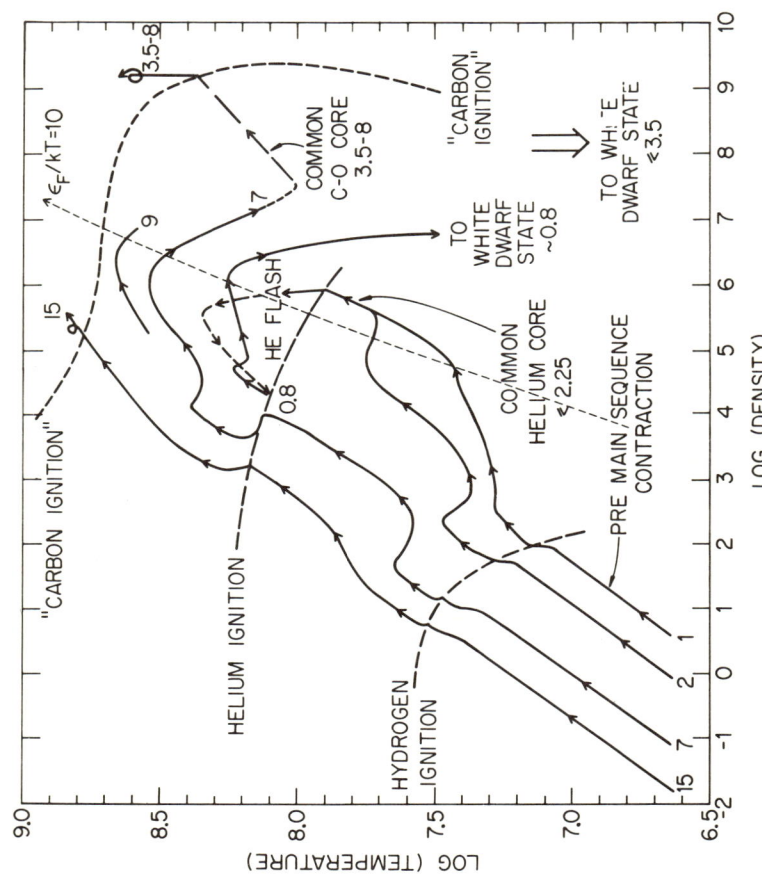

Fig. 3. Density and temperature at the centers of stars. Time is the parameter which varies in the direction of the arrows along each curve. The hydrogen and helium ignition curves have been empirically derived. The carbon ignition curve is an approximation obtained by equating an assumed neutrino loss rate to an adopted carbon-burning rate.

Astrophysical arguments may be used to estimate the range within which θ_α^2 probably lies. If the abundance ratio of C^{12} to O^{16} in the cosmic abundance distribution is assumed to be similar to the abundance ratio formed at the centers of intermediate mass and high mass stars, then $\theta_\alpha^2 \approx 0.08$. However, after core helium exhaustion, the abundance ratio left behind by the helium-burning shell is larger than that left at the center. On the assumption that the carbon-to-oxygen ratio achieved by averaging over products of all helium-burning phases approximates the cosmic ratio we may estimate that θ_α^2 lies in the range $0.05 < \theta_\alpha^2 < 0.15$.

Another estimate of θ_α^2 may be made by comparing the location of the theoretical blue limit during core helium burning with the location of this limit suggested by the position of Galactic Cepheids in the H-R diagram. The result is $0.1 < \theta_\alpha^2 < 0.25$. Both astrophysical estimates are model dependent and are therefore no substitute for an experimentally based estimate, provided, of course, that the interpretation of experimental results is not itself strongly model-dependent.

After the exhaustion of central helium, stars of intermediate mass return to the giant branch for the second time, rising upward along this branch as a carbon-oxygen core grows in mass. At the edge of the carbon-oxygen core, helium burns in a shell. Hydrogen burns in a shell separated from the helium-burning shell by a thin layer of helium (a mass of about $0.001 M_\odot$).

(2) <u>Paths in the density-temperature plane</u>. Conditions at the centers of stars of several different masses vary along the curves shown in Fig. 3. Along the gravitational contraction tracks preceding hydrogen ignition, density ρ and temperature T are related by $\rho/T^3 \propto 1/M^2$, where M is the stellar mass. Most of the core hydrogen-burning phase is spent very near the hydrogen ignition locus, demonstrating that thermonuclear reactions act as sensitive thermostats.

Significant departure from the hydrogen ignition curve does not occur until hydrogen is exhausted over the inner ten percent of the star's mass. In stars more massive than about $2.25 M_\odot$, the hydrogen-exhausted core contracts gravitationally in such a way that ρ/T^3 again remains nearly constant. Helium is ignited at the center and contraction is halted when central temperatures reach approximately $10^8 °K$. The ensuing core nuclear burning phase takes place under conditions that remain close to those prevailing on the helium ignition locus. Because of the much higher temperature sensitivity of helium-burning reactions relative to the temperature sensitivity of hydrogen-burning reactions, the slope of the helium ignition locus is much less steep than that of the hydrogen ignition locus.

In the hydrogen-exhausted core of stars less massive than about $2.25 M_\odot$, electrons become highly degenerate. The dashed line in Fig. 3 labeled $\epsilon_F/kT = 10$ indicates where the zero-temperature electron Fermi energy, ϵ_F, is ten times kT. Prior to helium ignition, central density and temperature for all stars less massive than about $2.25 M_\odot$ converge onto a common path where $\epsilon_F/kT \sim 20$. When the mass in the hydrogen-exhausted core reaches about $0.5 M_\odot$, helium ignites in an almost explosive fashion. Core temperatures rise until degeneracy is lifted. Central density and temperature then adopt a new position on the helium ignition curve. The ensuing evolution is very similar to that of stars more massive than $2.25 M_\odot$. The transition from the first helium ignition point to the second "initial" helium burning point is indicated schematically in Fig. 3 for a model of mass $0.8 M_\odot$.

In all stars, helium burning produces a convective core. The larger the mass of the star, the larger the core. When helium vanishes at the center, it vanishes also over the entire region defined by the convective core at its maximum extent. In stars more massive than about $10 M_\odot$, the helium-exhausted core contracts and heats until carbon is ignited at the center.

Stars somewhere in the mass range $8-10 M_\odot$ ignite carbon at some distance from the center of the helium-exhausted core. This is possible because neutrino losses (if they occur at a rate predicted by several theories of weak interactions) retard the rate at which temperatures rise at high densities and bring about a temperature distribution in the core that manifests a peak at some distance from the center. Only a portion of the central ρ-T track for a $9 M_\odot$ star is shown. Note that, had we ignored conditions away from the center, we would have predicted that carbon burning would not begin until much higher densities and somewhat lower temperatures were reached at the center. In fact, an off-center portion of the $9 M_\odot$ star reaches the carbon ignition curve as central density and temperature reach the end of the $9 M_\odot$ track shown in Fig. 3. Carbon ignition has occurred over a broad region about $0.5 M_\odot$ from the center. Although the calculations have not yet been performed, it is expected that a carbon "flash" (analogous to the helium "flash" that follows helium ignition in the electron-degenerate cores of lighter stars) will partially lift degeneracy in the carbon-oxygen core and that quiet carbon burning will thereafter continue in a fashion similar to that in more massive stars that ignite carbon at the center. If this is the case, then the central values of ρ and T in the $9 M_\odot$ star will jump to a position on the carbon ignition curve closer to that occupied by central ρ and T in the $15 M_\odot$ star.

For stars in the mass range $3.5-8M_\odot$, it is expected that paths of central ρ and T will converge onto a common path that strikes the carbon ignition curve when $\epsilon_F/kT \sim 200$. The long-dash curve in Fig. 3 labeled "common C-O core" is taken from Paczynski's calculations, which suggest that carbon burning is ignited when central ρ and T reach the carbon ignition curve. The short-dash curve following the last portion of the solid $7M_\odot$ curve is a conjecture, which anticipates the results of further time-consuming calculations.

It has not yet been established that the thermal instability, which occurs in the nuclear burning region outside of the carbon-oxygen core (and which has been ignored in Paczynski's calculations), does not influence the value of central ρ and T when carbon ignition occurs. Nor has it been established that, even when the thermal instability is ignored, carbon ignition will occur first at the center. This is due to the fact that the temperature profile in the carbon-oxygen core is sensitive to uncertainties that still remain in electron conductivity and in neutrino loss rates. Exploration of the possibilities is still underway.

To complete the story in the ρ-T plane, we must remark on the fate of stars of mass less than $3.5M_\odot$. Such stars will follow a curve similar to that one shown in Fig. 3 for $0.8M_\odot$ except that, the larger the initial mass, the larger will be the carbon-oxygen core presumably become before descent to the white dwarf stage begins and the larger will be the central density which is approached assymptotically as the central temperature drops. It is assumed that, in such stars, mass in excess of $1.4M_\odot$ will be lost before the carbon-oxygen core reaches a mass of $1.4M_\odot$.

(3) <u>Details of structure and consequences of the thermal instability</u>. The next three figures describe overall characteristics of a typical red giant of intermediate mass ($7M_\odot$) that consists of a large carbon-oxygen core ($\sim 0.945M_\odot$), a thermonuclear region where both hydrogen and helium are burning, and an extensive convective envelope.

The distribution of matter (ρR^2) and the distribution of particle kinetic energy (kT) are shown in Fig. 4 as functions of distance from the center (R). The entire radial extent of the carbon-oxygen core is only one-one hundredth of our Sun's radius (about one earth radius). The quantity ρR^2 is a measure of the mass contained in a spherical shell of unit thickness. The variation of ρR^2 with R reminds one of the probability distribution of matter in an atomic nucleus. The analogy with an atomic nucleus is enhanced by the relative flatness of the distribution of thermal kinetic energy per particle (temperature);

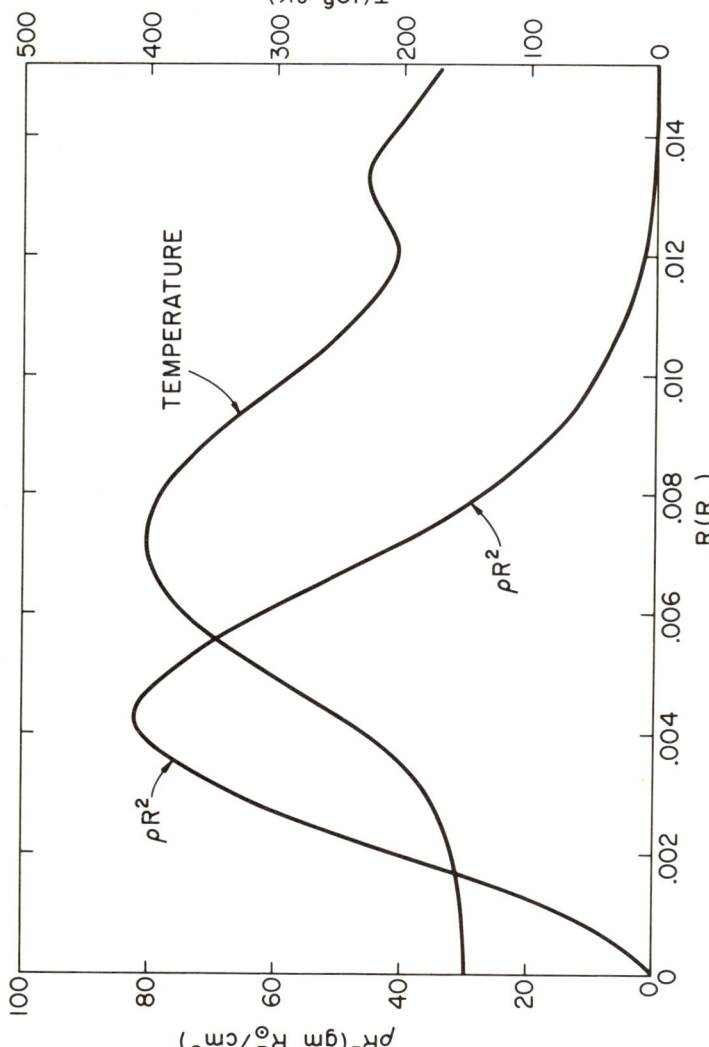

Fig. 4. "Shell density" (= ρR^2) and temperature in carbon-oxygen core of mass $M_{core} \sim 0.95 M_\odot$. Here ρ is the density and R is the distance from the center.

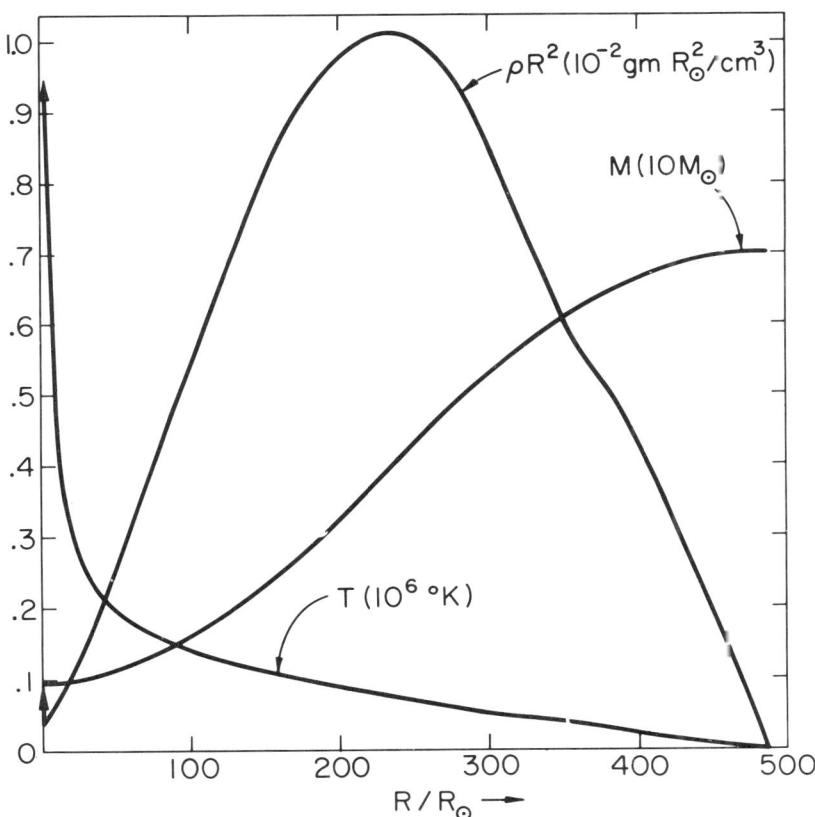

Fig. 5. "Shell density", temperature, and mass in the envelope of a 7M$_\odot$ star with the carbon-oxygen core described in Fig. 3.

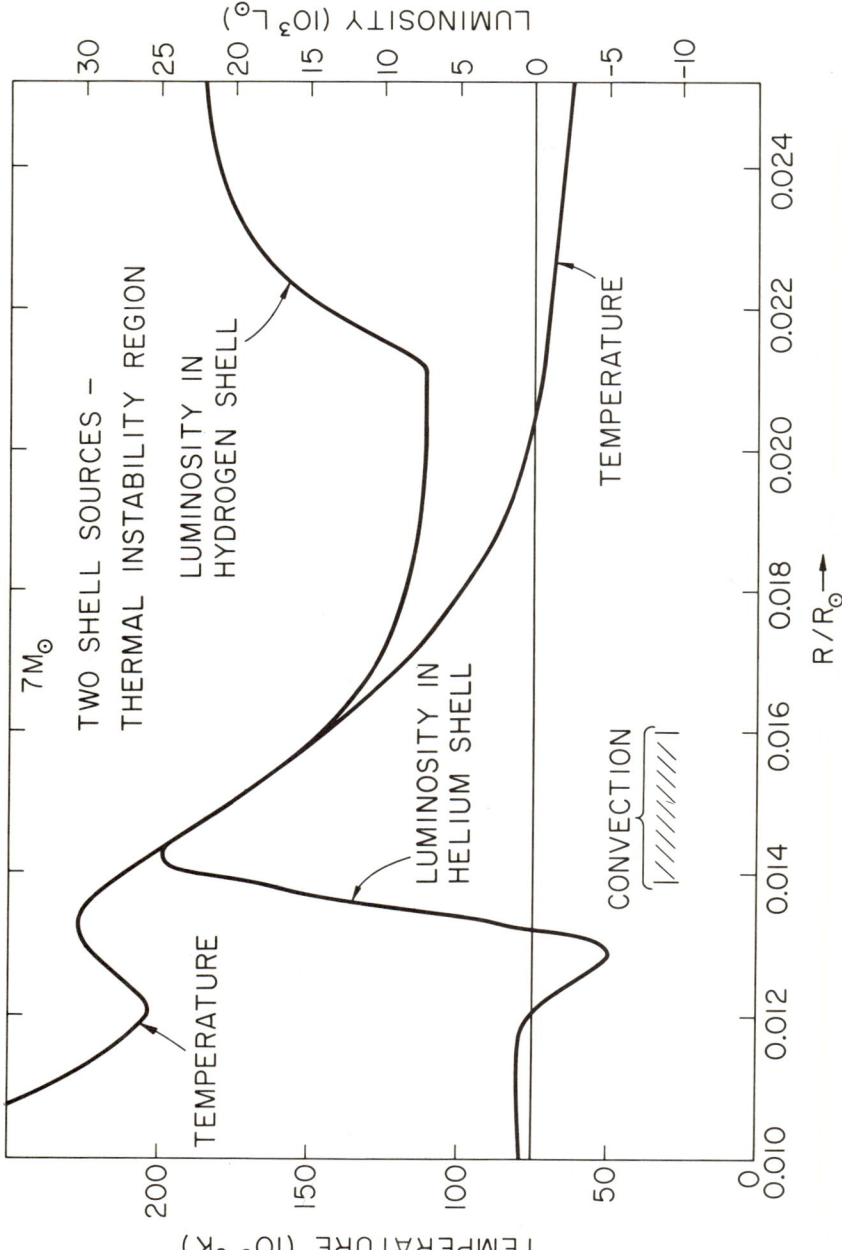

Fig. 6. Temperature and luminosity versus radius in the double shell region. Same model star, same time as in previous figures.

this distribution is characteristic of a tightly coupled system of particles interacting via strong, short-range forces. The strong coupling agent in the carbon-oxygen core is electron conductivity.

In contrast with the atomic nucleus, however, the carbon-oxygen core is a continuously evolving system in which the properties of the surface play a vital role. The kinetic energy in the core is maintained at a high value because of the release of gravitational potential energy as more carbon and oxygen are deposited in the core by the helium burning shell and more helium is added by the hydrogen burning shell. The depression in temperature near the stellar center is due to postulated neutrino losses. Finally, the secondary maximum in temperature beyond the primary maximum is due to helium burning reactions. As a consequence of the thermal instability, this secondary maximum appears and disappears at intervals of a few thousand years.

Fig. 5 describes conditions in the stellar envelope outside of the nuclear burning region. Both the matter distribution and the temperature (kinetic energy per particle) distribution are reminiscent of the electron cloud in the ground state of the hydrogen atom.

In Fig. 6 are shown temperature and luminosity distributions in the nuclear burning, thermally unstable region between the carbon-oxygen core and the diffuse envelope. Helium burning reactions are occurring at peak intensity during an early, relatively mild helium "flicker". Most of the energy produced in the flicker is used up in heating and expanding the matter within and on either side of the shell. As time progresses, the amplitude of the flicker increases. For example, in the next two flickers, the peak rate of energy production by helium burning reactions increases to $5.0 \times 10^4 L_\odot$ and then to $1.4 \times 10^5 L_\odot$. However, since the energy released per gram by helium burning reactions is roughly 10 times smaller than that released by hydrogen burning reactions and since the helium shell and the hydrogen shell remain separated by a roughly constant amount of helium ($\sim 0.001 M_\odot$), the average rate of energy production by helium burning reactions is roughly 10 times smaller than that by hydrogen burning reactions.

At the peak of the helium flicker, a convective zone extends from somewhere within the helium-burning region to somewhere just below the hydrogen-helium discontinuity. The minimum separation between the outer edge of the convective region and the hydrogen discontinuity decreases with time. Abundances by mass in and near the nuclear burning region are shown in Fig. 7 at a time corresponding to the peak of an early, weak flicker.

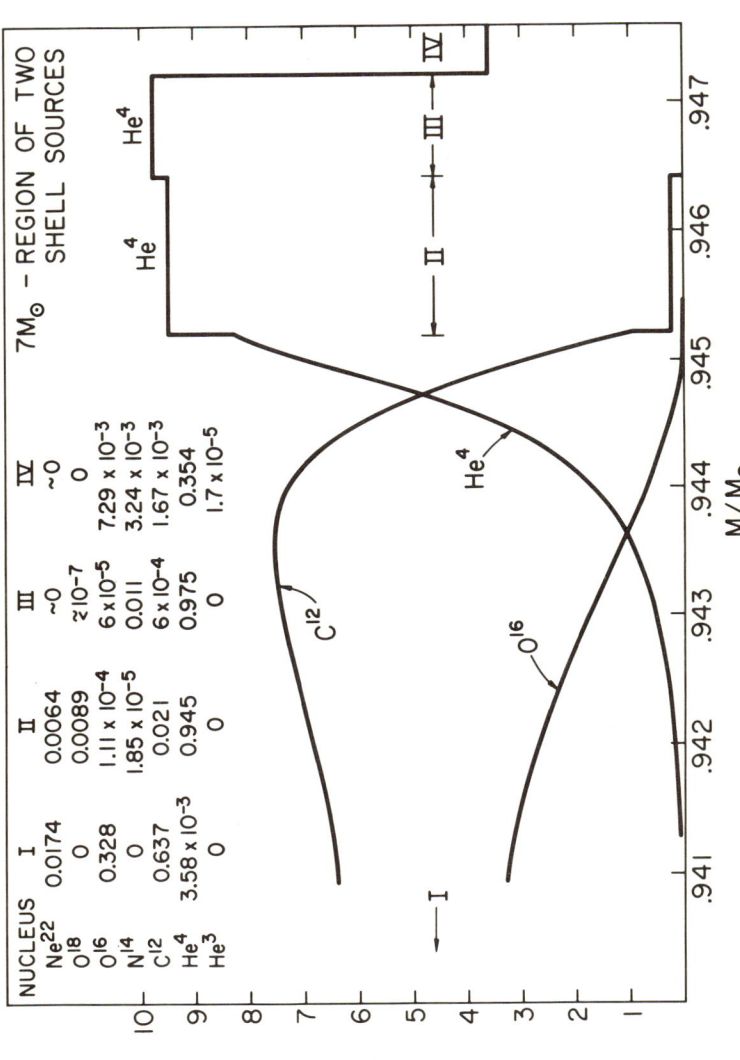

Fig. 7. Composition variables versus mass, in double shell region. Same star, same time as in earlier figures.

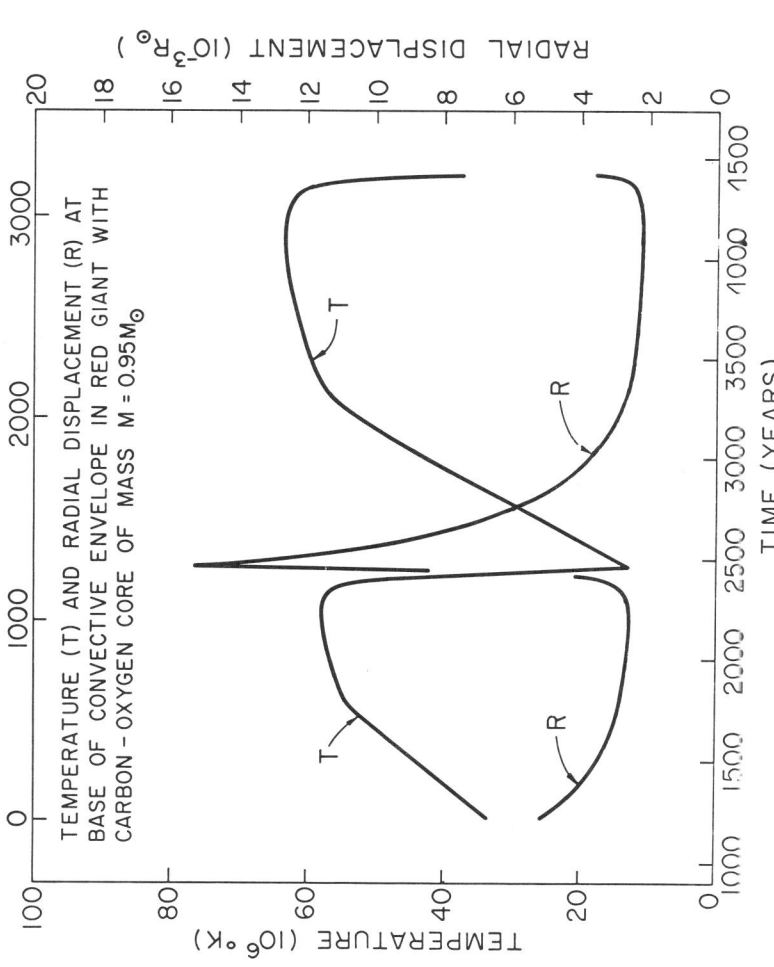

Fig. 8. Temperature at the base of the convective envelope and the radial position of this base versus time.

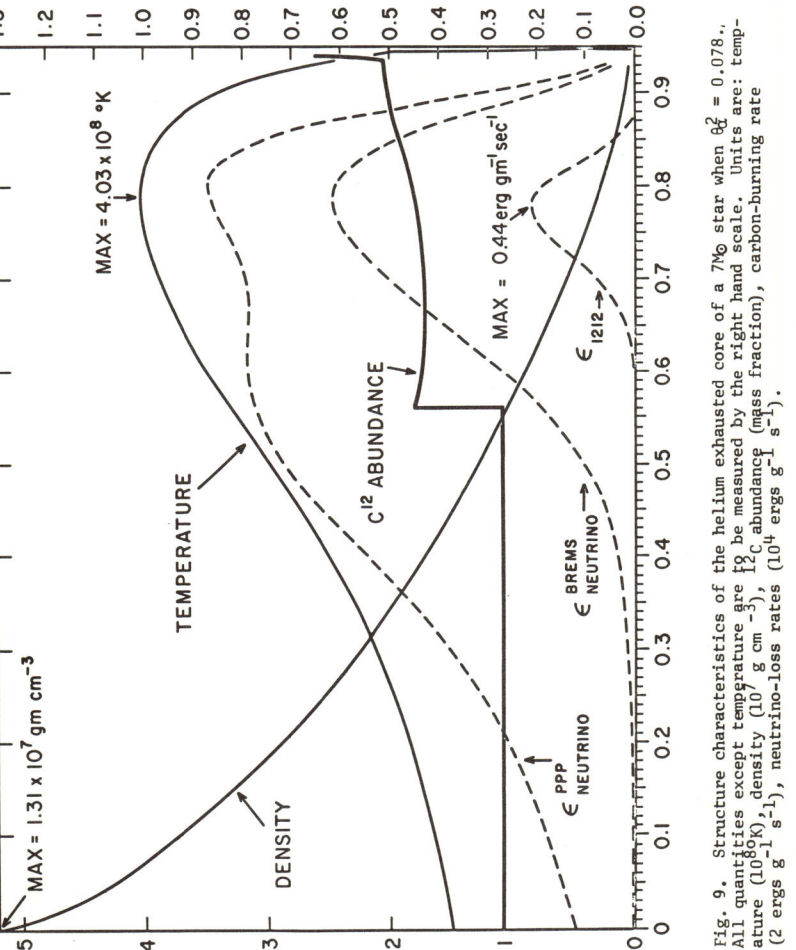

Fig. 9. Structure characteristics of the helium exhausted core of a 7M☉ star when $\theta_n^2 = 0.078$. All quantities except temperature are to be measured by the right hand scale. Units are: temperature (10^8 K), density (10^7 g cm^{-3}), C^{12} abundance (mass fraction), carbon-burning rate (2 ergs g^{-1} s^{-1}), neutrino-loss rates (10^4 ergs g^{-1} s^{-1}).

In subsequent flickers, all of the O^{18} in the inter shell region is converted into Ne^{22}.

It seems likely that, at the peak of some subsequent, sufficiently violent helium flicker, convection will extend beyond the hydrogen-helium discontinuity. Then, hydrogen will mix into regions which later reach temperatures considerably in excess of those at which hydrogen burning normally occurs. At the same time, newly formed products of helium burning will mix into hydrogen-rich regions. The ensuing reactions are sure to create a rich variety of new nuclei that are not normally formed under pure hydrogen-burning or under pure helium-burning conditions. In particular, it is certain that, following the formation of C^{13} by the $C^{12}(p,\gamma)N^{13}(\beta^+\nu)C^{13}$ reactions, neutrons formed by the $C^{13}(\alpha,n)O^{16}$ reaction will act on seed nuclei to form s-process elements.

These newly formed s-process elements, along with newly formed carbon, will be convected to the surface during the quiescent phase of each thermal oscillation, when the rate of helium burning is near minimum and the base of the convective envelope reaches down <u>into</u> the edge of the hydrogen-burning region.

Other interesting phenomena occur in the envelope. Shown as a function of time in Fig. 8 are the temperature and radial displacement at the base of the convective envelope. The high-temperature, small-radial-displacement phases are associated with low energy production by helium-burning reactions. During the high-temperature phases, some He^3 built up during the main sequence phase is converted into Be^7. In the first few months of the low-temperature phases, Be^7 is converted entirely into Li^7 by electron capture. During the high temperature phases, Li^7 in the envelope is destroyed completely.

The possibility presents itself that, every several thousand years, the Li^7 abundance at the surface of red giant of intermediate mass varies from zero to a possibly large value. A quantitative investigation shows that, at least during the early, low-amplitude flickers, the maximum Li^7 produced is less than 10^{-10} by mass, certainly nothing to write home about. However, a projection of current results suggests that the maximum surface Li^7 abundance increases with each flicker. A substantial surface abundance of Li^7 may develope before a supernova event occurs.

(4) <u>Carbon ignition in the electron-degenerate core</u>. The distributions of density, temperature, carbon, and neutrino loss rates in a carbon-oxygen core of mass $0.95M_\odot$ are shown in Fig. 9. Shown separately are the loss rate, ϵ_{ppp}, due to pair annihilation neutrinos, photoneutrinos, and plasma neutrinos and the loss

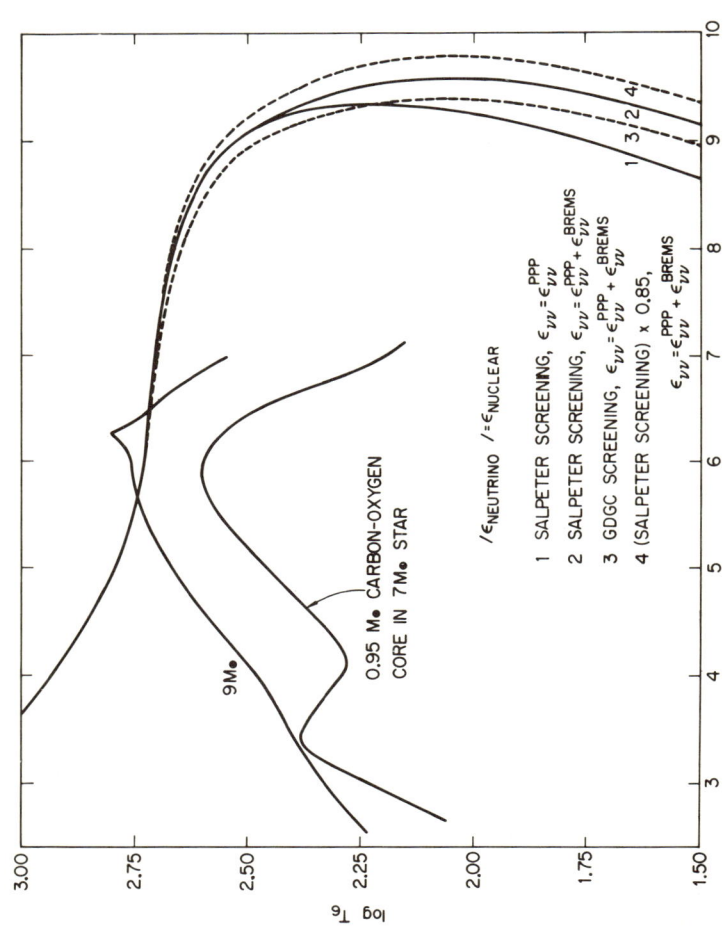

Fig. 10. Density and temperature within the carbon-oxygen cores of $7M_\odot$ and $9M_\odot$ stars at times corresponding to the points labeled 7 and 9 on the central ρ,T paths for $7M_\odot$ and $9M_\odot$ stars in Fig. 3.

rate, ε_{BREMS}, due to bremstrahlung neutrinos. The temperature distribution is sensitive to the neutrino rates adopted. Since there is no guarantee that the chosen rates (which are <u>guessed</u> by theoreticians and which cannot be detected in the terrestial laboratory) are even remotely near the truth, we should be cautious in accepting the temperature distributions found with the chosen set. For example, correct rates might reduce the central temperature considerably relative to the off-center temperature maximum. Such a redistribution would of course increase the probablity of off-center carbon ignition.

The distribution of carbon is also uncertain. The carbon abundance shown in Fig. 9 for a $7M_\odot$ star is the consequence of (a) attributing a value of 0.08 to the reduced width θ_α^2 of the $C^{12}(\alpha,\gamma)O^{16}$ reaction and (b) of neglecting semiconvection. During the core helium-burning phase, the formal convective zone reaches out to $0.55M_\odot$ from the center. Semi-convection might increase the size of the effective mixing region with the consequence that the abundance of carbon left at the end of core helium burning is less than shown in Fig. 9. Choice of a larger value of θ_α^2 has the same effect. For example, choice of $\theta_\alpha^2 = 0.25$ leads to a final central carbon abundance of less than 0.01. It is expected that the contrast between the final central carbon abundance and the final abundance left behind by the helium-burning shell, after the hydrogen shell is reignited, will increase with increasing θ_α^2 and/or with an enlargement of the effective mixing region during core helium burning. Both effects will therefore enhance the probability of off-center carbon ignition relative to the probability of carbon ignition at the center.

In the final figure, Fig. 10, are shown density-temperature values <u>within</u> the carbon-oxygen core of a $7M_\odot$ model in which hydrogen has just been reignited and in which a thermal instability of large amplitude has just been established. Also shown are ρ-T values within the core of a $9M_\odot$ model that has begun to burn carbon off center.

Approximate carbon ignition curves are also shown in Fig. 10. These are obtained by equating a carbon burning rate with a neutrino loss rate. The bend-over of all curves at high density is due to the enhancement of the carbon burning rate by strong-screening. Curves 1 and 3 differ only in that energy loss by neutrino bremstrahlung is omitted in deriving curve 1. Curves 2 and 4 differ from curve 3 due to the choice of screening potential. The strong screening potential estimated by Salpeter in 1954 is used in deriving curve 3. This potential is multiplied by 1.17 and 0.85 in deriving curves 2 and 4, respectively. Curve 2 is appropriate if one adopts the potential derived recently by Graboski <u>et al</u>.

It is expected that, in the $9M_\odot$ model, carbon burning will spread from the point of off-center ignition into the center and that the subsequent development will be somewhat analogous to what occurs in the helium flash in less massive stars. Electron degeneracy will be partially lifted and the major core carbon-burning phase will occur under only weakly degenerate conditions, with central ρ and T remaining near the carbon ignition curve in Fig. 3, perhaps to the left of the line $\epsilon_F/kT = 10$.

As time progresses, the entire ρ,T profile for the $7M_\odot$ star will deform and move to the right in Fig. 10. Since no calculations have yet been made with neutrino bremsstrahlung processes included, it is not established that central ρ and T will touch the relevant carbon ignition curve before the ρ and T at some off-center points. If off-center ignition does occur, it is not yet clear whether detonation will be a consequence of whether a flash phenomenon will ensue. In the latter case, an explosive event will be deferred until after central carbon is exhausted and an electron-degenerate core of mass near $1.4M_\odot$ and containing Mg^{24} and other products is reestablished.

THE s-PROCESS IN STARS

Roger K. Ulrich

Astronomy Department
University of California
Los Angeles
and
California Institute of Technology
Pasadena

Introduction

The basic form of the process of slow neutron addition as a mechanism for converting iron to heavier elements was first discussed by Cameron (1954, 1955). Later analyses, especially those of Burbidge, Burbidge, Fowler and Hoyle (1957, hereafter referred to as B^2FH), Clayton, Fowler, Hull and Zimmerman (1961, hereafter referred to as CFHZ) and Seeger, Fowler, Clayton (1965) have shown how the products of this process can be disentangled from the products of other processes and treated quantitatively. The basic argument concerning the time scale of the s-process has been given by B^2FH and has led to the consensus that s-process element building occurs on a time scale of thousands of years. I note, however, that a contrary position favoring a short time scale has been adopted by Cameron (1959). The presence of high abundances of magic neutron number elements in N, S, and BaII stars and the presence of Tc in some N and S stars has generally supported the idea that stars are able to form heavy elements by the s-process and bring those elements to their surfaces. Fowler, Burbidge and Burbidge (1955) showed that S stars are capable of synthesizing a large portion of the solar system s-process nuclei. Sanders (1957) has studied the s-process in the models of Schwarzchild and Harm (1967) and concluded that these are a likely site for the s-process. A <u>detailed</u> model for accomplishing both s-process nucleosynthesis and mixing to the surface has not heretofore been available. I would like to briefly mention possible mixing models and then turn to a detailed discussion of the s-process.

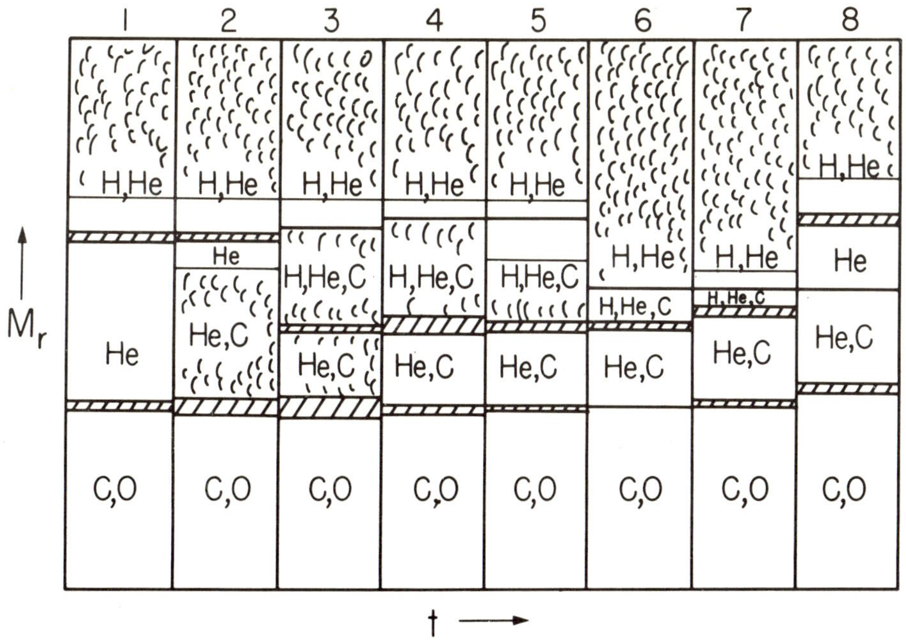

Figure 1: Schematic diagram of indirect mixing in a low mass star. Energy generating layers are denoted by cross-hatching. Convective zones are denoted by ⸨⸨⸨. Chemical composition boundaries are marked by heavy lines. Radiative - convective boundaries are marked light lines. The quiescent helium shell at time 1 becomes unstable at time 2 and mixes with hydrogen at time 3. A secondary flash occurs between times 3 to 5 and cools off the helium shell. The hydrogen shell becomes quiescent at time 6 and causes the convective envelope to mix a portion of the former convective shell to the surface. Between times 6 to 8 the hydrogen shell burns out to the point where the helium shell is again unstable and the cycle can repeat. This process is not based on any calculations but is based on the general tendency for energy generating layers and convective boundaries to maintain a constant mass separation.

Mixing Models

John Scalo and I have recently developed a model for mixing (Ulrich and Scalo 1972a; Scalo and Ulrich 1973) based on a model of a $5M_\odot$ model published by Weigert (1966). Our model is probably appropriate for stars in the mass range ~3 to $10M_\odot$. This mixing process can bring matter from the vicinity of the He burning shell to the surface of the star in a time period of 1 to 2 years. Smith and Sachmann (1973) have proposed a mechanism for mixing for stars in a mass range nearer $1M_\odot$. Both mixing models rely on a He shell flash to bring carbon-rich matter in contact with hydrogen-rich matter, and the mixing in both models is driven by energy from the first step of the CN cycle: $^{12}C(p, \gamma)^{13}N$. The differences between the two models arise from the fact that Scalo and I have made an attempt to treat the energy release during the mixing process in detail, whereas Smith and Sachmann have assumed that the rate of diffusion of hydrogen into the high temperature region is not influenced by the energy release. We have suggested (Ulrich and Scalo 1972b) that the inward diffusion of hydrogen in the Smith-Sachmann mode will be stopped by the energy release and that a disconnected secondary shell flash will occur. This suggested sequence of events is shown in Figure 1. The model which Scalo and I have developed suggests that the inward diffusion of hydrogen in the $5M_\odot$ model is stopped essentially at the point where hydrogen and carbon first come in contact. The energy release from the $^{12}C(p, \gamma)^{13}N$ reaction is distributed unevenly over the contact surface which we call the mixing interface. This assymetric energy release leads to mixing currents which we have treated with a standard meteorological model for a maintained plume (Morton, Taylor and Turner 1956). The most important property of the mixing interface is that it prevents all but a very small quantity of hydrogen from entering the convective shell. An overview of our model is shown in Figure 2.

The major portion of all carbon and hydrogen mixed together at the mixing interface becomes a part of a plume and is swept into the outer convective envelope. Plume mixing, therefore, causes the mass of the He-C convective shell to decrease and the position of the hydrogen shell source to migrate inwards. When the thermal pulse from the helium shell source has died down, the hydrogen rich matter will be too close in mass to the helium shell source for both to burn at the same rate. For both shells to burn at the same rate, the helium shell must produce about one tenth the energy produced by the hydrogen shell source. The flow of this energy requires a temperature gradient. Prior to the thermal instability the two shells were separated in mass by just the right amount such that both could burn simultaneously. After the hydrogen shell has migrated inwards, the He shell cannot immediately ignite because then the hydrogen shell would be too hot and produce too much energy. Therefore, the helium shell source must remain cool until enough hydrogen has been converted to

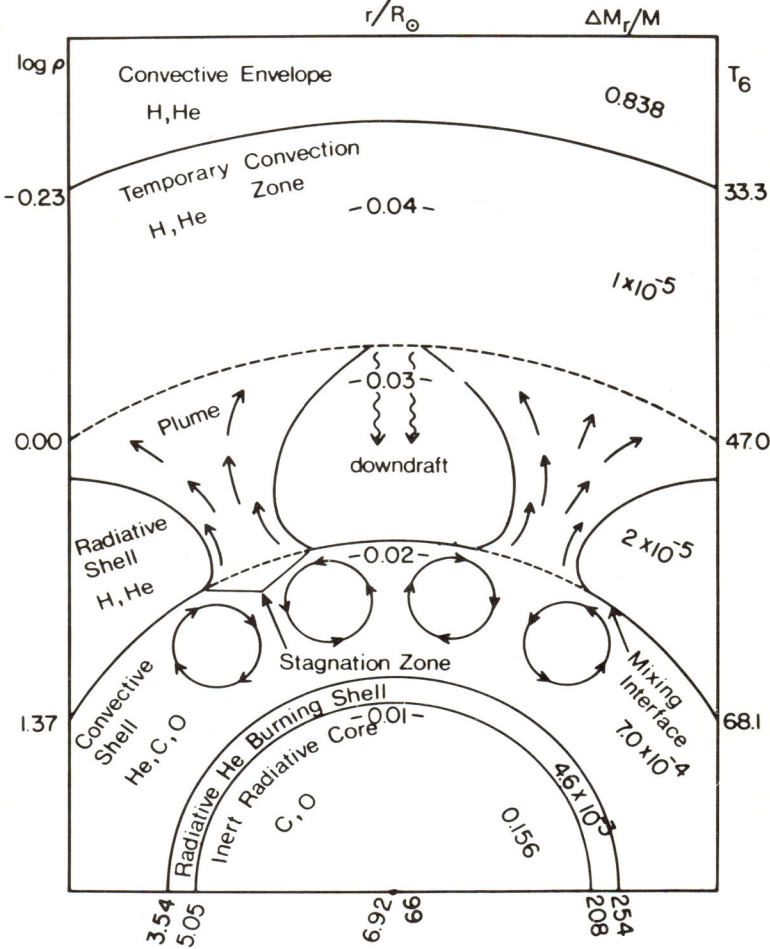

Figure 2: Scale drawing of Weigert's model during plume mixing. The scale in terms of $r/R_\odot$ is shown by the number in a vertical column in the center of the figure. Other quantities from left to right are $\log \rho$, the descriptive name of each zone with the principle chemical constituents, the mass fraction contained in each zone and T_6, the temperature in $10^6 \,^\circ K$. The flow pattern of convection in the He, C, O convective shell is indicated schematically. The thickness of the stagnation zone is not to scale and is about one tenth as thick as indicated. The normal convective envelope boundary is shown to be at $T_6 = 33$ although according to Weigert's data, this boundary could also be at $T_6 = 15$.

helium to restore the appropriate mass separation of the two shell sources. It is during this stage of evolution that unexposed seed nuclei are added from the outer envelope. These nuclei pass through the hydrogen burning shell unaltered.

The mass of the region between the two shell sources remains roughly constant during the evolution of the star. I denote this roughly constant mass separation by M_S. For the case of Weigert's model shown in Figure 2, $M_S = 7.0 \times 10^{-4} M$ where M is the total mass of the star. At the beginning of each instability phase, some of the matter between the shells was present during the previous flash and some was added through the hydrogen shell source. I denote the quantity of mass which has remained between the shell sources for two successive instability periods as $r M_S$ where r is a number less than unity. The quantity of mass added through the hydrogen shell source is therefore $(1-r)M_S$. The probability that any particular nucleus can remain between the shell sources for N flashes is r^N. I denote by $f M_S$ the quantity of mass mixed from the convective shell region into the outer envelope. Clearly $f<1-r$.

Neutron Source

The small quantity of hydrogen which crosses the mixing interface is swept down to the base of the convective region where the temperature is 2.5×10^8 °K. On the way down the hydrogen is converted to ^{13}C by being captured by a ^{12}C followed by a β^- decay. Virtually none of the ^{13}C is further converted to ^{14}N because the abundance of ^{13}C is very much less than the abundance of ^{12}C, and therefore most of the available protons are captured by ^{12}C. This feature of our model is crucial because ^{14}N in large quantities could act as a neutron poison. When the ^{13}C reaches the inner boundary of the convective shell, the temperature is high enough to cause 80-95% of all ^{13}C nuclei to capture an α particle and yield ^{16}O and a neutron within a single convective turnover time. The first few neutrons catalytically convert the small amount of ^{14}N from the initial CNO-Cycle into ^{14}C through the sequence of reactions pointed out by Caughlin and Fowler (1964): ^{14}N (n,p) ^{14}C, and ^{12}C (p,y) $^{13}N(\beta^+)^{13}C(\alpha,n)^{16}O$. The final step in the sequence is sufficiently slow that it might be hard to convert a large quantity of ^{14}N to ^{14}C by this cycle during a single He shell flash. After this conversion is completed the neutrons are then free to be captured by heavier elements such as iron.

Repeated Flashes

The primary goal of the work I am reporting on has been to study the s-process in Weigert's (1966) model. While this model may be uncertain in some of the details, the basic character of the evolution as described above is undoubtedly correct. The most important properties are the following:
1) the neutron exposures are brief and occur repeatedly --

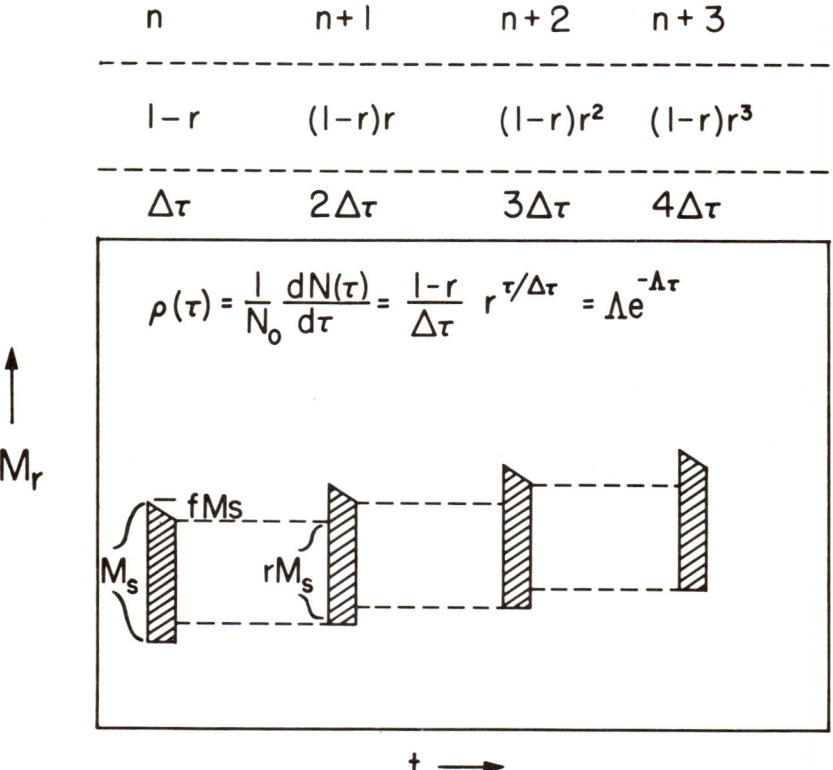

Figure 3: Schematic diagram of the repeated flash model for the s process. Neutron exposure occurs within the cross-hatched regions of the mass-time diagram. The steady advance of the helium burning layer causes the inner boundary of the region to move outward between flashes. Mixing of matter into the envelope causes the outer boundary of the region to move inward. The top line above the drawing labels the number of each flash. The second line gives the fraction of matter in the convective shell region which entered at flash N. The third line gives the neutron exposure of the matter which entered at flash N. The equation within the figure gives the relative abundance of nuclei which have received a neutron exposure of τ after a large number of flashes.

the rest periods between flashes are considerably longer than
the exposure periods; 2) the neutron exposure per flash is
about the same for all matter contained in the convective shell
zone of mass M_s. The neutron exposure is denoted by $\Delta\tau = pv_t N_n \Delta\tau$,
is the thermal velocity; 3) the shell zone loses mass during
the flash because the plumes transport the matter to the surface and between the flashes because some ^{4}He is burned to ^{12}C.
Thus, only a mass $r M_s$ remains in the convective shell zone
between successive flashes. Those neutron enriched nuclei which
leave the shell zone are effectively destroyed for s-process
nucleosynthesis because they are no longer able to capture
neutrons; 4) at the beginning of each flash a mass $(1-r) M_s$ of
matter from the outer envelope is incorporated into the convective shell zone. This matter contains seed nuclei which have
not previously been exposed to neutrons.

The repeated cycles are illustrated schematically in Figure 3.
The number of nuclei per gram in the shell, dN, which have neutron
exposures between τ and $\tau + d\tau$ relative to the number of seed
nuclei in the incoming matter, N_o, is seen to be an exponential
function. Previous treatments of the s-process have obtained
a neutron exposure distribution like the exponential function
as a semi-empirical fit. In contrast the above argument shows
that this distribution arises as a natural consequence of the
repeated flash model and occurs within a single star. Also,
this model gives a definite abundance for the enriched nuclei
in the shell so that the quantity of shell matter in the solar
system can be calculated. A derivation based specifically on
the repeated flash model is given below. The resulting equations
are of wider validity than the repeated flash model so a more
general derivation of these equations is given in the appendix.

As long as $\Delta\tau$ is small then the abundances of the nuclei
with small σ which govern the s-process do not change substantially during the flash. After a sufficient number of flashes,
a steady state will develop in which all abundances are constant:
Each heavy nucleus can be destroyed either by further neutron
capture or removal from the mass region between the shell
sources, and each heavy nucleus can be replaced when its progenitor nucleus captures a neutron. The ultimate source of
progenitor nuclei is the matter brought in from the outer envelope.
Let N_A be the number of nuclei of atomic mass A per gram in the
convective shell and N_{56}^o be the number of iron nuclei per gram
in the matter in the outer envelope. Let $N_H M_s$ be the total
number of hydrogen nuclei added during the flash. It is convenient to measure the neutrons which result as if they were
added along with the ^{56}Fe. Let this hypothetical neutron abundance be $N_n^o = N_H/(1-r)$. The steady state balance of iron nuclei
gives

$$(1-r) N_{56}^o = (1-f/2)\, \sigma_{56} \Delta\tau N_{56} + (1-r) N_{56} \quad . \tag{1}$$

Ulrich

Balance of heavy nuclei gives

$$\sigma_{A-1} \Delta\tau N_{A-1} = (1-f/2) \sigma_A \Delta\tau N_A + (1-r) N_A \qquad (2)$$

Since all neutrons are captured rapidly, $\Delta\tau$ is constrained by

$$(1-r) N_n^o = \sum_{A=56}^{209} (1-f/2) \sigma_A \Delta\tau \, N_A \qquad (3)$$

If each of the above equations were multiplied by M_s, the left-hand side would be the total number of each species added to the shell each flash, the first term on the right-hand side would be the number of each species destroyed by neutron capture, and the second term on the right-hand side of Equations (1) and (2) would be the number of each species which does not remain in the region between the two shell sources. The terms involving $\sigma_A \Delta\tau$ are multiplied by a factor $1-f/2$ in order to account for the fact that the shell region shrinks during the flash due to plume mixing. The overall distribution of heavy element abundances is governed by the ratio N_{56}^o / N_n^o. Through Equation (3) this ratio governs the parameter Λ defined to be

$$\Lambda = \frac{1-r}{(1-f/2)\Delta\tau} \qquad (4)$$

Equations (1) and (2) immediately yield the solution

$$\frac{\sigma_A N_A}{\Lambda N_{56}^o} = \prod_{A=56}^{A} (1+\frac{\Lambda}{\sigma_A})^{-1} \qquad (5)$$

which may be expressed as

$$\log \frac{\sigma_A N_A}{\Lambda N_{56}^o} = -\frac{\Lambda}{2.303} \sum_{A'=56}^{A} \frac{C(\Lambda/\sigma_{A'})}{\sigma_{A'}} \qquad (6)$$

where $C(x) = x^{-1} \ln(1+x)$. Equation (3) then gives N_n^o/N_{56}^o as

$$\frac{N_n^o}{N_{56}^o} = \sum_{A=56}^{209} \prod_{A'=56}^{A} (1+\frac{\Lambda}{\sigma_{A'}})^{-1} \qquad (7)$$

Some representative values of N_n^o/N_{56}^o as a function of Λ are given in Table 1. The cross sections used were mostly those given by Allen, Gibbons and Macklin (1972) at an energy of 30 Kev. Exceptions are the $^{136}B_a$ cross section given by Stroud (1972) and cross sections for unstable nuclei which I have estimated.

Table 1

Λ	0.5	1.0	1.5	2.0	2.5	3.0
N_n^o/N_{56}^o	72.	37.	21.6	13.7	9.4	6.9

Table 1 (cont'd.)

Λ	3.5	4.0	5.0	6.0	8.0	10.0
N_n^o/N_{56}^o	5.3	4.3	3.1	2.52	1.79	1.42

The value of Λ required to fit the solar system abundances is between 4 and 5. Thus between 3 and 4 neutrons per seed nucleus can be added during the flash. In Weigert's model $r = 0.7$ and $M_s = 7 \times 10^{30}$ gm. Adopting an iron abundance of 7.3 on the long $N_H = 12$ scale and a hydrogen mass fraction of 0.75 for the primordial matter gives $(1-r)M_s N_n^o = (3.2 \times 10^{25}$ gm$)/m_H$ as the number of seed nuclei added between flashes. The number of neutrons added during the flash must therefore be about $(10^{26}$ gm$)/m_H$. If these are added to the shell in the form of hydrogen over a span of roughly five years, the energy released in converting them to neutrons will increase the luminosity of the shell by roughly 5%. This energy release is sufficiently small that it is not likely to modify the dynamics of convection near the mixing interface. The addition of this amount of hydrogen in about five years requires an average flux across the mixing interface of 2.5×10^{-2} (gm-H) cm^{-2} s^{-1}. Since the density is about 25 gm cm^{-3} and a typical convective velocity is .5 km/sec, the typical mass fraction of a hydrogen in a descending convection current must be on the order of 2×10^{-8}. This hydrogen mass fraction is a factor of 5000 less than the value of 10^{-4} estimated by Scalo and Ulrich (1973) from crude considerations of the hydrodynamics near the mixing interface. This discrepancy may indicate either than the catalytic conversion of ^{14}N to ^{14}C is too slow to be effective and the excess protons mixed in are stored in the form of ^{13}C which never captures an α particle, or that the filtering process at the mixing interface is considerably more efficient at keeping out protons than we estimated. I favor the latter alternative because ^{13}C appears too short-lived to store many excess protons. In a continuous s-process in which both seed nuclei and protons are mixed into the shell region at the same time, it would be necessary to keep the hydrogen mass fraction to about 5×10^{-5} in the incoming matter.

Time Scale

In the traditional slow process, the time scale is set by three decays discussed by B^{2}FH:
$$^{79}\text{Se}\,(\gamma,\beta^-)\,^{79}\text{Br}, \quad ^{85}\text{Kr}\,(\beta^-)\,^{85}\text{Rb} \text{ and } ^{154}\text{Eu}\,(\beta^-)\,^{154}\text{Gd}.$$

The half-lives are respectively 35 years at the base of the convective shell in Weigert's model, 12 years and 16 years. The

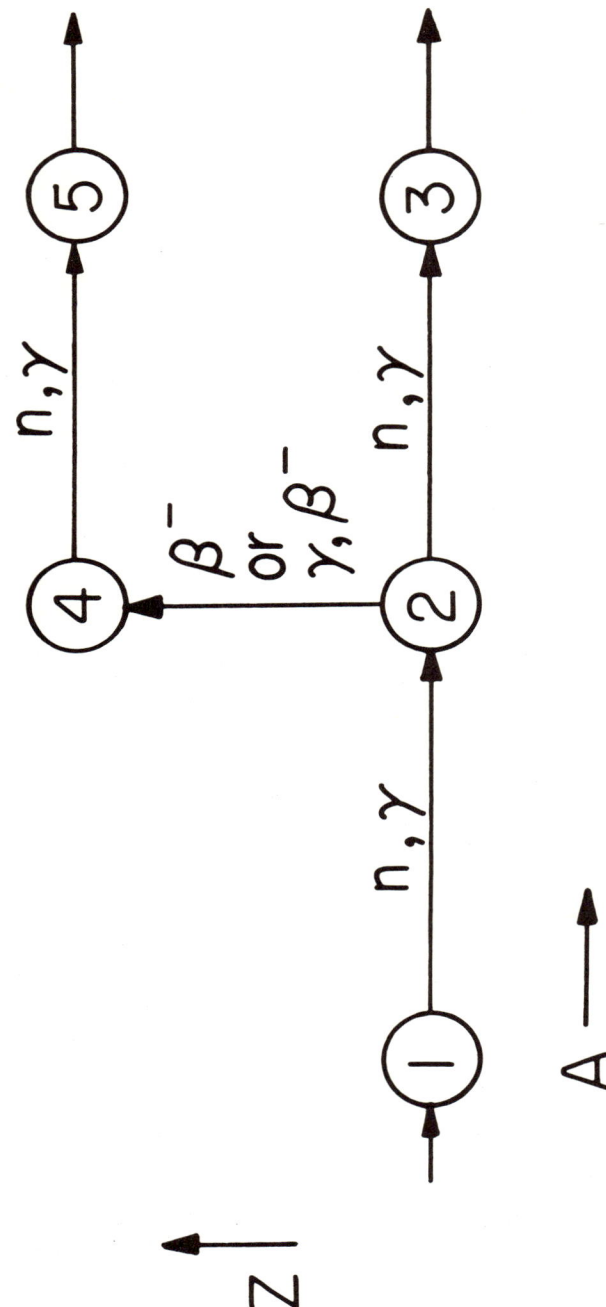

Figure 4: Flow diagram for an archetype branch at nucleus 2. Some of the nuclei 2 formed during the flash can decay during the flash; all nuclei 2 are assumed to decay between the flashes. This general scheme also applies to electron capture decays. The abundance of nucleus 1 is assumed constant.

available evidence from meteoretic abundances suggests that the first decay occurred about 1/4 of the time, the second decay, about 80% of the time or more and the third decay, more than half the time. When the neutron exposure occurs continuously, then the above decays are sufficient to determine the flow of a nucleus through the valley of β stability. There are a few difficulties with the continuous s-process which serve to put us on guard even though they are not major flaws. 1) The simultaneous production of the correct branching ratios at ^{79}Se and ^{85}Kr requires fairly special physical conditions since the decay of ^{79}Se occurs through an excited state and the rate is very temperature sensitive. 2) The abundance of ^{152}Gd is about typical of p-process abundances while the abundance of ^{164}Er is only slightly less than typical s-process abundances and nearly a factor of 10 greater than p-process abundances. If essentially all the ^{154}Eu with a 16 year half-life is assumed to decay to ^{154}Gd, then it is likely that some of the ^{151}Sm with an 87 year half-life will also decay. This latter decay could easily lead to overproduction of ^{152}Gd. The branching at ^{152}Eu between electron capture producing ^{152}Sm, and β$^-$ decay producing ^{152}Gd could help with this problem providing the electron capture is favored by a high density. A high density however would favor the electron capture by ^{163}Ho rather than the possible (γ, β$^-$) decay of ^{163}Dy which leads to the production of ^{164}Er on the s-process.

In an intermittent s-process as is implied by Weigert's model, the possibility arises for nuclei to decay between periods of neutron irradiation. While most unstable nuclei will decay during the 3,000 - 4,000 year interval between flashes, this decay can only change the branching ratio if the neutron exposure during the flash is sufficiently small that few of these unstable nuclei capture a neutron after they are formed. The archetype branch in Figure 4 illustrates the decay of the unstable nucleus 2. As any nucleus continues to capture neutrons it will eventually enter the branch at nucleus 1. An additional neutron capture converts nucleus 1 to nucleus 2. At this point there are the following three alternatives: 1) nucleus 2 can decay during the neutron irradiation period to form nucleus 4; 2) nucleus 2 can capture a neutron to form nucleus 3; 3) nucleus 2 can remain until the end of the irradiation period and then decay to form nucleus 4. If either alternative 1) or 2) occurs, then the daughter nucleus can undergo further neutron captures. If alternative 3) occurs most of the time then the decay of nucleus 2 is independent of its half-life. This alternative is important for the long-lived species where alternative 1) is unlikely. In this case alternative 2) is unlikely as long as $\sigma_2 \Delta\tau < 1$. The time scale determining species ^{79}Se and ^{85}Kr are near magic neutron number nuclei and have relatively small cross sections (I have estimated these to be

Ulrich 150

200 and 50 m-barns respectively). Choice of $\Delta\tau \simeq 10^{-2}$ then permits both nuclei to decay between flashes.

All decays with $t^{1/2} > 20$ days have been studied quantitatively by assuming the abundance of nucleus 1 is constant and integrating the equations

$$\frac{dN_2}{d\tau} = \sigma_1 N_1 - \sigma_2 N_2 - \frac{a}{t_{1/2}} N_2$$

$$\frac{dN_3}{d\tau} = \sigma_2 N_2 - \sigma_3 N_3 \qquad (8)$$

$$\frac{dN_4}{d\tau} = \frac{a}{t_{1/2}} N_2 - \sigma_4 N_4$$

$$\frac{dN_5}{d\tau} = \sigma_4 N_4 - \sigma_5 N_5$$

from $\tau = 0$ to $\tau = \Delta\tau$ subject to the conditions

$$N_2(0) = 0$$
$$N_3(0) = r\, N_3(\Delta\tau) \qquad (9)$$
$$N_4(0) = r\, N_2(\Delta\tau) + N_4(\Delta\tau)$$
$$N_5(0) = r\, N_5(\Delta\tau).$$

The quantity a is an adjustable parameter related to the neutron density:

$$a = \frac{\ln 2}{v_t N_n} \qquad (10)$$

All abundances are then averaged over $\Delta\tau$. Neutron capture by nucleus 4 for several branches of interest forms a short-lived isotope which decays to nucleus 5. In this case nucleus 5 should be shown with Z one unit greater than nucleus 4. I have neglected this additional decay in all cases since the rapid decay prevents the formation of a significant quantity of these nuclei.

Results for branching at ^{79}Se and ^{85}Kr are shown in Figure 5. The branching ratio B_3^{-1} is defined to be $(\sigma_3 + \Lambda) < N_3 > / (\sigma_1 N_1)$ where $<N_3>$ is the average of N_3. The observed branching ratios were obtained from the relative abundances of ^{80}Kr and ^{80}Se for the ^{79}Se decay and from the abundances of ^{86}Kr relative to an average of ^{82}Kr and ^{84}Kr for the ^{85}Kr decay. Both decays can be accounted for by $\Delta\tau = 0.01$ to 0.02 independent of temperature and neutron density. The remaining branches indicate a $\simeq$ 350 milli barn years. Use of $\Lambda = 4.8$ as indicated from the fit to meteoretic data given below allows a complete determination of the flash properties. The best fit is with $\Delta\tau = 0.016$, $r_{-3} = 0.927$, Δt(flash) = 8 years, and at T = 22 Kev, $N_n = 2 \times 10^8$ cm^{-3}.

Application to Meteoritic Abundances

The abundances given in Mason (1971) have been resolved into s-process and r-process components. Table 2 gives the adopted elemental abundances. All abundances in this section are on the log N_{Si}=6 scale. The rare earth abundances were taken from Cameron (1968) since those given in Mason's book were not normalized to Cl meteorites. The s-process abundances were multiplied with the cross sections given by Allen, Gibbons and Macklin (1971) and the resulting σN_S curve is shown in Figure 6. The computed values of σN_S which are shown in this figure are based on $\Lambda = 4.8$. The shell region abundances given by Equation (5) have been diluted by a factor of 150 in order to obtain the computed points. All branching corrections have been applied using the flash parameters given above. It is evident that the existence of branching is important for nuclei between the Ba peak and the Pb peak and that the branching improves the fit to the data. Of special interest is the fact that branching permits a choice of a large value of σN_S in this region and this in turn reduces the gradient of σN_S near Ba.

Some nuclei are produced predominantly by the s-process but have a small contribution from the r-process. For these nuclei an r-process abundance must be subtracted from the total abundance in order to obtain an s-process abundance. Conversely, nuclei produced predominantly but not entirely by the r-process must be obtained by subtracting the computed s-process abundances. The resolution must be done in such a way that N_r and σN_s are both smooth. The adopted r-process abundances are shown in Figure 7. The small dots are abundances which are interpolated in such a way as to produce the most consistent s-process abundances. The large dots represent abundances which are essentially the low interpolated abundances between A = 83 and A = 95. In particular, ^{87}Rb is formed mostly from neutron capture on the predominantly s-process nucleus ^{86}Kr. Thus the ^{87}Rb - ^{87}Sr chronometer is primarily an s-process chronometer.

TABLE 2

Adopted Elemental Abundances

Z - Element	Suess Urey	Cameron (1968)	Mason (1971)	Notes
26 Fe	6.0×10^5	8.9×10^5	9.01×10^5	
27 Co	1800	2300	2300	
28 Ni	2.74×10^4	4.57×10^4	4.8×10^4	
29 Cu	212	919	540	
30 Zn	486	1500	1600	
31 Ga	11.4	45.5	48.	
32 Ge	50.4	126	133.	
33 As	4.0	7.2	6.6	
34 Se	67.6	70.1	69.	
35 Br	13.4	20.6	17.5	
36 Kr	51.3	64.4	45.	1
37 Rb	6.5	5.95	7.1	
38 Sr	18.9	58.4	27.	
39 Y	8.9	4.6	4.8	
40 Zr	54.5	30	28.	
41 Nb	1.00	1.15	1.15	2
42 Mo	2.42	2.52	4.0	
44 Ru	1.49	1.6	1.9	
45 Rh	0.214	0.33	0.36	3
46 Pd	0.675	1.5	1.3	4
47 Ag	0.26	0.5	0.96	
48 Cd	0.89	2.12	2.4	
49 In	0.11	0.217	0.21	
50 Sn	1.33	4.22	3.6	
51 Sb	0.246	0.381	0.36	
52 Te	4.67	6.67	5.8	
53 I	0.80	1.41	1.09	
54 Xe	4.00	7.10	7.1	5
55 Cs	0.456	0.367	0.37	
56 Ba	3.66	4.7	4.8	
57 La	2.00	0.36	0.36	10
58 Ce	2.26	1.17	1.17	10
59 Pr	0.40	0.17	0.17	10
60 Nd	1.44	0.77	0.77	10
62 Sm	0.664	0.23	0.23 10	10
63 Eu	0.187	0.091	0.091	10
64 Gd	0.684	0.34	0.34	10
65 Tb	0.0956	0.052	0.052	10.
66 Dy	0.556	0.36	0.36	10
67 Ho	0.118	0.090	0.090	10
68 Er	0.316	0.22	0.22	10
69 Tm	0.0318	0.035	0.035	10
70 Yb	0.220	0.21	0.21	10
71 Lu	0.050	0.035	0.035	10
72 Hf	0.438	0.16	0.47	
73 Ta	0.065	0.055	0.019	6
74 W	0.49	0.71	0.16	6

Table 2 (continued)

Z - Element	Suess Urey	Cameron (1968)	Mason (1971)	Notes
75 Re	0.135	0.055	0.052	
76 Os	1.00	0.71	0.72	
77 Ir	0.821	0.43	0.61	
78 Pt	1.625	1.13	1.2	
79 Av	0.145	0.20	0.19	
80 Hg	0.284	0.75	0.5	7
81 Tl	0.108	0.182	0.121	
82 Pb	0.47	2.9	3.2	8
83 Bi	.144	0.164	0.16	
90 Th	- -	0.034	0.043	
92 U	- -	0.0234	0.036	9

Notes:

1. The Kr abundance is reduced from that given by Cameron in order to follow the decrease in Sr abundance.

2. The Nb abundance is essentially unknown. Cameron's value has been adopted.

3. The Rh abundance is from the Abee meteorite, type E1.

4. The Pd abundance omits the discordant data of Greenland (1967) Geochim. Cosmochim Acta 31, 1733.

5. The Xe abundance is established by the r-process peak at ^{129}Xe and the x-only isotope ^{128}Xe. The r-only isotopes ^{134}Xe and ^{136}Xe appear overabundant and can be processed by fission cycling on the r-process.

6. The W and Ta abundance is for the Murray meteorite type C2.

7. The Hg abundance is based on an r-process extrapolation for ^{199}Hg and an s-process interpolation for ^{202}Hg.

8. The Pb abundance for the Orgueil meteorite (type C1) has been reduced 20% to account for contamination.

9. Abundances 4.7 billion years ago. 25% was ^{235}U at that time.

10. Rare earth abundances taken from Cameron (1968) since those in Mason's book were not normalized to C1 chondrites.

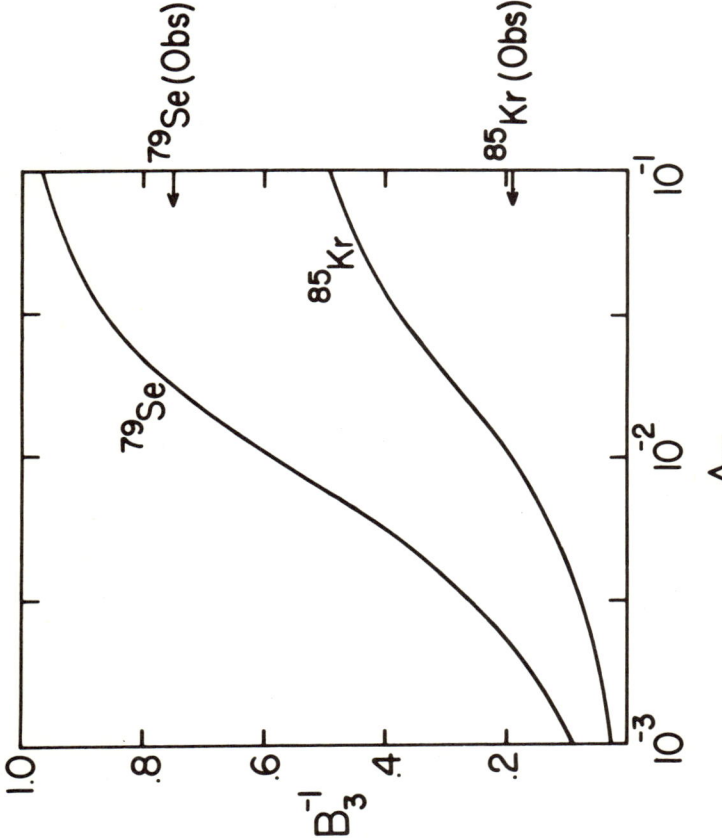

Figure 5: Comparison of calculated and observed branching ratios at ^{79}Se and ^{85}Kr. Changes in the halflife of ^{79}Se or the neutron density by a factor of two from the values given in the text change B_3^{-1} by less than 10%.

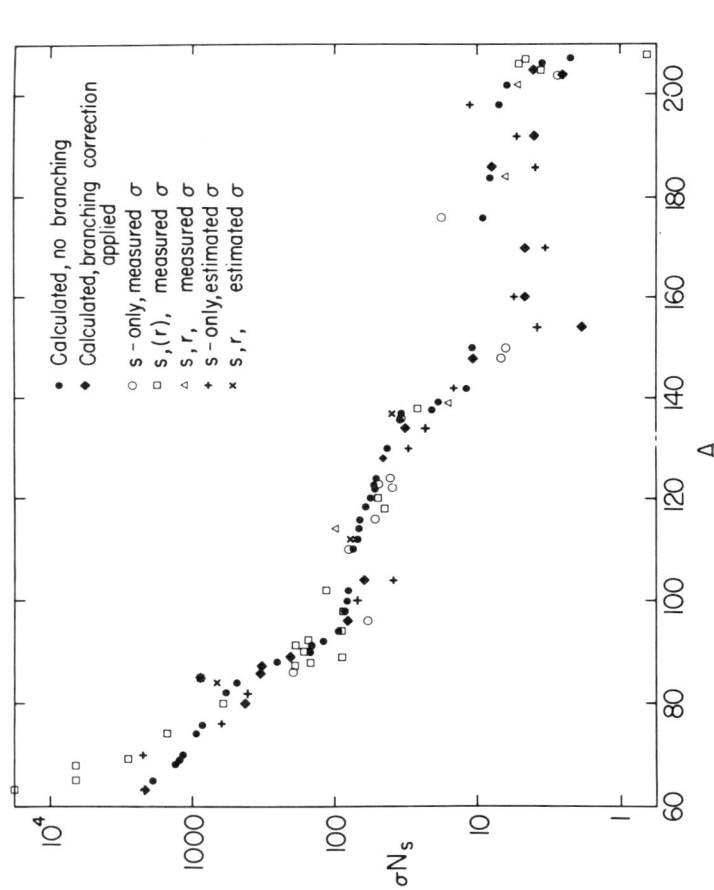

Figure 6: Comparison between calculated and observed σN_s points. The calculated points are based on $\Lambda=4.8$, $\alpha=350$ millibarn-years and $\Delta\tau=0.016$. The shell matter has been diluted a factor of 150 in the meteorites to obtain this level of abundance. The breaking point between □ and △ is at $N_r/N_s = 0.25$. No cases with $N_r/N_s > 0.7$ are included and only two cases ^{98}Mo and ^{102}Ru have $N_r/N_s > 0.4$.

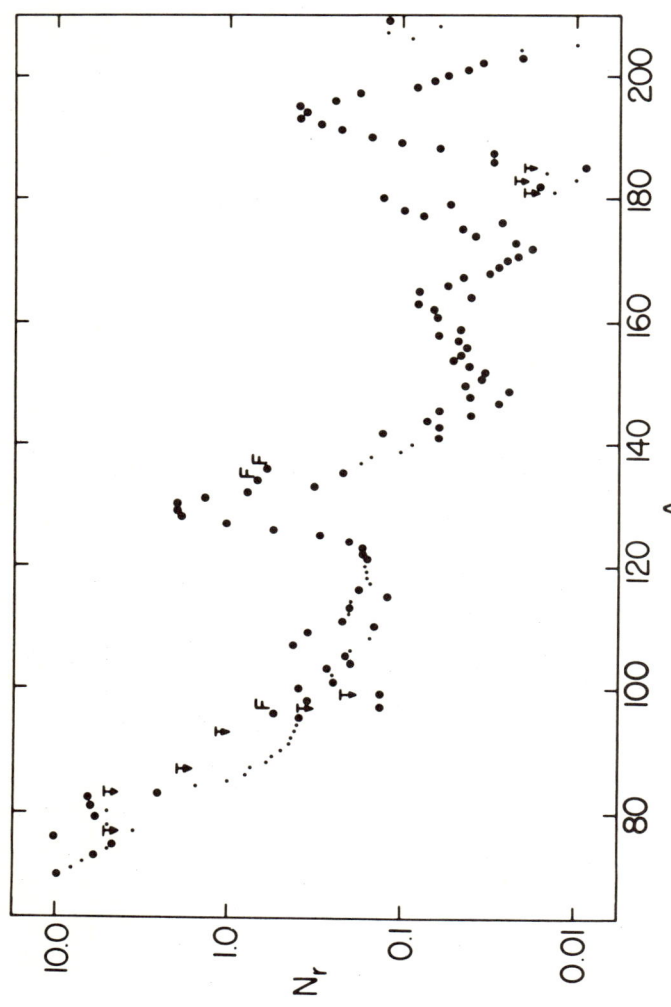

Figure 7: The adopted r-process abundances vs. A. The large dots are abundances which are nearly independent of the s process abundances. The small dots have been interpolated so as to yield a smooth distribution of N_r and a smooth distribution of σN_s. The value of N_r between $A = 85$ and $A = 96$ is smaller than previous resolutions because ^{86}Kr and ^{87}Rb are primarily synthesized on the s process according to the model discussed in the text. The peak between $A = 177$ and $A = 180$ is due to a high abundance of Hf. This peak is probably mostly spurious. Probable fission products are labeled with an F.

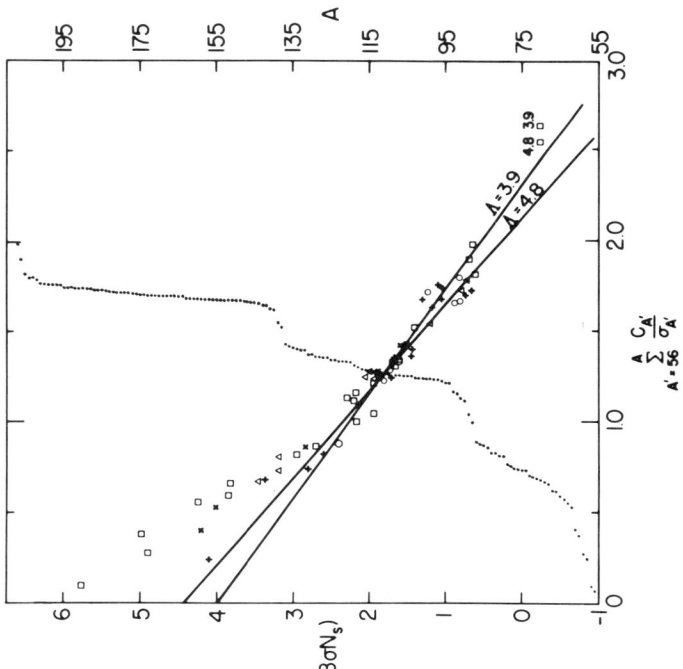

Figure 8: The σN_c values plotted against $\sum_{A'=56}^{A} C_{A'}/\sigma_{A'}$. The values of $\Gamma(y) \equiv y^{-1} \ln(1+x)$ were computed for $x = 4.8/\sigma$. For reference A is shown as a function of this variable by the dots and the values of A are given on the right hand side. The position of the ^{208}Pb point is a sensitive function of Λ and positions are shown for $\Lambda = 4.8$ and $\Lambda = 3.9$.

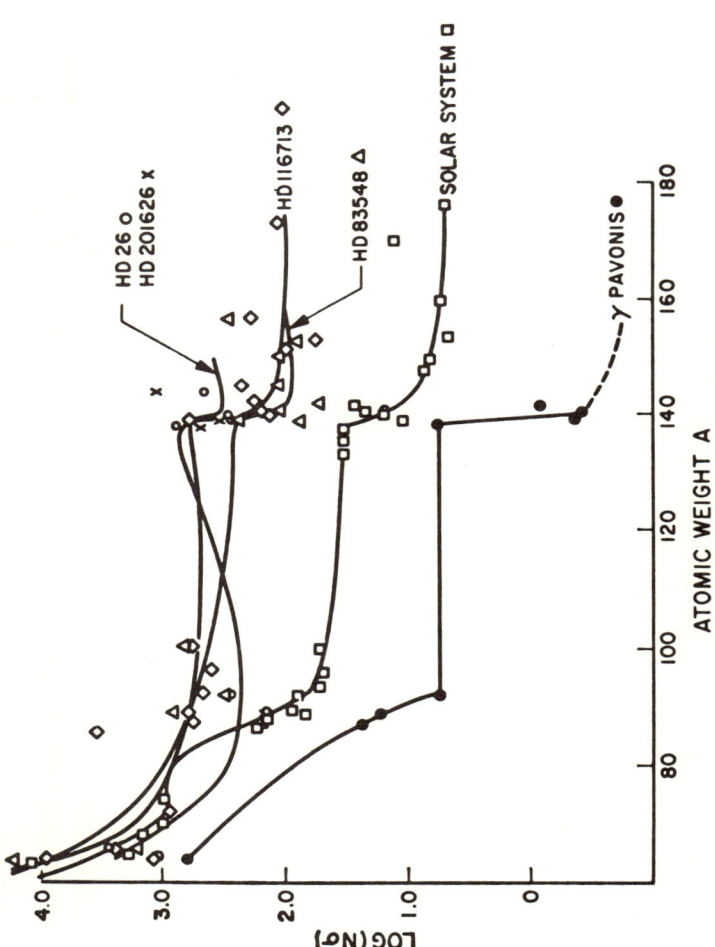

Figure 9: The distribution of σN_s vs. A observed on the surfaces of four evolved stars as given by Danziger (1966).

Figure 8 shows a test of Equation 6. It is seen that the curve is linear between A = 70 and A = 207. The nuclei between A = 57 and A = 69 also lie along a straight line with a slope corresponding to Λ = 9.5. The dilution factor for these nuclei is about 6 to 10. The very low degree of dilution suggests that these nuclei were synthesized at about the same time as the ^{56}Fe.

Evolution of Stellar Envelopes

Figure 9 is a reproduction of a figure published by Danziger (1965) which shows the observed abundances in several evolved stars compared to solar system abundances and abundances in Y Pav. The evolved stars are BaII stars (HD116713, HD83548) and CH stars (HD26 and HD201626). Note especially that the overabundance of the elements is a relatively uniform function of A and does not show the peaked structure found by CFHZ to be characteristic of a single neutron exposure. <u>The stars are clearly able to produce internally a distribution of neutron exposures</u> and the mechanism discussed above seems to be a plausible explanation. The steady state analysis given above is not appropriate for the earliest flashes, and R.B. Carlos and I have applied the formalism given by CFHZ to study these flashes. We have assumed that each flash adds an invariant number of neutrons and a set of fresh seed to the shell. We keep track of when each set of nuclei was added to the shell. After each flash we determine the total neutron exposure of each set of nuclei and the fraction of those which still remain in the shell region. We then add up the abundance distributions for all nuclei which are still present in the shell and determine the total abundance distribution for the shell region. A portion of the shell matter is then added to the outer envelope. Because the outer envelope is much more massive than the shell region in Weigert's model, the heavy element abundances are greatly diluted for the first flashes.

Figure 10 shows the evolution of a model envelope when the flash parameters are chosen to conform with the solar system abundance requirements. The fraction of the shell matter deposited in the envelope, f, must be smaller than 1-r. Consequently, the low value of 1-r of 0.073 required by the ^{85}Kr and ^{79}Se branches results in a very slow build up of heavy element abundances in the envelope. This set of parameters requires the inner edge of the shell zone to advance less than 5% of the shell mass between flashes. According to Weigert's calculations this advance should be more like 20 - 30% of the shell mass. I believe a likely cause for this discrepancy is the fact that plume mixing causes the outer edge of the shell zone to move inward. This deposits hydrogen at a smaller fraction than before plume mixing. This additional hydrogen must be burned

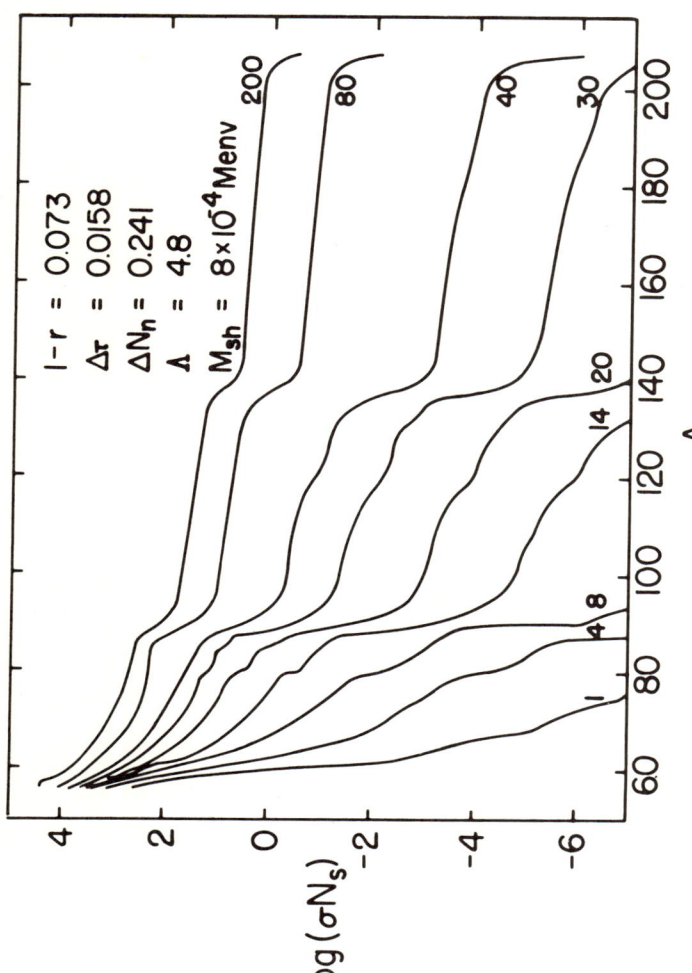

Figure 10: The evolution of the envelope as a function of the number of flashes. Each flash deposits 5% of the shell mass in the envelope. This set of parameters matches the ^{85}Kr and ^{79}Se branching ratio. The parameter ΔN_n is the number of neutrons captured per heavy nucleus in the shell region.

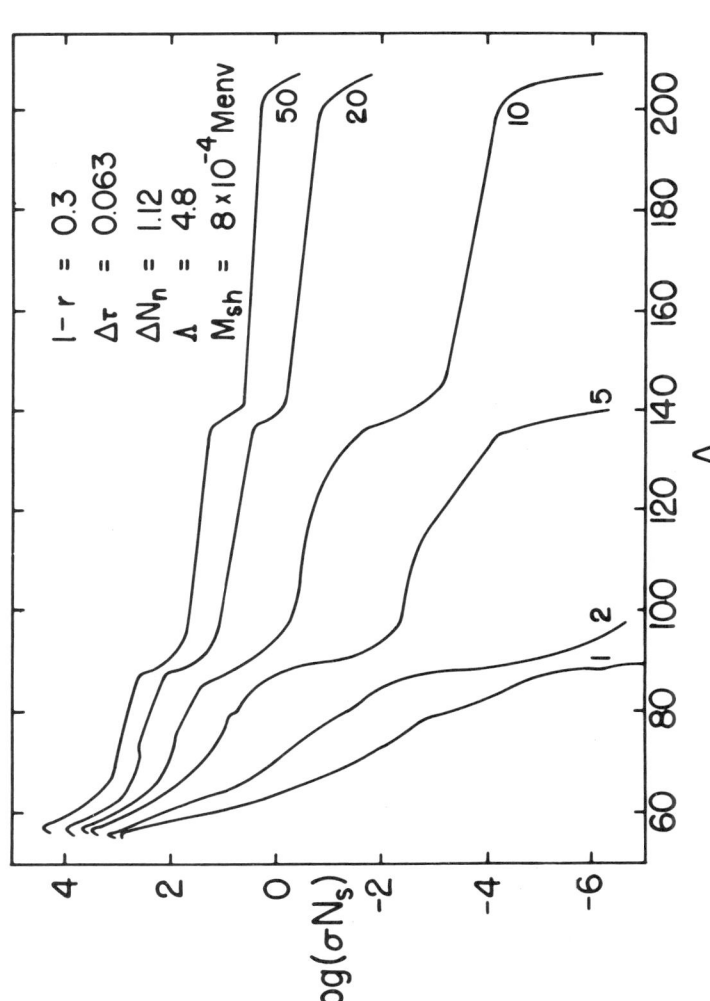

Figure 11: Same as figure 10 only with $\Delta\tau$ and $1-r$ both increased a factor of about four.

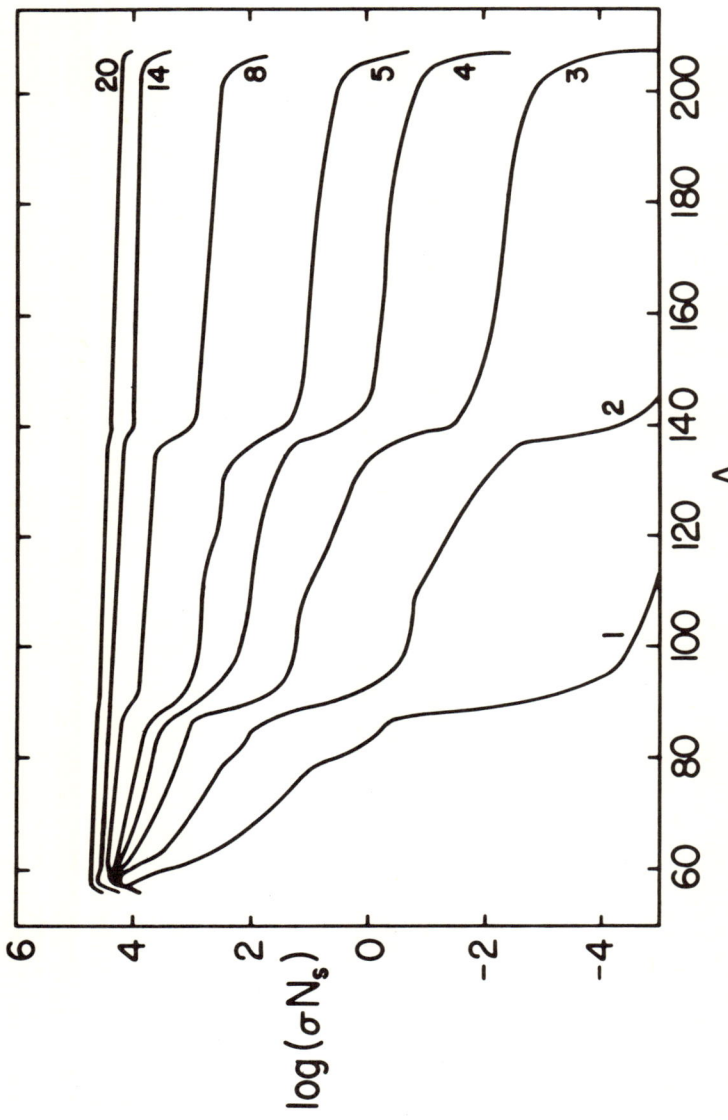

Figure 12: Same as figure 10 only with $\Lambda = 1.0$, $\Delta\tau = 1-r = 0.062$, $\Delta N_n = 2.2$, $M_{sh} = 0.1$ Menv.

before the He shell can re-ignite. The additional time required to consume the extra hydrogen permits thermal transients introduced by the preceeding flash to die away before the He shell ignites. Therefore the instability of the shell can set in immediately when He burning begins. This effect may change the lifetime of the star in the two-shell phase of evolution. The evolution of the envelope is much more rapid when a value of $1-r = 0.3$ is used as indicated by Weigert's calculations. This result is shown in Figure 11. After about the 80th flash in the first case and the 20th flash in the second, the shell abundances have reached a steady state and further flashes merely enrich the envelope. The steady state abundances are in good agreement with the simple theory given above.

The final application of the model shown in Figure 12 is prompted in part by the observation by Kraft (1973) that FG Sge has increased its surface abundances of s-process elements by a factor of about 6 in the last eight years. This star is at the center of an old planetary nebula and is currently decreasing its temperature rapidly. It probably ejected a portion of its outer envelope some time ago as a result of a He shell flash but enough matter remains for the star to once again become a red giant. It is now cool enough for the heavy elements synthesized in the latest flash to be swepth to the surface by a deepening convective envelope. Both elements in the Y - Zr s-process peak and in the La s-process peak appear to be enhanced the same amount. The model shown in Figure 12 is appropriate to a low mass star ($1-1.5 M_\odot$) in which the envelope is only 10 times as massive as the shell. The final abundance distribution could be added to the solar system abundances and produce ^{208}Pb alone. At the 5th flash the surface abundances all increase by about a factor of 6 as required by Kraft's observations of FG Sge. The existence of a planetary nebula around this star would also seem to favor the low mass interpretation proposed here.

Discussion and Summary

The following are the important conclusions of this work:
1) It is possible to synthesize all s-process nuclei between $A = 70$ and $A = 205$ within a single star by means of a repeated series of He shell flashes. About one gram out of 150 grams of iron peak nuclei in the solar system has been processed through such a shell flash region. For the optimal parameters used in Figure 10, at least 80 flashes are required to come to the final distribution of abundances. Also for this set of parameters about 5% of the mass between the hydrogen and helium shell sources can be enriched in s-process elements and mixed into the outer envelope each flash. Surface abundances of heavy elements in carbon stars obtained by Utsumi (1970) and

Kilston (1973) suggest that about 10% of the outer envelopes of these stars has been processed through the shell region. This processing fraction is compatible with the plume mixing model of Scalo and Ulrich (1973).

2) ^{87}Rb is primarily an s-process chronometer. Therefore, the majority of the s-process nucleosynthesis occurred at about the same time as the r-process nucleosynthesis since the ^{87}Rb - ^{87}Sr chronology indicates an age similar to the ^{187}Re - ^{187}Os and the U-Th - Pb chronologies.

3) Some s-process with mixing probably occurred immediately after the formation of the iron peak nuclei. This s-process synthesis is needed to explain the nearly linear distribution of σN_s vs $\Sigma C/\sigma_A$, shown on Figure 8 between A = 57 and A = 75. One gram out of six of ^{56}Fe has apparently been processed this way. Because of the probable presence of protons there is no reliable information from β decays which could establish a time scale for the process.

4) The increase in heavy element abundances observed by Kraft to occur in FG Sge can best be explained as an event in a low mass star. This is primarily because of the favorable ratio of shell mass to envelope mass and because of the limitation on the fraction of the shell mass which can be processed in each flash. The factor 1-r used in constructing Figure 12 could be increased from 0.06 to 0.3 without substantially changing the shape of the σN curves. The ratio of shell mass to envelope mass could then be decreased by a factor of five and still permit a 6-fold increase of surface heavy element abundances in a single flash. No set of parameters in Weigert's 5M$_\odot$ model would permit such a change to occur in a single flash.

The solution given in Equations 5 - 7 is valid under conditions considerably less restricted than the repeated shell flash model and the plume mixing model. It requires only than the s-process occurs in a convective or mixed region and that fresh seed nuclei and neutrons are added to the mixed region at the same rate. The resultant s-process abundances depend rather sensitively on the relative rates of addition of seed nuclei and neutrons. Other models for s-process nucleosynthesis may exist which possess the required properties to yield the abundance distribution in Equations 6 - 8. It is important to show for any such model that the correct ratio of neutrons to seeds is at least plausible. While the dynamics of the mixing interface are complex, it is at least plausible that the correct ratio can be produced in the plume mixing model.

I would like to thank Prof. Kraft for informing me of his observations of FG Sge and for permission to quote these observations prior to publication. Numerous discussions with W. A. Fowler about this work have been extremely helpful. L. H. Aller, J. M. Scalo and S. Kilston have kindly read the manuscript and provided me with several helpful suggestions. The computations by Mr. Carlos and myself were supported by funds granted by the UCLA Campus Computer Network. A major portion of this work was done while I was supported by the Kellogg Radiation Laboratory through grants GP-28027 and GP-36687X.

REFERENCES

Allen, B. J., Gibbons, J. H., and Macklin, R. L. 1971, Adv. in Nucl. Phys., 4, 205.
Burbidge, E. M., Burbidge, G. R., Fowler, W. A. and Hoyle, F. 1957, Rev. Mod. Phys., 29, 547.
Cameron, A. G. W. 1954, Phys. Rev., 93, 932.
Cameron, A. G. W. 1955, Ap. J., 121, 144
Cameron, A. G. W. 1968, "A New Table of Abundances", in Origin and Distribution of the Elements, Ahrens, ed. (New York: Pergamon Press).
Caughlin, G. R. and Fowler, W. A. 1964, Ap. J., 139, 1180.
Clayton, D. D., Fowler, W. A., Hull, T. C. and Zimmerman, B. A. 1961, Ann. Phys., 12, 121. (CFHZ)
Danziger, I. J. 1966, Ap. J., 143, 527.
Fowler, W. A., Burbidge, G. R. and Burbidge, E. M. 1955, Ap. J., 122, 271.
Kraft, R. B. 1973, in preparation.
Mason, B. 1971, Handbook of Elemental Abundances in Meteorites, (New York: Gordon and Breach).
Morton, R. B., Taylor, G. I. and Turner, J. S. 1956, Proc. Roy Soc. London A., 234, 1.
Sanders, R. H. 1967, Ap. J., 150, 971.
Scalo, J. M. and Ulrich, R. K. 1973, Ap. J., July 1, in press.
Schwarzschild, M. and Harm, R. 1967, Ap. J., 150, 961.
Seeger, P. A., Fowler, W. A. and Clayton, D. D. 1965, Ap. J. Supp., 97, 121.
Smith, R. L. and Sachmann, I. J. 1973, preprint.
Stroud, D. B. 1972, Ap. J., 178, L93.
Ulrich, R. K. and Scalo, J. M. 1972a, Ap. J., 176, L37.
Ulrich, R. K. and Scalo, J. M. 1972b, paper given at the Red Giant Star Conference, Bloomington, Ind.
Utsumi, K. 1970, Pub. Ast. Soc. Japan, 22, 93.
Weigert, A. 1966, Zeits F. Ap., 64, 395.

Ulrich

Appendix

The s-Process in a Convective Region

Consider a convective region of Mass M_S and suppose that the matter within this region has been previously exposed to neutrons. Let a small blob of unexposed matter with a mass δM be added to the convective region. The abundance per gram of an element of atomic mass A is N_A in the convective region and is either zero for A=56 or N_{56}^o for ^{56}Fe in the blob. (For the purposes of this derivation I am considering ^{56}Fe as the only seed nucleus. Similar arguments apply to each additional seed nucleus and the total abundance of any nucleus is the sum of the contributions from all seed nuclei.) Associated with the addition of the blob is the release of a quantity of free neutrons. It is simplest to consider these neutrons as resulting from the conversion to free neutrons of some species with an abundance N_n^o in the blob. In fact, the only necessary feature of the model is that the neutrons, seed nuclei and matter deficient in heavy nuclei are added to M_S simultaneously or nearly simultaneously. As a result of the release of the neutrons, all nuclei in M_S receive a neutron exposure of $\delta\tau = \rho v_t \langle N_n \rangle \delta t$. The convective region must be sufficiently well mixed that all nuclei receive the same average neutron exposure. The average free neutron abundance $\langle N_n \rangle$ and δt are determined by the condition that all the neutrons are captured.

Consider first the balance of species A where A≠56. The number of A before δM is added to M_S is $M_S N_A$. As δM is added to M_S the number of A formed by neutron capture is $M_S N_{A-1} \sigma_{A-1} \delta\tau$ and the number of A destroyed by neutron capture is $M_S N_A \sigma_A \delta\tau$. After δM is added and the region is completely mixed the number of A is $(M_S+\delta M)(N_A+\delta N_A)$. Balance of A gives

$$(M_S+\delta M)(N_A+\delta N_A) = M_S N_A + M_S N_{A-1}\sigma_{A-1}\delta\tau - M_S N_A \sigma_A \delta\tau. \quad (A1)$$

For ^{56}Fe the term $M_S N_{A-1}\sigma_{A-1}\delta\tau$ is replaced by $\delta M N_{56}^o$. After dividing by $M_S \delta\tau$ canceling $M_S N_A$ the following equations result

$$\frac{\delta N_A}{\delta\tau} = \sigma_{A-1}N_{A-1} - \sigma_A N_A - \frac{\delta M}{M_S \delta\tau} N_A \quad (A2)$$

$$\frac{\delta N_{56}}{\delta\tau} = \frac{\delta M}{M_S \delta\tau} N_{56}^o - \sigma_{56} N_{56} - \frac{\delta M}{M_S \delta\tau} N_{56} \quad (A3)$$

The number of neutrons added is $\delta M N_n^o$ while the number captured is $M_s \sum_{A=56}^{209} N_A \sigma_A \delta\tau$. The value of $\delta\tau$ is therefore determined by

$$\frac{\delta M}{M_s \delta\tau} N_n^o = \sum_{A=56}^{209} \sigma_A N_A \qquad (A4)$$

These correspond to Equations (1) - (3) of the text as long as $\delta N_A = 0$, δM is identified with $(1-r)M_s$ and M_s of this appendix is identified with $(1-f/2)M_s$ of the text. Removal of mixed matter only determines the rate of change of M_s and does not alter the relative abundance.

NUCLEOSYNTHESIS IN RED GIANTS

Richard L. Smith, I-Juliana Sackmann and Keith H. Despain

California Institute of Technology, Pasadena, California

I. Direct Convective Mixing

The stage of stellar evolution considered here is the red giant phase during which the star contains both a hydrogen-burning and a thermally unstable helium-burning shell source. Schwarzschild and Härm (1967) have shown that during some of the helium shell flashes in a 1 $M_\odot$ star, a convective zone extends from the helium burning shell and barely penetrates into the s-shaped hydrogen profile at the inactive hydrogen burning shell. Although the molecular weight gradient at the hydrogen-burning shell produces a situation which is highly stable against convection, we suggest, as have Cameron and Fowler (1971), that the penetration just described, and perhaps in combination with convective overshooting, may transport sufficient hydrogen inward to a temperature $T \simeq 1 \times 10^8$ °K where the $\rho(^{12}C,\alpha)^{13}N$ reaction and subsequent reactions can provide a rapid release of energy to extend the convective zone into hydrogen-rich layers.

Should the flash-driven convective zone penetrate into hydrogen-rich layers, the mixing of elements produced in the interior to the surface might be imagined to occur either directly or indirectly. Detailed stellar evolution calculations are required to determine the extent of the convective regions as a function of time. Here, we make the simplest assumption. We assume that the flash-driven convective zone temporarily joins the convective envelope, resulting in a convective zone which extends from the surface down to a temperature $T \simeq .5 - 2 \times 10^8$ °K.

II. Combined Mixing and Nucleosynthesis

Given a deep, temporary convective envelope as described above, it can not be assumed that the abundance of each element is spatially constant, since at the base of the convective envelope, the lifetimes of some elements against reactions are of the same order of magnitude, or less than, the convective mixing time across the envelope (~10^8 sec). We

Smith et al.

describe simultaneous nucleosynthesis and convective mixing by means of a diffusion equation with source terms. For each species i, and for a convective envelope with time-independent runs of density and temperature, this equation is

$$\rho \frac{\partial X_i}{\partial t} = \frac{1}{r^2} \frac{\partial}{\partial r} (r^2 \rho D \frac{\partial X_i}{\partial r}) + \sum_{j,k} X_j X_k T_{jk} - \sum_{\ell} X_i X_\ell T_{i\ell} \quad (1)$$

X_i is the mass fraction of species i, ρ is the local mass density, r is the radial coordinate, t is the time, and spherical symmetry is assumed. The first term on the right-hand side gives the mass flow rate at each point for element i, and D is the appropriate diffusion constant. The second and third terms on the right describe the sources and sinks for species i due the nuclear reactions. A derivation of Equation (1), without the source terms, is given by Landau and Lifshitz (1959). The source term for the reaction $j + K \to i + \ldots$ is given by

$$X_j X_k T_{jk} = n_i \frac{\rho^2 X_j X_k}{1+\delta_{jk}} \frac{A_i}{A_j A_k} N_A <\sigma v>_{jk}, \quad (2)$$

where $<\sigma v>_{jk}$ is the average reaction rate per particle pair, A_i is the atomic weight of element i, and n_i is the number of i-particles produced in each reaction. The sink term T for the reaction $i + \ell \to k + \ldots$ is given by

$$- X_i X_\ell T_{i\ell} = - \frac{\rho^2 X_i X_\ell}{1 + \delta_{i\ell}} \frac{1}{A_\ell} N_A <\sigma v>_{i\ell}, \quad (3)$$

where N_A is Avogadro's number. The source and sink terms can be suitably modified for beta decay and electron capture reactions. The reaction rates $N_A <\sigma v>_{jk}$ were obtained from the tables of Fowler, Caughlan, and Zimmerman (1972).

A network of 13 nuclear reactions for 11 elements was followed here in detail. The elements considered were

^3He, ^7Be, ^7Li ;
^{12}C, ^{13}C, ^{14}N, ^{15}N;
^{16}O, ^{17}O;
^4He, H.

The reactions included were

^{3}He (α,γ) ^{7}Be (e^-, ν) ^{7}Li, ^{3}He $(^3$He, $2p)\alpha$
^{7}Li (p,γ) ^{8}Be (2α)
^{7}Be (p,γ) ^{8}B (e^+, ν) ^{8}Be (2α)
^{12}C (p,γ) ^{13}N (e^+, ν) ^{13}C (p,γ) ^{14}N (p,γ) ^{15}O
$\qquad\qquad\qquad\qquad\qquad\qquad$ ^{15}O (e^+,ν) ^{15}N (p,α) ^{12}C
^{15}N (p,γ) ^{16}O
^{16}O (p,γ) ^{17}F (e^+, ν) ^{17}O (p, α) ^{14}N
^{13}C (α,n) ^{16}O.

The local diffusion constant D was approximated by the expression

$$D \approx v_c \cdot l_{mix}.$$

v_c, the convective velocity, and l_{mix}, the convective mixing length, were determined from the model for the envelope and the mixing length description of convection. D was typically 10^{14} cm^2/sec near the bottom of the convective zone and 10^{17} cm^2/sec near the surface.

Deep convective envelopes were constructed by integrating the usual stellar structure equations inward from the surface, with a chosen value for the effective temperature and luminosity. In order to construct a convective envelope with a temperature $\sim 10^8$ °K at the base, the local luminosity L_r was increased sufficiently at each layer during the course of the integration so that the condition $\nabla_{rad} > \nabla_{ad}$ was barely maintained.

The results of numerical solutions of Equation (1) will be described here for envelopes based upon two models: (1) the 1 $M_\odot$ model of Schwarzchild and Harm (1967) at the ninth helium-shell flash and (2) the 5 $M_\odot$ model of Weigert (1966) at the sixth shell flash. The mass fractions adopted as initial conditions for Equation (1) were:

	1 $M_\odot$ (Z = .001)		5 $M_\odot$ (Z = .044)	
H	0.,	.9	0.,	.56
He	.5,	.1	.5,	.40
^{12}C	.25,	.14Z	.48,	.14Z
^{16}O	.25,	.42Z	.02,	.42Z
^{14}N	0.,	.05Z	0.,	.05Z
^{3}He	0.,	.002	0.,	.0001

The first number in each column refers to the inter-shell region and the second refers to the region exterior to the hydrogen-burning shell.

III. Surface Enrichment of Carbon, Nitrogen and Lithium

The numerical solutions which we have obtained for Equation (1) show that the parameter which most strongly influences the time behavior of the surface abundances is the temperature at the base of the temporary convective zone (T_b), i.e., the depth of deep mixing.

a. Eruptive Carbon Stars

For the 1 $M_\odot$ model, and for $T_b = 2 \times 10^8$ °K, we find that hydrogen burns quite rapidly as it mixes inward, such that an amount of energy equal to the gravitational binding energy of the envelope is released within a time interval (t_H) of .02 years. Similarly, for $T_b = 1 \times 10^8$ °K, this time interval is about one-tenth of a year. Since the dynamical timescale of a red giant envelope is $10^6 - 10^7$ sec, we would expect this situation to lead to a dynamical event and mass loss, provided the star does not react hydrostatically in a manner to stop the mixing. Should most of the envelope be ejected, the surface abundances of the remnant would be similar to the abundances observed in R Coronae Borealis stars, i.e., very little hydrogen and much helium and carbon. Hydrodynamic calculations are required to determine the exact time and amount of mass loss. However, if mass is lost down to a point interior to the hydrogen-burning shell, the abundances of ^{12}C, ^{16}O, and ^4He in the remnant will be nearly those existing in the inter-shell region prior to mixing. The hydrogen mass fraction X_H predicted by the computed models is low, $X_H = 10^{-6} - 10^{-2}$, depending on the time of envelope shedding. The abundances of ^{13}C, ^{14}N, and ^7Li also depend critically upon the time of mass ejection.

Looking at the composition of the ejected envelope, we find that the calculations show a slight enrichment of carbon ($^{12}C/^{16}O \simeq 1.3$), some enhancement of ^{12}C relative to ^{14}N ($^{14}N/^{12}C$ variable, between .001 and .1), and an enhancement of ^7Li ($X_{^7Li} + X_{^7Be} \simeq 10^{-8}$). There is practically no increase in ^{13}C in the ejected envelope.

b. Non-eruptive Mixing

For envelopes with $T_b = 5 \times 10^7$ °K, hydrogen burns suffi-

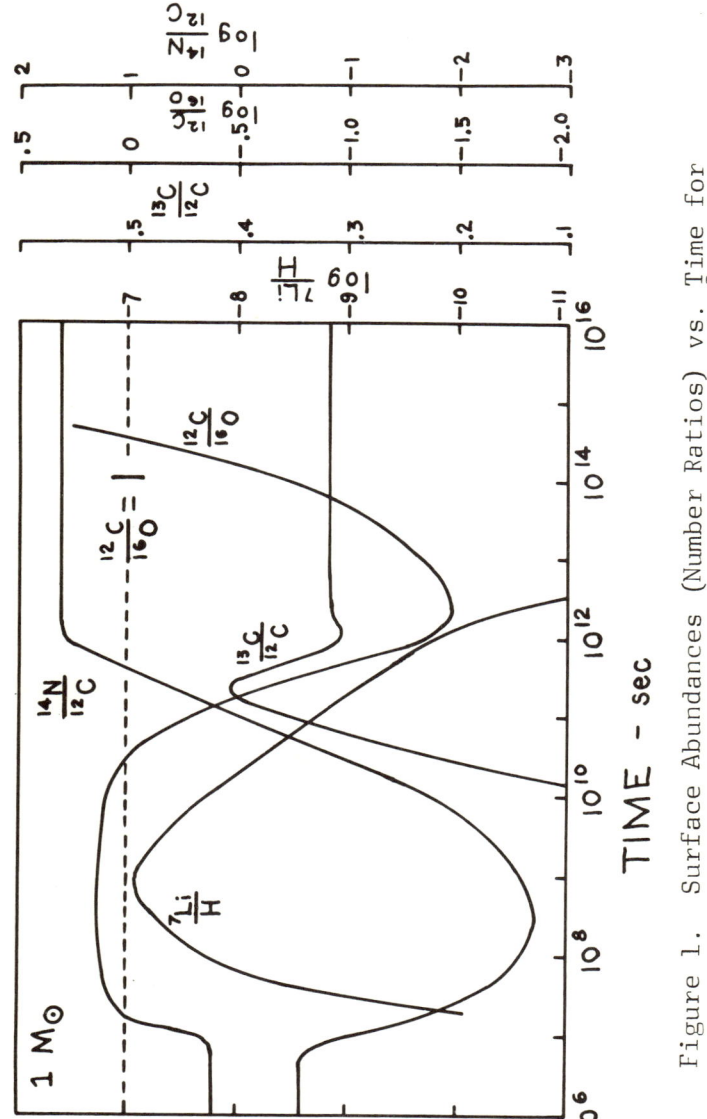

Figure 1. Surface Abundances (Number Ratios) vs. Time for the 1 M$_\odot$ model. Model parameters: L=10^{3}L$_\odot$, log T$_{eff}$=3.519, temperature at base of convective zone ≡ 5.02 × 10^7 °K.

ciently slowly so that $\bar{L}_H$, the time-averaged luminosity produced by hydrogen burning, is close to the surface luminosity adopted for construction of the model envelope, and rapid mass loss is not expected. The behavior of the surface abundances with time (measured from the beginning of deep mixing) is shown in Figures 1 and 2 for the 1 $M_\odot$ model and the 5 $M_\odot$ model. The curves displayed here (except for the $^7Li/H$ curve) can be interpreted as showing the abundances resulting from a single deep mixing phase of a time duration chosen along the horizontal axis, or, roughly, as showing the abundance resulting from a repetitive series of deep mixing flashes, where now the cumulative time of deep mixing is measured along the horizontal axis. As we have solved Equation (1) in the context of a <u>fixed</u> stellar structure, the results are only representative, but could be improved upon by following, simultaneously, the evolution of the star during the mixing process.

We find that ^7Li is produced according to the mechanism proposed by Cameron and Fowler (1971). The maximum lithium content, $^7Li/H = .7 \times 10^{-7}$ and $.5 \times 10^{-7}$, for the 1 $M_\odot$ and the 5 $M_\odot$ models, respectively, is in remarkable agreement with the super-rich lithium stars. WZ Cas has the highest known lithium content with $^7Li/H = (.8 + .2) \times 10^{-7}$ (Wallerstein and Conti 1969). However, the maximum Lithium abundance can be varied by nearly a factor of 10 by varying T_b within the range $4 - 10 \times 10^{-7}$ °K. Except for the reaction ^3He (^3He, 2p)^4He, the equations describing ^7Li production are nearly linear, so that the maximum ^7Li abundance scales nearly linearly with the adopted initial ^3He abundance.

Considering the $^{12}C/^{16}O$ ratio for the 1 $M_\odot$ model in Figure 1, we can divide the time axis into three regions. For short duration (t ≤ 3 × 10^{11} sec) or long duration mixing (t ≥ 3 × 10^{14} sec) $^{12}C/^{16}O > 1$. For intermediate mixing intervals $^{12}C/^{16}O < 1$. On the other hand, for the 5 $M_\odot$ model $^{12}C/^{16}O > 1$ only for t ≥ 3 × 10^{15} sec, due to the relatively low mass of the intershell region.

When $^7Li/H$ is maximum, there is very little ^{13}C. But ^{13}C builds up very rapidly, so it is possible to have a fairly large ^7Li abundance and a large $^{13}C/^{12}C$ ratio. Our calculations predict that the $^{13}C/^{12}C$ ratio in lithium-rich carbon stars to be highly variable. The peak in the $^{13}C/^{12}C$ curve occurs when ^{13}C has attained equilibrium with ^{12}C, but ^{12}C is not in equilibrium with ^{15}N. The final $^{13}C/^{12}C$ equilibrium value is reached as ^{12}C comes to equilibrium with ^{15}N.

We wish to thank Professor William A. Fowler for many helpful suggestions concerning this problem. This work was supported in part by the National Science Foundation (GP-27304, GP-28027).

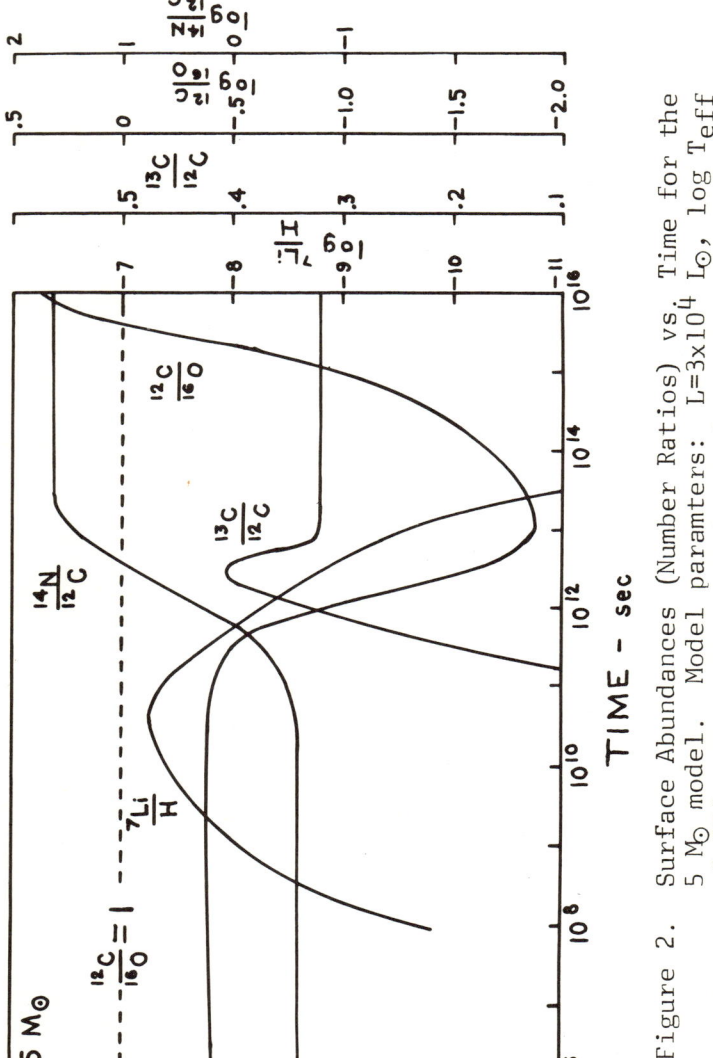

Figure 2. Surface Abundances (Number Ratios) vs. Time for the 5 $M_\odot$ model. Model paramters: $L=3\times10^4\ L_\odot$, log T_{eff} =3.515, temperature at base of convective zone = 5.02×10^7 °K.

REFERENCES

Cameron, A.G.W., and Fowler, W. A. 1971, Ap. J., 164, 111.
Fowler, W. A., Caughlan, G. R. and Zimmerman, B. A. 1972, private communication.
Landau, L. D. and Lifshitz, E. M. 1959, Fluid Mechanics (Reading, Mass: Addison-Wesley).
Schwarzchild, M. and Härm, R. 1967, Ap. J., 150, 961.
Wallerstein, G. and Conti, P. S. 1969, Ann. Rev. Astron. and Astrophys., 7, 99.
Weigert, A. 1966, Z. Astrophysik., 64, 395.

CHAPTER IV. The Carbon Detonation Model

One of the more uncertain aspects of the evolution of stars in the mass range $4 \leq M/M_\odot \leq 8$ is what happens when $^{12}C + ^{12}C$ ignites. If the ignition occurs first at the center it apparently occurs under conditions of extreme electron degeneracy. If the rate of energy generation form the $^{12}C + ^{12}C$ reaction is not compensated by some process of energy loss, a thermal runaway and thermonuclear detonation can result. If the density is not extremely high the star will be disrupted completely, ejecting about $1.4 M_\odot$ iron-group nuclei.

This model has the following problems: (1) it is inconsistent, (2) it makes too much iron, and (3) it makes no pulsar (neutron star remnant). Different theoreticians give weight to these problems to differing degrees, but no one seems too happy with the model. It was, however, the first supernova model to derive directly from stellar evolutionary calculations. Probably the most important problem with the model is its inconsistency; as Paczynski first pointed out in detail, neglect of Urca processes is invalid. Couch and Arnett discuss their attempt at a direct numerical attack with a realistic nuclear reaction network. Graboske reviews recent work at Livermore which has improved our understanding of electron screening. The screening factor has a vital influence upon the ignition point and thereby the question of remnant formation. We note that an equally important question, that of uncertainty in the neutrino emission rates, has not been explored. The abundance of ^{12}C at ignition depends upon the amount of ^{12}C left after helium burning; Dyer discusses new experimental and theoretical work on the crucial reation $^{12}C (\alpha,\rho) ^{16}O$.

Wheeler, Bruenn and Buchler all discuss hydrodynamic behavior after the thermal runaway: under what conditions do we get a remnant? There is not yet any great satisfaction for those who wish to make pulsars from this mass range. The Couch-Arnett models become hydrodynamic at a density which may be too low to avoid complete disruption according to the new hydrodynamic results.

Chapter IV   178

The critical transition stage from hydrostatic to hydrodynamic
evolution needs further study as does the nature of the convective flow in the Urca-damped core. There is still no unambiguous
answer in sight (except solution by decree).

References to Chapter IV

Early Work

    Arnett, W. D. 1969, Astrophys. and Space Sci., 5, 180.
    Paczynski, B. E. 1970, Acta Astron., 20, 47.

The Urca question

    Paczynski, B. E. 1972, Astrophys. Lett., 11, 53.

Nucleosynthesis

    Arnett, W. D., Truran, J. W. and Woosley, S. E. 1971, Ap. J., 165, 87.

For more recent references, the bibliographies following
the papers are reasonably complete.

DEGENERATE CARBON BURNING

Richard G. Couch and W. David Arnett

The University of Texas, Austin, Texas 78712

Carbon burning at high densities appears to be relevant only for stars having a main sequence mass in the 4-8 $M_\odot$ range. In these stars the neutrino losses are great enough to cause a temperature inversion following core helium burning. This drop in central temperature delays the initiation of carbon burning until much higher densities are attained ($\rho_c \approx 2 \times 10^9$ g/cm^3, $T_c \approx 2 \times 10^8$ °K). In particular, a core of approximately 1.4 $M_\odot$ is formed with the remainder of the mass contained in an extended envelope. This structure is independent of the total mass of the star. (Paczynski, 1970).

The interest in this phase of stellar evolution was motivated mainly by the desire to form pulsars from stars in this mass range (Gunn and Ostriker 1971). However, the possible formation of pulsars is only one aspect of the problem. Since these stars are relatively numerous, their ultimate fate has enormous implications for theories of galactic nucleosynthesis. The work of Arnett (1969) indicated that the most likely fate of these objects was a supernova explosion and complete dispersal. No satisfactory resolution of the apparent discrepancy between this result and the interpretation of pulsar statistics was put forth until Paczynski (1972) pointed out that large scale convective mass flow could generate Urca neutrinos at a rate sufficient to stabilize the carbon burning and prevent detonation. This suggestion by Paczynski provided the impetus for the present work. It was apparent that the subsequent evolution of these objects would depend on the particulars of the nuclear processing.

A computer code was written which accurately follows the evolution of abundances during carbon burning, while at the same time describing the evolution of the core as the mass and composition change. There are 25 nuclei in the reaction matrix. For 12 of these nuclei the beta decay or electron capture rates, and the associated neutrino losses are explicitly calculated. The core evolution is followed by using a simple algorithm which gives the rate at which mass is added to the core (Arnett 1971), and then searching for the central density which reproduces the core mass. This is accomplished by

TABLE 1

NUCLEI FOR WHICH BETA RATES ARE CALCULATED EXPLICITLY

Beta Decay	Electron Capture	ρ_F
^{31}Si	^{31}P	4.01(7)
	^{27}Al	6.54(8)
^{25}Na	^{25}Mg	8.01(8)
^{23}Ne	^{23}Na	1.20(9)
^{21}F	^{21}Ne	2.65(9)
	^{24}Mg	3.09(9)
	^{24}Na	3.09(9)
	^{20}Ne	6.56(9)

integrating the equations of hydrostatic equilibrium and mass continuity from the center outward for an assumed density and then performing a series of iterations of the density until the correct mass is obtained. A zero temperature equation of state (Salpeter 1961) is used in this phase of the calculation and general relativistic corrections are included.

The calculation begins at the ignition point for carbon burning. From the equation of state, the net change in internal energy due to nuclear burning, gravitational contraction, and neutrino losses, and a knowledge of the structure of the core, the temperature behavior and the extent of convection can be determined. The convective region is divided into radial mass zones, and the net reaction rates and the neutrino losses are determined by integration over these zones.

Figure 1 gives ignition curves for various initial mass fractions of ^{12}C, and for the screening corrections of Salpeter and Van Horn (1969) and those from the more recent work of DeWitt et al. (1973). The effect of the new screening corrections is to decrease the ignition density by about 30%. The pair, plasma, and photoneutrino rates are from Beaudet et al. (1967), and the bremstrahlung rates are from Festa and Ruderman (1969). The carbon rate is that of Patterson, Winkler, and Zaidins (1969).

The beta decay, electron capture, and associated neutrino loss rates were obtained by numerical integration of the expressions presented in Hansen (1966). The rates were stored as functions of the temperature and density again using the method of Hansen. Table 1 lists the nuclei for which neutrino loss rates were calculated and the densities at which the fermi energy equals the transition energy. These densities cover a reasonably broad range and include Urca shells both above and below the ignition densities. The four pairs which can contribute to the Urca losses are all reasonably abundant, their interactions proceed by allowed transitions, and their abundances are likely to be enhanced during carbon burning. Consequently, the inclusion of other possible seed nuclei is not expected to alter the results.

Solar system values were assumed for the initial abundances with the exception of ^{12}C, ^{16}O, and ^{22}Ne. The ^{22}Ne was enhanced by buildup from ^{14}N during the helium burning phase, and the remainder of the core consisted of equal parts of ^{12}C and ^{16}O by mass. The screening corrections used are from DeWitt et al. (1973), and the initial core model was from Paczynski (1971). On Figure 2 the central temperature and density and the ^{12}C mass fraction are plotted as a function of time. The central temperature rises very rapidly once the ignition point is reached. This rise in temperature is halted

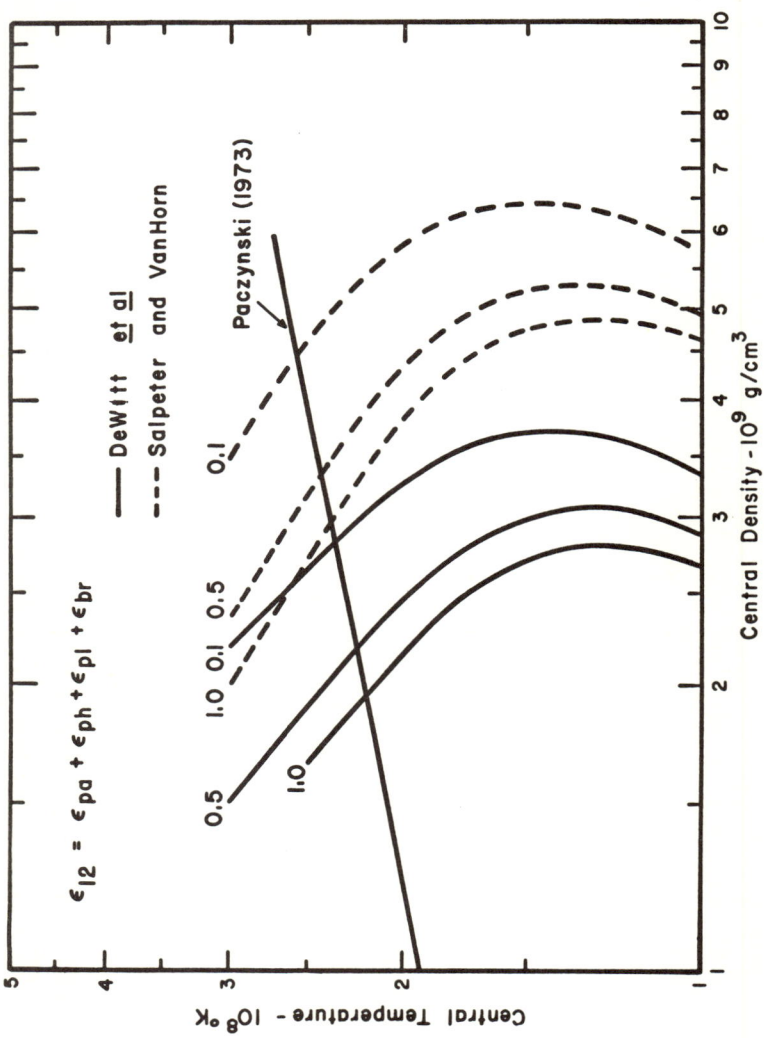

Figure 1. Ignition curves for ^{12}C mass fractions of 0.1, 0.5, and 1.0 for the indicated sources of screening corrections.

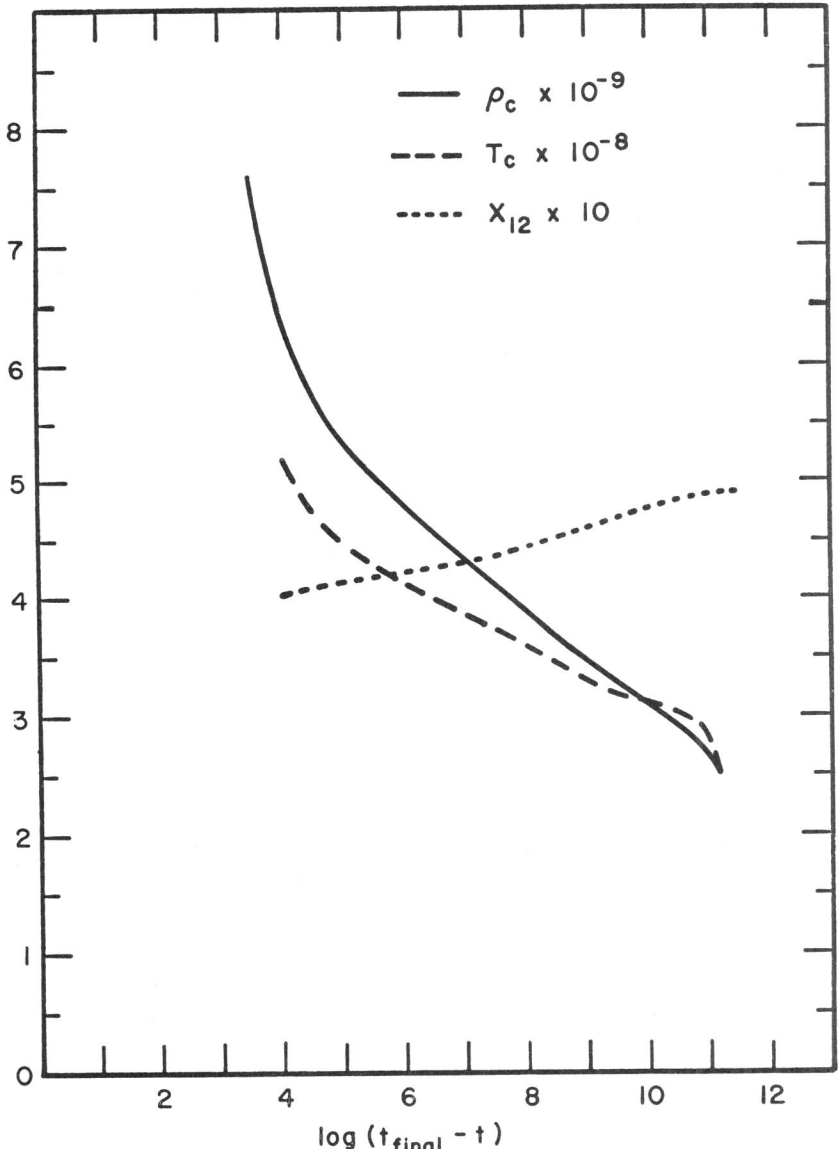

Figure 2. Behavior of central density, central temperature and the mass fraction of ^{12}C in the convective region as a function of time. The quantity t_{final} is an arbitrary time choosen in order to expand the time scale in the region where evolution is rapid.

when the convective region begins to extend to densities at which ^{23}Ne decay becomes significant. At this point the abundance of mass-23 nuclei has been enhanced by about a factor of 100 over the solar system abundance of ^{23}Na. Consequently, the ^{23}Na-^{23}Ne pair are a copious source of neutrinos. The core then proceeds through a quasi-hydrostatic phase. The existence of this hydrostatic phase seems quite well established. If convection is not inhibited by some unusual mechanism, then the core should be well mixed on a time scale short compared to the beta lifetimes (Paczynski 1972). The ^{23}Na-^{23}Ne pair is such an efficient producer of neutrinos that order of magnitude variations in the ^{12}C reaction rate and in the screening corrections have negligible effect. Since the carbon burning alone produces enough ^{23}Na to stabilize the core, the result applies to both population I and population II objects.

Once the burning is stabilized the central temperature required for equilibrium is determined by a Newton-Raphson iteration method. The central density increases as the core grows and the composition changes moving the Urca shell further from the center and allowing the temperature to rise. The evolution soon becomes very rapid. At this stage the increase in central density is being driven by the increase in μ_e resulting from electron captures on the products of carbon burning. Consequently, the higher the temperature the more rapid is the approach to the singular state. Eventually the temperature and density are high enough that the Urca shell can no longer contain the burning region. However, this occurs at about the same time that the core begins dynamic collapse. It is not possible within the context of the present calculation to predict the ultimate fate of the object. A mass fraction of only about 0.1 in the form of carbon had been consumed when the core became dynamic.

There are several uncertainties in this type of calculation which must be discussed. In the case described above, the ^{21}Ne-^{21}F Urca shell very nearly contained the burning. The uncertainties in the relevant reaction rates are such that the effect of this pair could be enhanced to the extent that the convective region would begin to contract and the central temperature drop once the central density increased beyond this Urca shell. A calculation was made with an initial ^{12}C mass fraction of only 0.10. In this case the contraction of the convective region was observed, but the calculation has not yet been followed beyond that point. It seems likely that as the central density continues to increase the temperature will rise to the point at which the ^{21}Ne-^{21}F shell will be overwhelmed and the convective region will again move out to the ^{23}Na-^{23}Ne shell mixing in a fresh supply of ^{12}C. Subsequent evolution will probably resemble that of Figure 2.

The most important and least understood rates are those of the $^{12}C + ^{12}C$ and $^{12}C(\alpha,\gamma)^{16}O$ reactions. The evolution will certainly depend on the amount of carbon available to burn. It appears that at most a ^{12}C mass fraction of 0.1 is required to be consumed in order to make the core dynamic. If the initial ^{12}C abundance is lower than the value 0.5 assumed here, then the ignition will take place at a higher density and with a more massive core. The net result is that the less ^{12}C available the less that is needed. In addition, the constant mixing with unprocessed material as the convective region grows makes it difficult to deplete the ^{12}C. The two ^{12}C mass fractions considered here are probably reasonable upper and lower limits to the actual abundance. Iben (1973) finds that with his models X_{12} is in the range 0.0 and 0.27. The lower value is consistent with the normally quoted limits on θ_α^2, but seems somewhat unreasonable. The actual uncertainty in θ_α^2 allows a value for X_{12} slightly larger than 0.27.

Perhaps more important than the actual rate of the ^{12}C reaction is the branching ratio between the proton and alpha particle channels which determines the relative amounts of ^{23}Na and ^{20}Ne produced. At these low temperatures a narrow region in the ^{24}Mg compound nucleus is being sampled and the likelihood of large variations in this branching ratio is increased. This ratio was assumed to be unity.

This research was supported in part by NSF grants GP-23282 and GP-32051 at The University of Texas at Austin.

REFERENCES

Arnett, W. D. 1969, Astrophysics and Space Science, 5, 180.
Arnett, W. D. 1971, Ap. J. 169, 113.
Beaudet, G., Petrosian, V., and Salpeter, E. E. 1967, Ap. J., 150, 979.
DeWitt, H. E., Graboske, H. C., and Cooper, M. S. 1973, to be published.
Festa, G. C., and Ruderman, M. A. 1969, Phys. Rev., 180, 1227.
Gunn, J. E., and Ostriker, J. P. 1971, Ap. J., 160, 979.
Hansen, C. J. 1966, Ph.D. Thesis, Yale University.
Iben, I. 1972, Ap. J., 178, 433.
Paczynski, B., 1970, Acta Astronomica, 20, 47.
_____., 1971, Acta Astronomica, 21, 271.
_____., 1972, Astrophysical Letters, 11, 53.
_____., 1973, to be published.
Patterson, J. R., Winkler, H. and Zaidins, C. S. 1969, Ap. J., 157, 367.
Salpeter, E. E. 1961, Ap. J., 134, 669.
Salpeter, E. E., and Van Horn, H. M. 1969, Ap. J., 155, 183.

THE INFLUENCE OF SCREENING EFFECTS ON CARBON IGNITION[*]

Harold C. Graboske, Jr.

Lawrence Livermore Laboratory, University of California

I. Screening Theory

Recent developments in the theory of screening of nuclear reactions have led to a general statistical mechanical formalism which, when combined with pair distribution functions of the electrons and nucleons, lead to a revision of the previous strong screening theory (DeWitt, Graboske and Cooper 1973). This generalization of the screening theory of Salpeter (1954) and Salpeter and Van Horn (1969, hereafter referred to as SVH) leads to the following form for the screening factor f:

$$f = e^{H(o)}$$

$$H(o) = \Gamma_o \{ \tfrac{9}{10} \bar{z}^{1/3}[(Z_1 + Z_2)^{5/3} - Z_1^{5/3} - Z_2^{5/3}]$$
$$+ c_1 \bar{z}^{2/3}[(Z_1 + Z_2)^{4/3} - Z_1^{4/3} - Z_2^{4/3}] \quad [1]$$
$$+ d_1 \bar{z}^{-2/3}[(Z_1 + Z_2)^{2/3} - Z_1^{2/3} - Z_2^{2/3}]$$

$$\Gamma_z \equiv Z^{5/3} \Gamma_o = 0.22747 \, Z^{5/3} (\rho_9/\mu_e)^{1/3}/T_9$$

This expression for the screening function $H(o)$ is valid for $\Gamma_z \gg 1$, arbitrary charges Z_1, Z_2, $\bar{z}$ (mean ionic charge), and a strongly degenerate electron gas. In the limit Z_1, $Z_2 \gg \bar{z}$ Eq. [1] reduces to the result given in Salpeter (1954) for strong screening.

For the specific case of the ^{12}C, ^{12}C reaction in a helium-exhausted core composed primarily of carbon and oxygen, $Z_1 = Z_2 \approx \bar{z}$, and the screening function simplifies to

---

[*] Work performed under the auspices of the U.S. Atomic Energy Commission.

$$H(o) = \Gamma_{Z_1}[1.0573 + 0.1478\,(\bar{Z}/Z_1)^{1/3}] - 0.1898\,(Z_1/\bar{Z})^{1/3} \quad [2]$$

The strong screening expression obtained by Salpeter for this case is $H(o) = 1.0573\,\Gamma_{Z_1}$. The new screening factor produces significantly enhanced screening, by a factor of $\exp(0.1478\,\Gamma_{Z_1} - 0.19)$ for pure carbon, and introduces a plasma charge dependence $(\bar{Z}/Z_1)$ which increases the screening function as the fraction of oxygen in the C-O mix is increased.

II. The Screening Transition Locus

Before investigating the effect of the new screening factors on the carbon ignition zone, a brief review of the transition from strong screening to pycnonuclear screening is necessary. This latter regime, lying toward higher densities relative to strong screening, is the region where strong density dependence replaces the strong temperature dependence of the thermonuclear regime. In SVH, an approximate general criterion was chosen to denote the boundary between these two regions:

$$\tau/\Gamma \sim 1$$
$$\tau = [38.348\,AZ^4/T_9]^{1/3} \quad [3]$$

For the ^{12}C, ^{12}C interaction where the unscreened energy generation rate (Arnett and Truran 1969) and the screening factor (Eq. [1]) are well-established, a more precise determination can be made. If the strong screening-pycnonuclear boundary is defined as the locus where a strong direct dependence on temperature changes to an inverse temperature dependence, the transition occurs where

$$\partial\varepsilon/\partial T = 0 \quad [4]$$

Solving this equation for the ^{12}C, ^{12}C reaction a set of transition lines is found for various values of X_c, the carbon abundance dependence, arising from the plasma charge dependence of the screening factor. The X_c dependence is very slight, and for the range $0.1 \leq X_c \leq 1.$ and $0.01 \leq \rho_g \leq 10$ a single relation,

$$\tau/T_6 = 7 \quad [5]$$

defines the screening transition. Eq. [5] predicts the transition locus with a maximum error of 2.5%, and is accurate to 1% or better over most of the range. For $\tau/T_6 > 7$, the nuclear

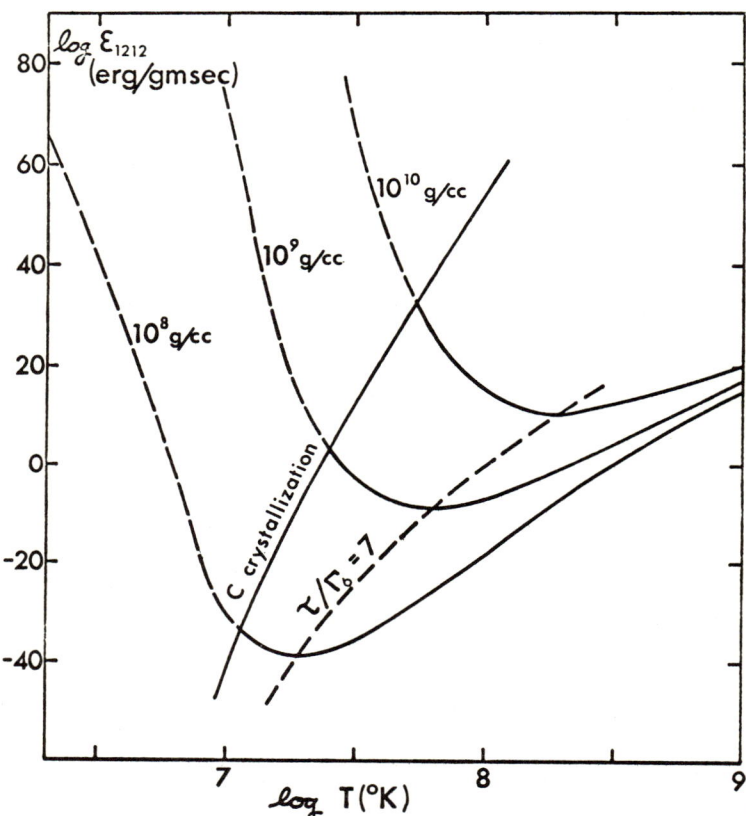

Figure 1. The screening transition locus ($\tau/\Gamma_6=7$) defines the minima of ε_{1212}. The thermonuclear region ($\tau/\Gamma_6>7$, toward higher T) has moderate ρ-dependence and very strong T dependence. The pycnonuclear region ($\tau/\Gamma_6<7$, toward lower T) has extremely strong density and inverse temperature dependence. For regions above the crystallization line pycnonuclear screening and reaction rates will have a very different form from the strong screening-fluid regime below $\Gamma_6=143$.

reactions are thermal, while for $\tau/\Gamma_6 < 7$, density effects dominate the rate.

To illustrate these effects, the screening transition locus is plotted in Figure 1, along with the isodensity energy generation rates for the $^{12}C, ^{12}C$ reaction. The region to the right of the transition locus is the thermonuclear region where most current C-O core models predict carbon ignition will occur. The energy rates here have a moderate density dependence and a strong temperature dependence, ranging from $T\sim 17$ at 10^{10} gm/cm^3, to $T\sim 25$ at 10^9 gm/cm^3 and $T\sim 40$ at 10^8 gm/cm^3. Along the transition locus, the energy rates reach their minimum and then increase rapidly to the left, in the pycnonuclear region. Here they attain very strong density dependence and very strong inverse temperature dependence. The evolution of a stellarmodel into this region would produce a pycnonuclear ignition which would probably be quite violent.

The source of the reversal of the energy rate curves is the screening factor: the same behavior occurs if the SVH strong screening factor is used, the only difference is that the transition locus lies at $\tau/\Gamma_6 \sim 5$. In the strongly screened thermonuclear region, the unscreened rate is generally many orders of magnitude larger than f, but for $\tau/\Gamma_6 \leq 7$ the screening factor begins to dominate the unscreened rate. At some point the strong screening factor must be replaced by the finite-temperature pycnonuclear screening factor, for example where $\tau/\Gamma_6 \sim 1$ as suggested in SVH.

The advent of pycnonuclear effects has previously been related to the existence of a rigid lattice for the ions. A brief review of the location of the phase transition from the strongly coupled coulomb fluid to the classical coulomb lattice is in order here, together with its relation to the screening transition. Numerous estimates have been made for the value of Γ_Z for crystallization, ranging from 18 to 170. Some have been based on the empirical Lindemann melting law, while others rely on Monte Carlo methods to calculate numerical models of the fluid-solid transition. A recent study by Hansen (1972) has produced the best value for the phase transition of the single species classical ion fluid. Using a 128 particle ion system, an accurate method for representing the ionic coulomb potential, with very large Monte Carlo chains (2×10^5 to 2×10^6 configurations), Hansen has significantly improved on the original Monte Carlo study (Brush, Sahlin and Teller, 1966), finding the classical ion fluid with uniform electron background crystallizes at $\Gamma_Z = 143$. The accuracy of Hansen's numerical pair distribution functions has been verified by an independent calculation at Livermore, using Hubbard's (1972) Monte Carlo program.

The location of the crystallization loci for pure carbon and oxygen systems is shown in Figure 2. An additional phase transition line is shown for a two-component plasma composed of C and O in equal numbers (Loumos and Hubbard 1973). The mixture transition line lies midway between the pure species transition lines, indicating a reasonable approximation for a C-O mixture crystallization temperature is achieved by scaling with the fractional carbon abundance. This set of crystallization lines defines the C-O crystallization zone, which lies well below the region where current models of C-O cores evolve.

Also plotted in Figure 2 is the screening transition line for the ^{12}C, ^{12}C reaction, which is seen to be in the region of the ion fluid. That density-dominant effects appear in the fluid region is not surprising, as the pair distribution functions for small separations of the nuclei are quite similar for the strongly coupled fluid exhibiting short range order to those for the fully ordered lattice.

III. The Carbon Ignition Zone and Non-central Ignition

The location of the carbon ignition zone is determined by the relation $\epsilon_{1212} + \epsilon_\nu = 0$, where ϵ_ν represents the sum of all neutrino emission processes (Beaudet, Petrosian and Salpeter 1967; Festa and Ruderman 1969). As discussed by Graboske (1973), the use of the new screening factor significantly increases ϵ_{1212} for large Γ_Z, reducing the density for carbon ignition. The low density end of the ignition lines, at the low Γ_Z end of the strong screening region ($1 \leq \Gamma_Z \leq 5$), are only slightly affected, but for $\Gamma_Z > 10$ the ignition densities found by Barkat, Wheeler and Buchler (1972, hereafter BWB) and Bruenn (1972) are reduced by factors of 2 to 3.

An additional feature of the strong screening factor is the production of doubled-valued carbon ignition lines. This effect is due to the reversal of the temperature dependence of ϵ_{1212}, and also occurs if the SVH screening factor is used. In Figure 2, the upper or thermonuclear branch, equivalent to the lines cited by Paczynski (1971), Bruenn and BWB, has a moderate X_c dependence due primarily to the X_c^2 term in the unscreened rate. The lower or pycnonuclear branch is the result of the rapid increase of ϵ_{1212} with decreasing temperature after the transition line is crossed, the extremely strong inverse-T dependence causing the pycnonuclear branch to lie much closer to the transition line. The virtual disappearance of the X_c-dependence along the pycnonuclear branch occurs because of the dominance of the screening factor with its weak dependence on $\bar{Z}/Z_1$.

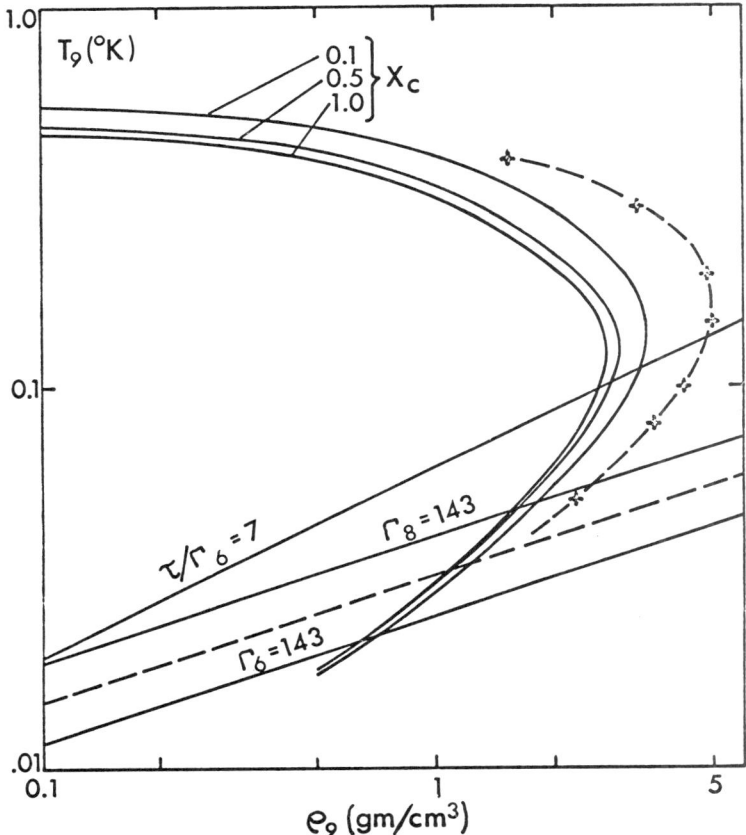

Figure 2. The C-O crystallization zone is the region lying between $\Gamma_6 = 143$ and $\Gamma_8 = 143$. The C-O mixture crystallization line (— — —) lies between the two pure species limits. The carbon ignition zone ($0.1 \leq X_c \leq 1.$) has a thermonuclear branch and a pycnonuclear branch, the latter lying close to the screening transition locus. The addition of 10^4-fold enhanced Urca shell neutrino losses to the plasma and bremsstrahlung rates shifts the ignition line for $X_c = 0.1$ (— — + —) to higher densities.

The double-valued form of the ignition zone results in a clearly defined maximum ignition density for specified X_c. For the plasma plus bremsstrahlung neutrino rates, these density maxima are ρ_g = 2.7, 2.9 and 3.4 for X_c = 1, 0.5 and 0.1. To achieve ignition at higher densities requires an additional strong neutrino emission process. The Urca process has been investigated by Bruenn (1972) using an averaging of several Urca shell emission rates. Using best current estimates for the abundances of Urca-active nuclei in post-helium-burning cores, he finds insignificant changes in the neutrino rate. With large enhancements, a higher density ignition zone is found; however the increased screening factors when combined with Bruenn's most extreme case (X_c = 0.1, 10^4-fold enhancement) yield a maximum ignition density of 5×10^9 gm/cm^3 (see Figure 2). It seems very unlikely, barring a major revision (downward) of the ^{12}C, ^{12}C reaction rate or (upward) of the neutrino rates, that degenerate carbon burning can be initiated at densities higher than the ρ_9 = 2-3 range.

One suggestion for a non-disruptive quasistatic ignition process is the possibility of non-central ignition (Iben 1972). A comparison of current quasistatic core models with the ignition lines is given in Figure 3. The original C-O core models of Paczynski are plotted in the form of the evolving center (ρ_c, T_c curve) and the structure line ($\rho(r)$, $T(r)$) for the radial dependence of the core at ignition. The structure line demonstrates the difficulty of achieving non-central ignition with such models; the center of the core clearly is the point of ignition, and with stronger neutrino losses, an even stronger decrease of T away from the center will decrease the probability of off-center ignition.

Also plotted in Figure 3 are the ρ_c, T_c values for the models of BWB and Bruenn. The varying core mass growth models of BWB remain essentially similar to Paczynski's results. Bruenn's 10^4-fold enhanced Urca model does ignite at a higher density (the corresponding ignition line for this model is that shown in Figure 2), and has the interesting feature that it achieves pycnonuclear ignition. For all these models a large restructuring of the core is required for non-central ignition to occur. For the Paczynski and BWB models, a strong increase in T away from the center must exist, while for Bruenn's model a marked drop in T away from T_c is required. Such significant structural alterations imply a different set of physical assumptions, such as the onset of convection. The combination of Urca shell emission driven by convection in a post-ignition core (Paczynski 1972; Couch and Arnett 1973) seems to satisfy the constraints of the screening effects. Ignition occurs in the $2-3 \times 10^9$ gm/cm^3 range and is central. The onset of convection steepens the temperature gradient but the center of the model remains the highest energy generating region.

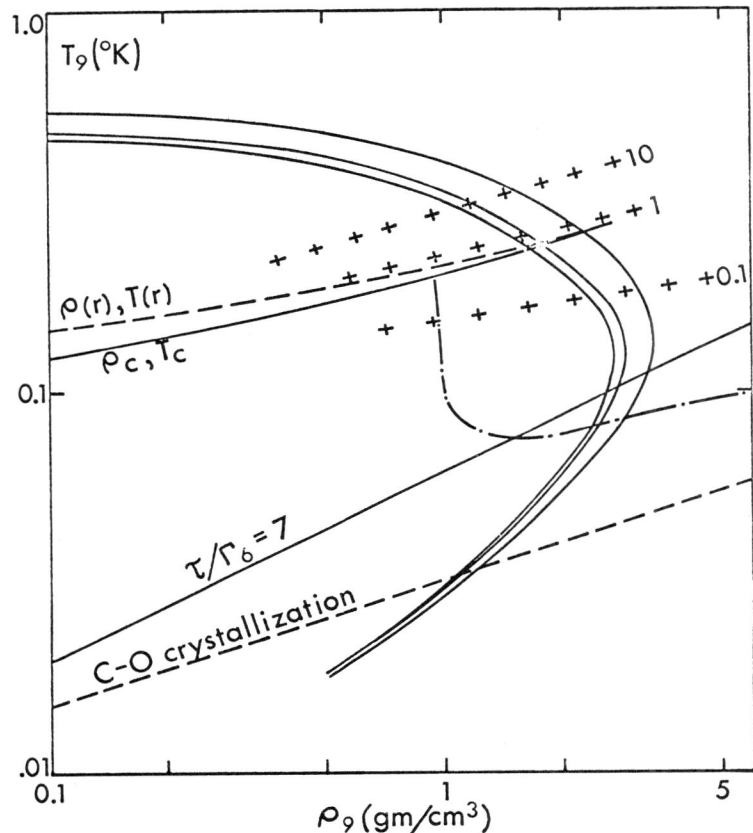

Figure 3. A comparison of C-O core models with the revised carbon ignition zone. Paczynski's models are represented by a time-dependent central point (ρ_c, T_c) and a structure line $\rho(r)$, $T(r)$. The BWB models (+ + +) are labelled 0.1, 1, 10 to indicate the scaled rate of core mass growth. Bruenn's 10^4-fold enhanced model (——·——·) undergoes pycnonuclear ignition.

Another possible mechanism for initiating non-violent quasistatic carbon burning in the highly degenerate C-O core is implicit in the behavior of the energy rate itself. The screening transition locus, corresponding to $\partial \epsilon / \partial T = 0$, is a ρ, T region where the screened energy rate is not only a minimum for any T at the chosen ρ, but is also essentially temperature independent. An examination of Figure 1 shows there is a finite region about $\tau / \Gamma_6 = 7$ where ϵ varies quite slowly with T. This region corresponds to a corridor of non-violent carbon ignition through the otherwise very strongly T dependent carbon ignition zone. In Figure 3 this corridor surrounds the $\tau / \Gamma_6 = 7$ screening transition line, so that if a given stellar core model evolved in such a manner as to cross the carbon ignition line at or near the transition line it would be quite possible that the carbon burning would, at least initially, be quite gentle. In this way quasistatic carbon burning, centrally ignited, could be achieved and if the model track were to continue evolving along the transition line, this weak temperature dependence would continue to obtain. The major drawback to this mechanism is that all previously calculated normal C-O core models lie in a well localized position, close to Paczynski's original model, which is well above the position of the transition line. It would still be interesting to do a parametric quasistatic model study to investigate the consequences of this type of high density ignition.

REFERENCES

Arnett, W. D. and Truran, J. W. 1969, Ap. J., 157, 339.
Barkat, Z., Wheeler, J. C. and Buchler, J. 1972, Ap. J., 171, 651.
Beaudet, G., Petrosian, V., and Salpeter, E. E. 1967. Ap. J., 150, 979.
Bruenn, S. W. 1972, Ap. J., 177, 459.
Brush, S. G., Sahlin, H., and Teller, E. 1966, J. Chem. Phys., 45, 2102.
Couch, R. G. and Arnett, W. D. 1973, Ap. J. Lett., 180, L101.
DeWitt, H. E., Graboske, H. C. and Cooper, M. S. 1973, Ap. J., 181, 439.
Festa, G. C. and Ruderman, M. A. 1969, Phys. Rev., 180, 1227.
Graboske, H. C. 1973, Ap. J., in press.
Hansen, J.-P. 1972, Phys. Letters 41A, 213.
Hubbard, W. B. 1972, Ap. J., 176, 525.
Iben, I. 1972, Ap. J., 178, 433.
Loumos, G. L. and Hubbard, W. B. 1973, Ap. J., 180, 199.
Paczynski, B. 1971, Acta Astr., 21, 271.
_____ 1972, Ap. Letters, 11, 57.
Salpeter, E. E. 1954, Austral. J. Phys., 7, 353.
Salpeter, E. E. and Van Horn, H. M. 1969, Ap. J., 155, 183.

MEASUREMENT OF THE $^{12}C(\alpha,\gamma)^{16}O$ CROSS SECTION*

P. Dyer

California Institute of Technology, Pasadena, California

I. Introduction

This paper is a report on the measurements of the $^{12}C(\alpha,\gamma)^{16}O$ cross section at low alpha-particle energies, carried out at the California Institute of Technology in collaboration with Professor C. A. Barnes, and Drs. D. C. Weisser and J. F. Morgan. The time-of-flight technique adopted for these measurements was explored first by Barnes, Adams, Shapiro, and Adelberger (Adams et al. 1968). The results given here should be regarded as preliminary; a more complete analysis will be published shortly.

The rate of the $^{12}C(\alpha,\gamma)^{16}O$ reaction appears to be the least well-known factor in the determination of the stellar core composition following helium burning. If the $^{12}C(\alpha,\gamma)^{16}O$ reaction rate is small compared with the rate of the $3\alpha \rightarrow {}^{12}C$ reaction, the main end product of helium burning will be ^{12}C; if large, the end product will be ^{16}O (in all but very massive stars). This uncertainty in the stellar core composition at the end of helium burning poses a serious problem for those who wish to perform detailed stellar model calculations for the later stages of evolution.

It is not presently possible to make laboratory measurements of the $^{12}C(\alpha,\gamma)^{16}O$ cross section at the effective stellar interaction energy of about 300 keV (c.m.), as the cross section is expected to be about 7 orders of magnitude too small for measurement by currently available techniques. Instead, the best that can be done is to measure the cross section over a wide range of energies which extends as low as possible, and then to extrapolate to 300 keV.

This extrapolation procedure is particularly difficult in the case of the $^{12}C(\alpha,\gamma)^{16}O$ reaction, because the major contribution to the cross section at 300 keV comes from the tail of a 1^- bound state in ^{16}O at 7.12 MeV excitation. The present

*Supported in part by the National Science Foundation GP-28027.

experiment can only attempt to observe the interference of the tail of this bound state with higher resonances. The rate of the $^{12}C(\alpha,\gamma)^{16}O$ reaction at astrophysically interesting energies is most conveniently parameterized in terms of the dimensionless reduced-alpha-width of the 7.12-MeV state in ^{16}O, a quantity denoted θ_α^2 (Fowler et al. 1967). In the present paper, the relation between θ_α^2 and the S-factor at 300 keV (c.m.) will be taken to be (Dyer 1973)

$$S(300 \text{ keV, MeV-barns}) = \theta_\alpha^2/0.70.$$

There have been many attempts to determine the rate of the $^{12}C(\alpha,\gamma)^{16}O$ reaction. Most of these have employed very indirect methods, and the results obtained by different investigators have varied widely. The present experiment, in which the gamma-ray yield of the $^{12}C(\alpha,\gamma)^{16}O$ reaction is measured directly, is the method which whould ultimately give the most reliable result. There has been one other attempt to measure the $^{12}C(\alpha,\gamma)^{16}O$ cross section at low alpha energies (Jaszczak et al. 1970a, 1970b). However, the statistical errors on these data are so large that the extrapolated cross section is indeterminate (Weisser 1970). In addition, the energy dependence of the cross section measured by Jaszczak et al. is not in agreement with that of the present measurement.

II. Experimental Method

The experimental configuration used for measuring the $^{12}C(\alpha,\gamma)^{16}O$ cross sections is conceptually quite simple: a beam of energetic alpha particles, an enriched ^{12}C target, and two large NaI(Tℓ) scintillation counters for detecting the emitted gamma rays. The difficulties in measuring the rate of the $^{12}C(\alpha,\gamma)^{16}O$ reaction are due to a very low yield (total cross sections of the order of nanobarns), and a relatively high background counting rate.

The major source of background arises from the ^{13}C contaminant in the target. This isotope is a prolific source of neutrons, when bombarded by alpha particles, from the $^{13}C(\alpha,n)^{16}O$ reaction. Some of these neutrons produce pulses in the detection system which are indistinguishable from those of the gamma rays of interest.

We have adopted two techniques to reduce this background. The first is to make a target which contains as little ^{13}C as possible. This target is then placed in a clean target chamber which is isolated from the accelerator by a liquid nitrogen cold trap and kept at high vacuum by ion-pumping. The second is to measure the difference in time between the occurrence of a reaction at the target and the detection of an outgoing particle by the NaI(Tℓ) detectors. If the detectors are placed 10 cm from the target, a gamma ray will require 0.3 nanoseconds to travel from the target to the detector, whereas

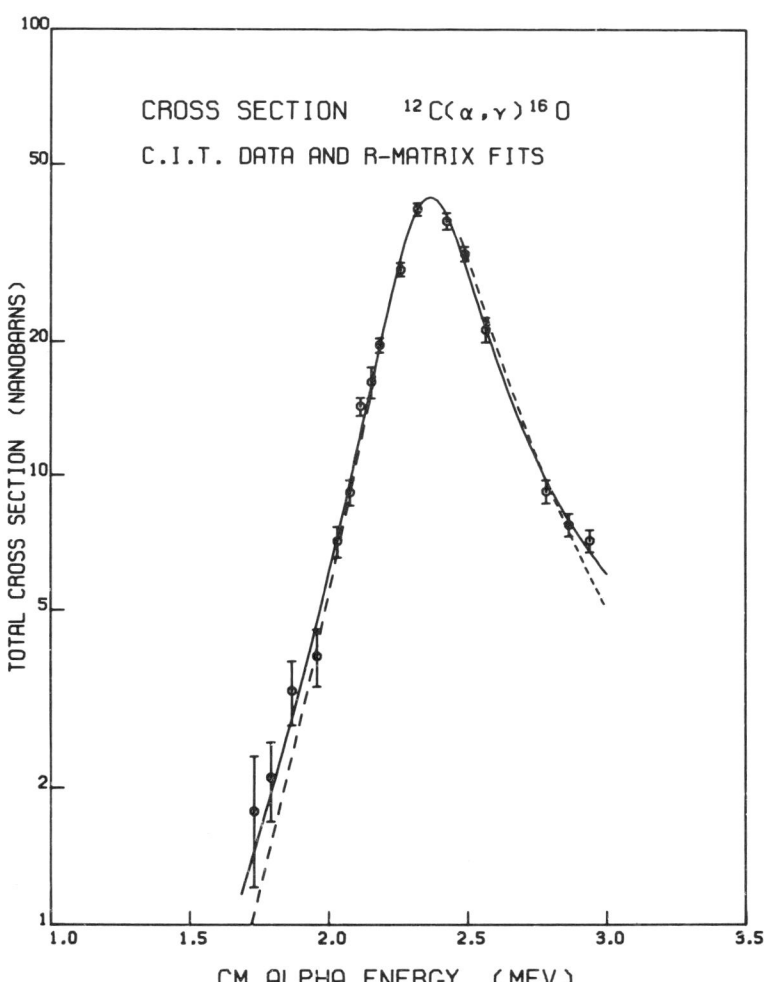

Fig. 1. The $^{12}C(\alpha,\gamma)^{16}O$ E1 contributions to the total cross sections are shown as a function of center-of-mass alpha energy. The resonance here corresponds to the 9.60-MeV state in ^{16}O. Circles with error bars indicate the experimental data. The solid curve shows the best fit by the 3-level R-matrix parameterization (three parameters varied); the dashed curve shows a fit with θ_α^2 constrained to zero.

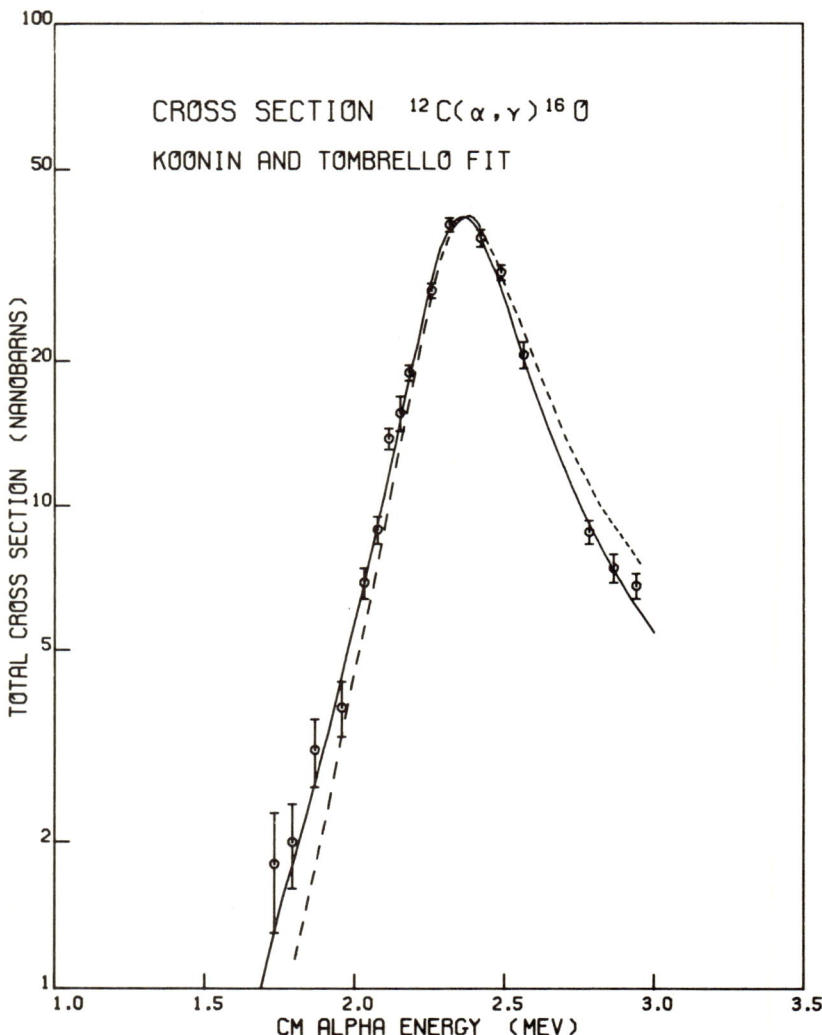

Fig. 2. This figure is the same as Fig. 1, except that the fits are those of the hybrid R-matrix optical-model parameterization. For the best fit (solid curve), two parameters were varied.

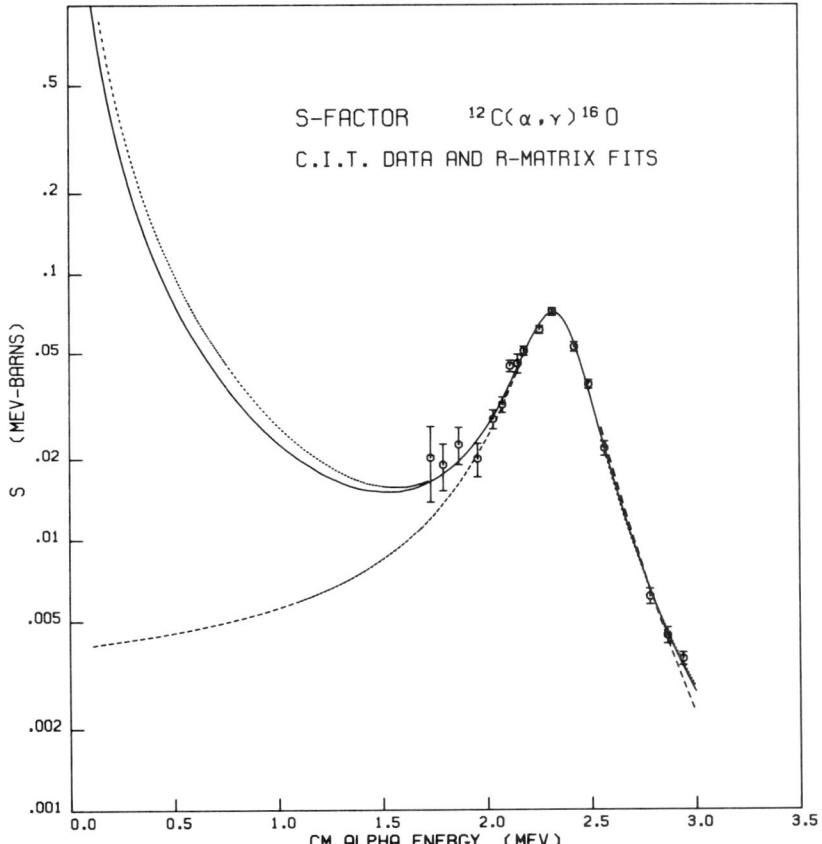

Fig. 3. The $^{12}C(\alpha,\gamma)^{16}O$ S-factor is shown as a function of center-of-mass alpha energy. The solid curve shows the best fit by the 3-level R-matrix parameterization; the dashed and dotted curves show fits with θ_α^2 constrained to 0 and 0.4, respectively.

a neutron of the relevant energy will require about 4 nanoseconds. A chopped and bunched beam from the ONR-CIT tandem accelerator was employed. The beam was on target for 1-1/2 nsec, and off for 285 nsec. Thus, with a detection system resolution of 2 nsec, it was possible to separate the neutrons and gamma rays by their different flight times to the detector.

An additional obstacle in obtaining measurements of the $^{12}C(\alpha,\gamma)^{16}O$ cross section is that high-energy gamma rays may be produced by the interaction of the alpha particles with low-Z impurities in the target. As an example, a target containing 0.5% ^{19}F would give a totally incorrect measurement, because of the relatively large cross section of the $^{19}F(\alpha,\gamma)^{23}Na$ reaction.

III. Results

In order to extrapolate the excitation function to lower energies, it is necessary to understand the reaction mechanism, both at 300 keV (c.m.) and at the energies where data have been obtained. It is necessary to take into account both resonant and direct capture processes, to consider all nearby states in ^{16}O, and to consider possible cascade radiative decays. Angular distributions have been measured both above and below the resonance corresponding to the 9.60-MeV state in ^{16}O, in order to take into account a small contribution from E2 capture and to assist in evaluating the contribution from higher resonances. The extracted E1 contribution to the total cross section is shown in Figure 1.

Thus far, two methods have been employed to parameterize the $^{12}C(\alpha,\gamma)^{16}O$ data in order to extrapolate to lower energies. A three-level R-matrix fit has been made by Weisser et al. (1970), where the three levels correspond to the bound 7.12-MeV state in ^{16}O, the 9.60-MeV state, and a fictitious background state at 20-MeV excitation. The preliminary result of this fit, shown in Figure 1, is that $\theta_\alpha^2 = 0.14^{+0.50}_{-0.08}$. Since θ_α^2 is expected to be a number between 0 and 1, it can be seen that this fit does very little toward putting an upper limit on θ_α^2. A second method, developed by Koonin and Tombrello (1972), is a hybrid R-matrix optical-model parameterization, in which the bound state is described by a single-level pole, and contributions from all higher states and from direct capture are given by a potential well. This method yields a preliminary value for θ_α^2 of $0.11^{+0.11}_{-0.07}$. The fit is shown in Figure 2. This method places a smaller upper limit on the value of θ_α^2. Figure 3 shows the yield of the $^{12}C(\alpha,\gamma)^{16}O$ reaction expressed in terms of the S-factor.

IV. Astrophysical Significance

It is of interest to consider what range of stellar core

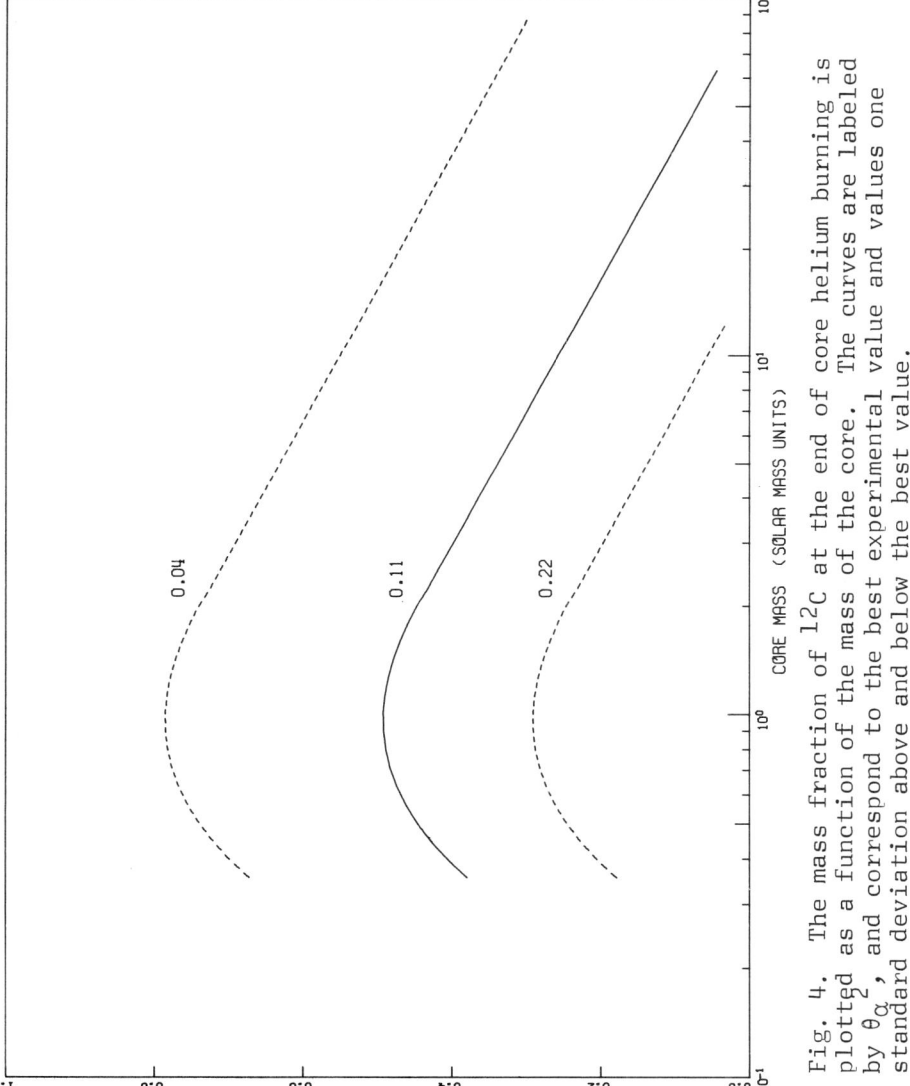

Fig. 4. The mass fraction of ^{12}C at the end of core helium burning is plotted as a function of the mass of the core. The curves are labeled by θ_{α}^2, and correspond to the best experimental value and values one standard deviation above and below the best value.

compositions at the end of helium burning corresponds to the range of θ_α^2 which is within one standard deviation of the value derived from the laboratory measurements. For stellar cores of mass greater than 2 $M_\odot$, Arnett (1972) has given an expression for X_c, the mass fraction of ^{12}C at the end of helium burning, as a function of the core mass M_α and of θ_α^2:

$$X_c = 0.31 - 0.667 \log (M_\alpha/M_\odot) - 0.267 \log \theta_\alpha^2.$$

Deinzer and Salpeter (1964) performed similar calculations, and considered lower masses as well. By combining these two analyses (normalizing at 2 $M_\odot$) it is possible to obtain a crude estimate of the mass fraction of ^{12}C for a range of values of θ_α^2. The result is given in Figure 4. As an example, it can be seen that for a 2 $M_\odot$ core, the value of X_c ranges from 25% to 75% for the range of θ_α^2 within one standard deviation of the best value. The experimental precision achieved up to this time does not determine the mass fraction of ^{12}C at the end of core helium burning with adequate precision.

Two developments within the past year should make it possible to extend the $^{12}C(\alpha,\gamma)^{16}O$ cross-section measurements to lower alpha energies, and thus to a better determination of the value of θ_α^2. A new chopping and bunching system constructed on the CIT 3 MV generator by Dr. M. R. Dwarakanath has made it possible to obtain up to 20 times the beam intensity previously available to us from the tandem accelerator. Also, methanol of ^{12}C isotopic purity much higher than previously available has been obtained from the Los Alamos Scientific Laboratory and has been successfully converted into an elemental ^{12}C target in a clean ultra-high vacuum system. It is hoped that additional measurements can be made shortly.

REFERENCES

Adams, A., Shapiro, M. H., Barnes, C. A., Adelberger, E. G., and Denny, W. M. 1968, Bull. Am. Phys. Soc., 13, 698.
Arnett, W. D. 1972, Ap. J., 176, 681.
Deinzer, W., and Salpeter, E. E. 1964, Ap. J., 140, 499.
Dyer, P. 1973, Ph.D. Thesis, California Institute of Technology.
Fowler, W. A., Caughlan, G. R., and Zimmerman, B. A. 1967, Ann. Rev. Astr. and Ap., 5, 525.
Jaszczak, R. J., Gibbons, J. H., and Macklin, R. L. 1970a, Phys. Rev., C2, 63.
Jaszczak, R. J., and Macklin, R. L. 1970b, Phys. Rev., C2, 2452.
Koonin, S. E., and Tombrello, T. A. 1972, to be published.
Weisser, D. C., Morgan, J. F., and Thompson, D. R. 1970, internal report.

THE CARBON DETONATION SUPERNOVA MODEL

J. Craig Wheeler
Harvard College Observatory, Cambridge, Mass.

I. Introduction

A supernova model based on a collapsing degenerate carbon/oxygen core may be of basic importance in gaining an understanding of a variety of supernova phenomena. For stars in the mass range $4 \lesssim M/M_\odot \lesssim 8$, strong neutrino loss mechanisms (Paczynski 1972, Couch and Arnett 1973) may prevent carbon ignition in the degenerate cores, $M \simeq M_{Ch} \simeq 1.4 M_\odot$, until the cores become unstable to collapse. In other situations stars could lose their envelopes leaving the core behind to become eventually dynamically unstable because of the effects of electron capture or of mass accretion from a companion in a binary system.

This notion can be exploited in the computation of numerical models without a detailed knowledge of the prior evolution of the core which led to its collapse. Assuming that the structure of the star as it begins free-fall collapse is not significantly different from its structure in hydrostatic equilibrium, one need only specify the central adiabat along which collapse begins. Since the prior evolution is not yet known exactly the collapse must be parameterized. This is done by starting the models at the central density at which collapse is thought to occur and then adjusting the initial temperature.

Wheeler, Hansen and Cox (1968) have estimated that carbon white dwarfs at zero temperature will become unstable at central density $\rho_c \sim 10^{10}$ g/cm^3. At this density the fundamental period becomes comparable to the timescale for fast electron-capture reactions (log ft $\sim$ 5) on some reasonably abundant seed nucleus, e.g. ^{23}Na. The two timescales at instability are both $\tau \sim 1$ sec. This calculation is being redone by Wheeler and Shapiro (1973), using a realistic temperature $\sim 5 \times 10^{8}$ °K.

II. The Model

The pre-supernova model adopted is a degenerate carbon core with central density $\rho_c = 10^{10}$ g/cm^3 and an isothermal temperature distribution. The model is spherically symmetric, ignores the effects of rotation and magnetic fields, and contains forty mass zones, typically. Collapse is initiated by increasing the mass of each zone by about a percent. This is meant to be reminiscent of mass accretion from a helium burning shell or a binary companion but is by no means thought to be in accord with the details of such a process.

The collapse is followed numerically by a hydrodynamic code which is a modified version of that described in Hansen and Wheeler (1969). The tabulated equation of state incorporates an arbitrarily relativistic and degenerate perfect electron gas, non-relativistic ions, and radiation. The ^{12}C-^{12}C reaction rates are based on the work of Patterson, Winkler and Zaidins (1969). The strong screening and pycnonuclear rates are those of Salpeter and Van Horn (1969) as modified by Hansen and Wheeler (1969) to be continuous with temperature in the high density regime. The influence of β-processes on the composition and neutrino loss rates of the detonated material, assumed to be in nuclear statistical equilibrium, given by Barkat, Buchler and Wheeler (1972) are incorporated. Other β-processes are negligible on dynamical timescales.

The initiation of detonation is perhaps the weakest point of these models. There is little hope that the model can reproduce accurately the deflagration phase preceding the detonation. Numerical results suggest that the deflagration phase is sensitive to the temperature even though a stationary Chapman-Jouguet detonation, once formed, would not be (cf. Buchler, Wheeler and Barkat 1971). As a result of the finite temperature steps between zones and the spreading of shock fronts over several zones, the distance required for a detonation to form spontaneously in the models is comparable to or larger than a density scale height even in favorable cases ($T \sim T_{ignition}$). The situation is thus prejudiced against the formation of a stable detonation front.

In a real star the shock front widths will be much smaller than other scales of interest, in particular the temperature scale height. The temperature will probably be near the ignition temperature over distances much greater than the width of the burning front. The most reasonable guess seems to be that a stable detonation forms at a radius much less than that of the first zone of the model, $\sim 10^7$ cm.

In these models a detonation is initiated by giving the first zone to flash an outward momentum corresponding to that contained in a Chapman-Jouguet detonation. Since the detonation velocity is about 3×10^8 cm/sec (cf. Buchler et al. 1971) and the momentum is mostly carried by three mass zones, the first zone is given an initial velocity of 10^9 cm/sec to impart the appropriate momentum. The numerics quickly spreads this momentum over several mass zones and a quasi-stable detonation then propagates. This process is performed in the central zone of the present models; for comparison see the discussion of off-center ignition by Buchler in this volume.

Three models have been run whose collapse adiabats correspond to temperatures 2.5×10^8, 4×10^8 and 6×10^8K at $\rho_c = 10^{10}$ g/cm^3. For details see Wheeler, Buchler and Barkat (1973). Figure 1 shows the evolution of the central conditions of the three models in the log T_c - log ρ_c plane. An ideal collapse adiabat would have slope $\Gamma_3 - 1 = (\delta \ln T/\delta \ln \rho)_S \lesssim 2/3$. The actual paths have slopes ~10 percent higher because of numerical inaccuracy. After the flash the temperatures are high enough that the degenerate electrons dominate, $\Gamma_3 - 1 \sim 1/3$.

The flash densities for the 2.5×10^8, 4×10^8 and 6×10^8K models are 2.7×10^{10}, 2.0×10^{10}, and 1.4×10^{10} g/cm^3, respectively. The leftward excursions in the post-detonation paths of the two cooler models in Figure 1 are due to the imposed initial velocity. Also given in Figure 1 are the contours where the burning rates are equal to 10^{19} ergs/gm/sec for carbon mass fractions 1.0, 0.5, and 0.1, respectively. This is essentially the condition for which a flash occurs. The presence of oxygen would thus affect these calculations significantly only if its mass fraction were rather large. The flash point contours in Figure 1 actually should turn over sharply at $\rho_c \sim 10^{11}$ g/cm^3 as temperature insensitive pycnonuclear reactions dominate the rates (see Wheeler 1973). Models collapsing on lower adiabats will thus all flash at $\rho_c \sim 10^{11}$ g/cm^3.

The 6×10^8K model formed a detonation very quickly even without an initial velocity and, because of the relatively low peak densities attained, it exploded completely, ejecting ~1.4 M$_\odot$ with ~2×10^{51} ergs. The behavior of the 4×10^8K model with induced detonation was qualitatively the same as that for the corresponding 2.5×10^8K model. The inner layers start to expand after the passage of the detonation wave but then collapse because of the overwhelming neutrino losses associated with elec-

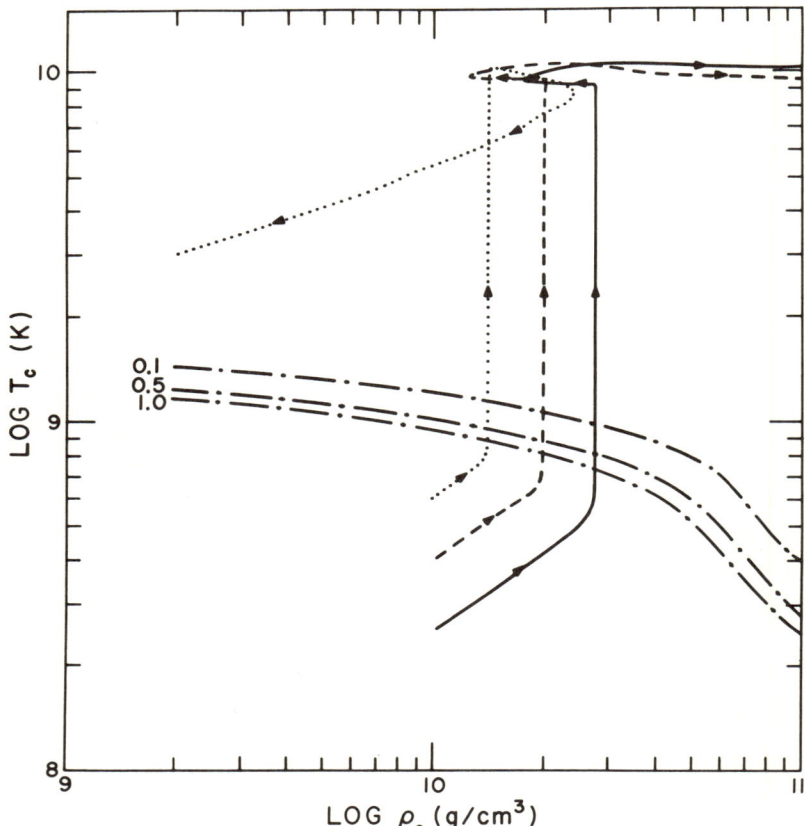

Fig. 1. The evolution of the central zone of the three models is depicted in the log T_c - log ρ_c plane. The initial density prior to collapse is 10^{10} g/cm^3. The initial temperatures are 2.5 x 10^8 K (solid line), 4 x 10^8 K (dashed line) and 6 x 10^8 K (dotted line). Arrows indicate the direction of the evolution. The dot-dash lines indicate the conditions for the energy generation rate being 10^{19} ergs/g/sec, roughly the criterion for the flashing of a zone, for carbon mass fraction x_{12} = 1.0, 0.5, and 0.1.

tron capture on the detonated material. The loss rates are typically $(5-10) \times 10^{19}$ erg/g/sec whereas the specific energies are $\sim 5 \times 10^{18}$ erg/g. Thus the characteristic dynamical time, $\tau \sim 0.1$ sec, is of the same order as that for the neutrino losses.

The final results of these calculations are that the 2.5×10^8 K model leaves a remnant of 1.37 $M_\odot$ with 0.07 $M_\odot$ (4 mass zones) ejected and the 4×10^8 K model leaves a remnant of 1.33 $M_\odot$ with 0.11 $M_\odot$ (7 mass zones) ejected both giving velocities in the range of $(2-3) \times 10^9$ cm/sec and associated energies of $\sim 4 \times 10^{50}$ ergs. The zones expelled are essentially only those which attain escape velocity with the original passage of the detonation wave.

The dynamical behavior is illustrated in Figures 2 (a-c) where the radii of selected mass zones of the models are shown as a function of time.

Since this code is unable to treat the details of neutron star formation, the collapse was handled by allowing each zone to fall freely when its density surpassed 10^{12} g/cm^3. This schematic treatment was checked by Wilson. His code incorporates a neutron matter equation of state and treats neutrino transport properly (see Wheeler, Wilson and Buchler 1973). Wilson found that the mass loss, in particular, is not sensitive to details of the remnant formation because of the decoupling of inner and outer regions evidenced in Figures 2a and 2b. Wilson's calculations also establish that if the remnant has a mass less than the static mass limit for the given neutron-matter equation of state there is no trouble forming a stable neutron star. The outflux of neutrinos cushions the infall and prevents overshoot to a black hole.

The models at detonation have an optical depth of about 100 for neutrinos and in fact Wilson's calculations indicate that a fair amount of energy from the core is deposited in the outer layers. In the model with initial temperature 2.5×10^8 K Wilson finds 0.12 $M_\odot$ ejected with 2.4×10^{51} ergs as compared with 0.07 $M_\odot$ with 4×10^{50} ergs for the original model. The velocities are about the same in either case, $\sim (2-3) \times 10^9$ cm/sec.

III. Type II Supernova

The model being discussed here corresponds to a Type II supernova inasmuch as it represents the core of a red-giant star with mass in the range 4-8 $M_\odot$. The energy liberated is at least 4×10^{50} ergs. This value does not include the energy deposited by neutrinos and any energy

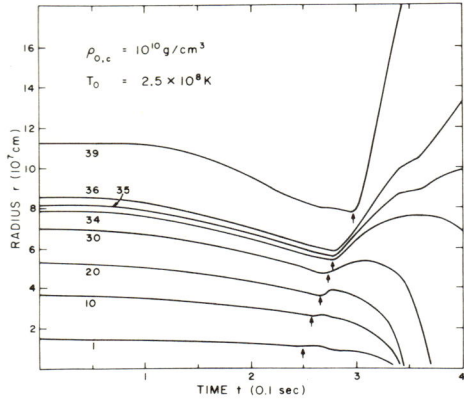

Fig. 2a. The radius versus time is plotted for selected mass zones of the model with initial temperature 2.5×10^8 K. The arrows indicate the flash points of the respective zones, hence the locus of the detonation front. Zone 36 is unbound and expelled from the star, Zone 35 is not. Note the decoupling of the interior of the star from the outer portions. This serves to define the mass expelled, somewhat independently of the interior dynamics.

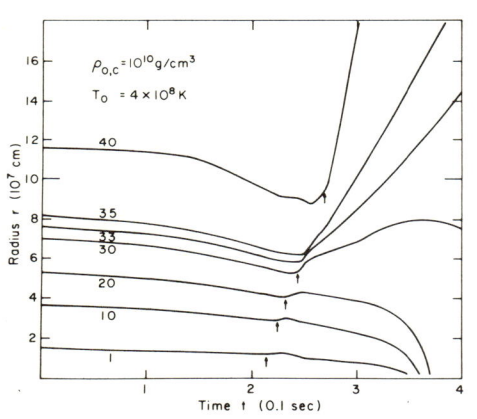

Fig. 2b. Same as figure 2a but for the model with initial temperature 4×10^8 K. Zone 33 is not ejected, zones exterior to it are.

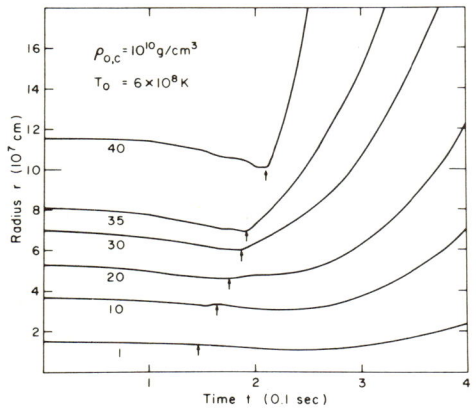

Fig. 2c. Same as figure 2a but for the model with initial temperature 6×10^8 K. The entire model is dispersed leaving no remnant.

stored in the remnant. The mass blown off should be the
3-7 $M_\odot$ in the hydrogen envelope. If radiation losses
from the envelope are neglected, energy conservation gives
an estimated lower limit for the velocity of the expelled
envelope of $\sim 4 \times 10^8$ cm/sec, a factor of two below estimated
Type II velocities. By assuming that neutrino deposition
brings the energy to 2.4×10^{51} ergs, one finds a velocity
$\sim 10^9$ cm/sec. There is thus no obvious need to involve the
energy stored in the remnant to account for Type II velocities.

The calculations reported here indicate that a core
collapsing on an adiabat lower than that passing through
$T_c \sim 4 \times 10^8$ K at $\rho_c \sim 10^{10}$ g/cm^3 will leave a remnant neutron
star. Determining the maximum adiabat for which a remnant
will be formed is subject to uncertainties associated with
the initiation and propagation of detonations, with the
nuclear reaction rates (mainly the screening factor), and
with the β-process rates in the detonated material.

If the present results are reasonably valid two main
projects suggested themselves: to determine the exact
collapse adiabat consistent with detailed evolutionary studies
and to calculate with two-dimensional hydrodynamics the same
model described here, but with rotation and magnetic field
effects included. The goal of showing that stars in the
mass range 4-8 $M_\odot$ give rise to Type II supernovae and pulsars
seems to be in view.

IV. Type I Supernova

Coincidentally, the model adopted here is nearly
identical to the one discussed by Hansen and Wheeler (1969)
as a possible model for a Type I supernova. One arrives at
this interpretation by assuming that the model is self-
contained, i.e. that the initial model is just a dense
white dwarf with no extended envelope. Finzi and Wolf (1967)
first suggested a collapsing white dwarf as a Type I model
because of the possibility of a stable lifetime as long as
10^{10} years, a condition which seems necessary in light of the
occurrence of Type I events in elliptical galaxies. Type I
supernovae also occur in spiral galaxies where their age is,
perforce, indeterminate.

The mass expelled and the velocities are similar in
the present calculation and in Minkowski's (1968) data on
Tycho's supernova of 1572, as shown in Table I. The mass
is somewhat model-dependent, both for the present results and
Minkowski's. The low value of the mass expelled occurs
naturally in the present model. Similar results might be

TABLE 1

Comparison of Model with Tycho's (1572) Supernova

	Model	Tycho
$M/M_\odot$	0.07	0.1
$V/10^9$ cm sec	2-3	2
$E_{kinetic}/10^{50}$ ergs	3	4 ($\sim\frac{1}{3} E_{total}$)
$E_{total}/10^{50}$ ergs	4	11 ($\times \frac{0.1}{n_H}$)†

†n_H is the hydrogen number density

obtained from other models involving hydrogen or helium explosions in degenerate stars, but on a seemingly ad hoc basis.

A white dwarf model should follow a collapse adiabat much lower than those calculated here. Such a model would probably detonate at the maximum density, $\sim 10^{11}$ g/cm^3 as discussed previously. Wheeler (1973) argues that such models are probably relatively insensitive to details of the nuclear reaction and β-process rates because the models would be well below the critical adiabat for remnant formation but not at such high densities that β-processes hinder propagation of the detonation.

For Type I supernovae, a reasonable scenario incorporating this model would picture a white dwarf accreting mass from a binary companion via an accretion disk. Except in the plane of the accretion disk one would then be observing the explosion of the bare core, in keeping with observed hydrogen depletion in Type I events (Kirshner et al. 1973). The pre-explosion time scale of the system can in principle be altered by adjusting the masses and orbital parameters of the binary system. The expelled matter is predominantly $^{56}N_i$, so perhaps the $^{56}N_i$ β-decay mechanism for the light curve (Colgate and McKee 1969) is appropriate.

This model predicts that Type I supernovae should also leave pulsar-like remnants. These pulsars might differ quantitatively from those generated in Type II explosions because of the possible differences in initial conditions, angular momentum, magnetic field density, etc. Wheeler (1973) summarizes the possible observational implications of this model. There is as yet no strong evidence one way or the other concerning pulsars in remnants of Type I events. There is some suggestion that the pulsar population comprises two components.

V. Conclusion and a Constraint

This work establishes a plausible evolutionary sequence leading to the formation of Type II supernovae and pulsars. It also suggests that the same physical mechanism, collapse followed by carbon detonation and neutron core implosion, is active in the formation of Type I supernovae.

Both of these processes may be intimately tied to the nature of stellar evolution in binary systems: the former because most massive stars are naturally found in binaries and the latter because accretion is a prime candidate for the instability mechanism. A fascinating question for the future concerns the influence of a binary companion on the stellar systems which can generate supernova explosions. Mass transfer onto an otherwise stable star may cause an explosion. Mass transfer from a star may prevent its evolution to an explosive phase. The evolutionary timescale, the observed properties of the explosion, the remnant produced and the products of nucleosynthesis may all be sensitive functions of the interactions of the stars in the system. Complete understanding of the final stages of stellar evolution will probably not come without a thorough study of the evolution of multiple stars.

BIBLIOGRAPHY

Barkat, Z., Buchler, J.-R., and Wheeler, J.C. 1972, Ap.J., 173, 183.
Buchler, J.-R., Wheeler, J.C., and Barkat, Z. 1971, Ap.J., 167, 465.
Colgate, S.A., and McKee, C.R. 1969, Ap.J., 157, 623.
Couch, R. and Arnett, W.D. 1973, Ap.J. (Lett.), 180, L101.
Finzi, A., and Wolf, R.A. 1967, Ap.J., 150, 115.
Hansen, C.J., and Wheeler, J.C. 1969, Ap. and Space Sci., 3, 464.
Kirshner, R.P., Oke, J.B., Penston, M.V. and Searle, L. 1973, submitted to Ap.J.
Minkowski, R. 1968, Nebulae and Interstellar Matter eds. B.M. Middlehurst and L.H. Allen University of Chicago Press, Chicago.
Paczyński, B.E. 1972, Ap. Letters, 11, 53.
Patterson, J.R., Winkler, H., and Zaidin, C.S. 1969, Ap.J., 157, 367.
Salpeter, E.E., and Van Horn, H.M. 1969, Ap.J., 155, 183.
Wheeler, J.C. 1973, submitted to Ap.J.
Wheeler, J.C., Buchler, J.-R., and Barkat, Z.K. 1973, submitted to Ap.J.
Wheeler, J.C., Hansen, C.J. and Cox, J.P. 1968, Astrophys. Letters, 2, 253.
Wheeler, J.C., and Shapiro, P. 1973, in preparation.
Wheeler, J.C., Wilson, J.R., and Buchler, J.-R. 1973, submitted to Ap.J.

THE CARBON DETONATION SUPERNOVA AND ASSOCIATED REMNANT FORMATION

S. Bruenn
Florida Atlantic University

Motivated by some rather suggestive observational and theoretical considerations, much recent effort has been devoted to establishing an evolutionary link between the late evolutionary stages of stars in the mass range $3.5 M_\odot \lesssim M \lesssim 8 M_\odot$ and supernova events which give rise to neutron stars. The main problem has been the explosiveness of C^{12} under the highly electron degenerate conditions characteristic of the cores of these stars (Arnett 1969) and the very high densities required for electron capture following detonation to be rapid enough to implode the core (Bruenn 1972a). A number of recent calculations (Arnett 1971; Barkat et al. 1972; Bruenn 1972b; Paczynski 1973) which explore the effect of additional neutrino emission mechanisms on the evolution of the degenerate core have not succeeded in sufficiently delaying C^{12} ignition for an implosion to occur. The new screening factors proposed by DeWitt et al. (1973) (see also Graboske et al. 1973) further aggravate the problem.

A way out of this dilemma was proposed by Paczynski (1972). He suggested that the convective mass flows that occur during the initial stages of a C^{12} burning runaway would give rise to greatly enhanced URCA neutrino losses, the overall process being sufficiently temperature sensitive to stabilize the C^{12} burning. A recent more detailed examination of this mechanism by Couch and Arnett (1973) has supported Paczynski's suggestion.

It is likely that still more detailed investigations of the evolution of degenerate cores undergoing hydrostatic C^{12} burning in the manner suggested by Paczynski will be forthcoming, and attempts will be made to link these with supernova events and neutron star formation. We have therefore been motivated to examine the dependence ρ_{crit} (the central density at detonation necessary for neutron star formation) on some of the relevant model parameters and input physics, and to place some constraints on the predetonation evolutionary stage in order that ρ_{crit} may be achieved or exceeded prior to detonation.

Predetonation Regime

The central density of a stellar core at which thermonuclear runaway causes the formation of a detonation wave is critical in determining whether or not the core will subsequently implode to form a neutron star. Let τ_n and τ_{HD} denote respectively the net energy generation and the hydrodynamical timescale of a given stellar core. By equating τ_n to τ_{HD} we can delimit a region of the ρ-T plane which is inacessible to the core prior to detonation, <u>i.e.</u>, the region in which $\tau_n < \tau_{HD}$.

Figure 1 illustrates some of the results obtained.

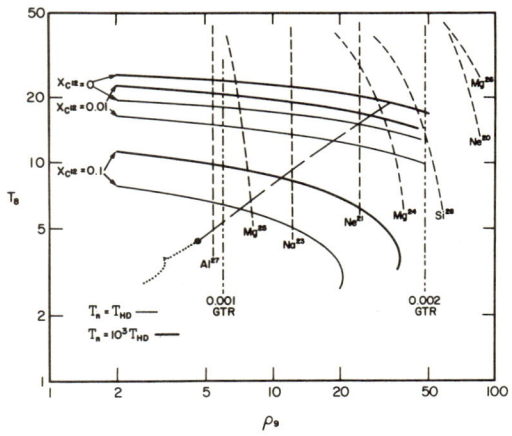

Figure 1
The loci of points representing the equality of various timescales relevant to the predetonation evolution and the central density at detonation of degenerate stellar cores. See text for details.

The heavy solid lines are the loci of points satisfying $\tau_n = \tau_{HD}$ for the indicated mass fraction of C^{12}. (Loci of points satisfying $\tau_n = 10^3 \tau_{HD}$ are also shown.) The assumed composition by mass was 0.39ζ O^{16}, 0.44ζ Ne^{20}, and 0.17ζ Mg^{24}, where $\zeta = 1-X_{c12}$, X_{c12} being the assumed mass fraction of C^{12}. This composition approximates that obtained by Arnett and Truran (1969) for quasistatic carbon burning at $6 \times 10^{(8)}$°K, and assumes that a photo-rearrangement of $Ne^{(20)}$ has not occurred. The energy generation timescale was determined from

$$\tau_n = \int_{T_o}^{T_1} \frac{C_v dT}{\dot{\varepsilon}(\rho,T,X_i)} \qquad (1)$$

where C_v is the isochoric specific heat, T is the temperature, and $\dot{\varepsilon}(\rho,T,X_i)$ is the net energy generation rate. (Neutrino emission was ignored since the neutrino cooling timescales are much greater than the timescales being considered here.) The temperature T_1 was chosen so that $\dot{\varepsilon}(\rho,T_1,X_i) = 2\dot{\varepsilon}(\rho,T_o,X_i)$, and the composition variables, denoted collectively by X_i, were assumed constant. This prescription for computing τ_n was checked against parallel computations utilizing a realistic nuclear reaction network, and was found to give results within a factor of two of the latter provided that the initial X_{c12}, when nonzero, was not less than about 0.07. Otherwise, the effect of the decrease in X_{c12} during burning could not be ignored. For $0 < X_{c12} \leq 0.07$ it was found that T_o in Eq. (1) required to satisfy $\tau_n = \tau_{HD}$ was given to good accuracy by

$$T_o = T_o(X_{c12} = 0) - qX_{c12}/C_v , \qquad (2)$$

where $q (\approx 4.7 \times 10^{17}$ ergs/g) is the energy released in the burning of one gram of C_{12}, and $T_o(X_{c12} = 0)$ is the temperature required to satisfy $\tau_n = \tau_{HD}$ if $X_{c12} = 0$. In the computations of τ_n, the $C^{12} - C^{12}$ and $O^{16} - O^{16}$ burning rates were taken respectively from Arnett and Truran (1969) and Truran and Arnett (1970), and the strong screening corrections wave those given by DeWitt et al. (1973). The hydrodynamical timescale was approximated by

$$\tau_{HD} = 1/\sqrt{(2\pi G_\rho)} \simeq 446/\rho^{1/2} \text{ sec.}, \qquad (3)$$

which for collapse probably represents a lower bound (Fowler and Hoyle 1964).

It is apparent from Figure 1 that the central density at which detonation is initiated in a core can be a sensitive function of X_{c12}. This figure will be used in the following sections in the discussion of results of some dynamical calculations.

Static Models

Suppose the evolution of a degenerate stellar core is quasistatic immediately prior to detonation. Then to very good approximation it can be assumed that the core is static and in complete hydrostatic equilibrium when detonation is initiated. A predetonation model satisfying these conditions will be referred to as a Static Model.

In a previous publication (Bruenn 1972a) we investigated the effect of the immediate predetonation central density, ρ_c, of Static Models on their postdetonation hydrodynamics. It was concluded that β-processes, which occur rapidly on the products of the detonation, would induce an implosion to neutron star densities only if $\rho_c > \rho_{crit}$ where ρ_{crit} was between 2.5×10^{10} and 3×10^{10} g/cm^3. We have reinvestigated the effect of ρ_c on the post-detonation hydrodynamics of Static Models for several reasons: (1) the development of an improved computer code, (2) finer zoning of the models, and (3) a more accurate and extensive tabulation of the β-rates and of the contribution of the nuclei to the equation of state (see Bruenn 1973 for details). The properties of the Static Models immediately prior to detonation were pretty much as described in Bruenn (1972a) (*i.e.*, homogeneous and isothermal at 3×10^{8} °K. The composition, however, was taken to be 0.54 O^{16}, 0.30 Mg^{24}, 0.02 Mg^{26}, and 0.14 Si^{28} (composition A of Truran and Arnett 1970), where the numbers denote mass fractions. This composition assumes a prior photo-rearrangement of Ne^{20}, and has somewhat less available energy than the compositions considered in Bruenn (1972a).

Some of the results of the computations are given in the first three rows of Table 1. Despite the increased accuracy of the computations, the basic result remains unchanged-- the value of ρ_{crit} lies between 2.5×10^{10} and 3×10^{10} g/cm^3. A few of the details of the postdetonation evolution were different, however. Whereas in the previous calculations the postdetonation temperature of a given mass zone decreased immediately after the passage of the detonation wave due to the intense neutrino emission associated with the β-processes, in the present calculations the postdetonation temperature began to increase. Not until the total neutron-proton ratio reached about 1.18 did the temperature level off and subsequently decrease. Because of the more than order-of-magnitude

TABLE 1

SOME RESULTS OF DETONATION-POSTDETONATION EVOLUTIONARY CALCULATIONS

Model	Detonation Aftermath
1. Static Model $\rho_{10}(d)=1.5$ M=1.440	Dwarf $M_r=0.14$
2. Static Model $\rho_{10}(d)=2.5$ M=1.446	Electron degenerate dwarf $M_r=1.26$
3. Static Model $\rho_{10}(d)=3.0$ M=1.447	Collapse to neutron star $M_c=1.37$
4. 3β Model $\rho_{10}(d)=1.5$ M=1.440	Electron degenerate dwarf $M_r=1.12$
5. 3β Model $\rho_{10}(d)=2.0$ M=1.443	Collapse to neutron star $M_c=1.35$
6. 10β Model $\rho_{10}(d)=1.0$ M=1.434	Electron degenerate dwarf $M_r=1.02$
7. 10β Model $\rho_{10}(d)=1.5$ M=1.440	Collapse to neutron star $M_c=1.36$

TABLE 1 (continued)

Model	Detonation Aftermath
8. Prior Collapse Model $\mu_e=2.003 \to 2.050$ $\rho_{10}(i)=0.6$, $\rho_{10}(d)=1.0$ $M=1.425$	No remnant
9. Prior Collapse Model $\mu_e=2.003 \to 2.050$ $\rho_{10}(i)=0.6$, $\rho_{10}(d)=1.5$	Collapse to neutron star $M_c=1.35$
10. Prior Collapse Model $\mu_e=2.003 \to 2.010$ $\rho_{10}(i)=1.5$, $\rho_{10}(d)=2.0$ $M=1.440$	Electron degenerate dwarf $M_r=1.25$
11. Prior Collapse Model $\mu_e=2.242 \to 2.250$ $\rho_{10}(i)=1.5$, $\rho_{10}(d)=2.0$ $M=1.150$	No remnant
12. Off-Center Detonation $\rho_{10}(d)=2.5$ $M=1.446$, $M_{det}=0.529$	Electron degenerate dwarf $M_r=1.24$
13. Off-Center Detonation $\rho_{10}(d)=3.0$ $M=1.447$, $M_{det}=0.538$	Collapse to neutron star $M_c=1.37$
14. Off-Center Detonation $\rho_{10}(d)=3.0$ $M=1.447$, $M_{det}=1.4466$	Electron degenerate dwarf $M_r=1.30$

TABLE 1 (continued)

DEFINITION OF SYMBOLS

M — mass (in $M_\odot$) of the predetonation model

M_r — mass (in $M_\odot$) of the remnant

M_c — mass (in $M_\odot$) of the material undergoing collapse to a neutron star

M_{det} — mass (in $M_\odot$) of the core enclosed by the detonating spherical shell in the Off-Center Detonation Models

$\rho_{10}(d)$ — central density (in 10^{10} g/cm^3) of the models at detonation

$\rho_{10}(i)$ — initial central density (in 10^{10} g/cm^3) of the Prior Collapse Models

$\mu_e = a \rightarrow b$ — a is the initial value of μ_e in the Prior Collapse Models, and b is the value to which μ_e was instantaneously switched in order to induce the predetonation collapse.

increase in the accuracy of the present calculations over those of Bruenn (1972a) we are inclined to give the present calculations more credence. Furthermore, there is a reasonable explanation for the postdetonation temperature behavior. The energy lost due to the neutrino emission which accompanies the rapid β-processes (primarily electron captures on free protons) immediately following detonation is primarily electron zero-point energy. The nuclear rearrangement which occurs following the free proton electron captures (the abundances of the nuclei are distributed in statistical equilibrium at the high postdetonation temperatures) shifts the nuclear abundance distribution to nuclei with greater binding energies per nucleon. Net thermal energy is therefore released until the abundance peak shifts to the neutron-rich side of Fe^{56}. (A more quantitative discussion of this is given in Bruenn 1973). The temperature-density histories of three Static Models detonating at $\rho_c = 1.5 \times 10^{10}$, 2.5×10^{10}, and 3×10^{10} g/cm^3, respectively, are given by the solid lines in Figure 2.

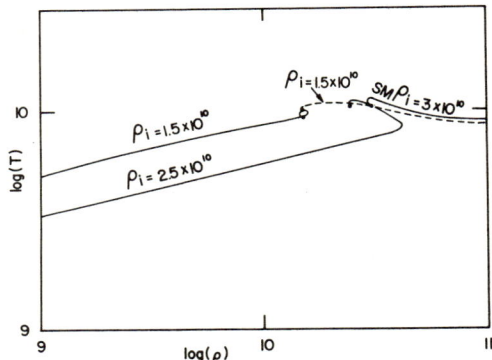

Figure 2

The postdetonation temperature-density histories of Static Models (solid lines) and a Prior Collapse Model (dashed line). The quantity ρ_i refers to the central density (g/cm^3) of the models at detonation.

Referring to Figure 1, the large value of ρ_{crit} characteristic of the Static Models makes it very unlikely that such a model with $X_{c12} = 0.1$ could evolve with a low enough temperature to avoid detonation before $\rho_c > \rho_{crit}$. In fact, the situation is much more severe. For quasistatic evolution the energy generation timescale, τ_n, must remain greater than the neutrino cooling timescale, τ_ν, and typically $\tau_\nu \gg \tau_{HD}$ for predetonation models. No calculation to date has indicated that a degenerate stellar core can evolve quasistatically to central densities exceeding ρ_{crit}.

Neutrino Rates

The rates of the β-processes used in the computations reported here and in Bruenn (1972a) were taken from the work of Hansen (1966, 1968). It is appropriate to consider the effect of possible future revisions of Hansen's (1966, 1968) rates of the value of ρ_{crit} for the Static Models, and the likelihood of a significant revision in the net of these rates.

The effect of possible revisions of the β-rates was examined by computing the postdetonation evolution of Static Models with these rates scaled by a factor of three and by a factor of ten. We will refer to the corresponding models as 3β and 10β Models, respectively. The results of the computations are given in the fourth to the seventh row of Table 1. It is apparent that a substantial upward revision in the net of Hansen's β-rates is necessary in order to significantly decrease ρ_{crit} for Static Models. In fact, scaling all rates upwards by a factor of ten only reduced ρ_{crit} by a factor of between two and three. A similar result was obtained by Bruenn (1972a).

To assess the likelihood of a substantial revision upwards in the net of the β-rates, we have computed the quantity $\Lambda(i) = n(i)\lambda_{e-cap}(i)$ for each of the nuclei (about 200) whose electron capture rates were considered in the hydrodynamic computations. The quantities $n(i)$ and $\lambda_{e-cap}(i)$ are the abundance (number/gram) and the electron capture rate (captures/nucleus/sec.) of nucleus i, respectively. (Electron capture is the predominant β-process for a significant portion of the postdetonation evolution.)

Figure 3 shows the eight largest $\Lambda(i)$ for each of a sample of densities, temperatures, and total neutron-proton ratios which approximate some of the postdetonation conditions of the Static Models. The $\Lambda(i)$'s have been normalized so that $\Sigma\Lambda(i) = 1$, where the sum is over all nuclei. The figure re-

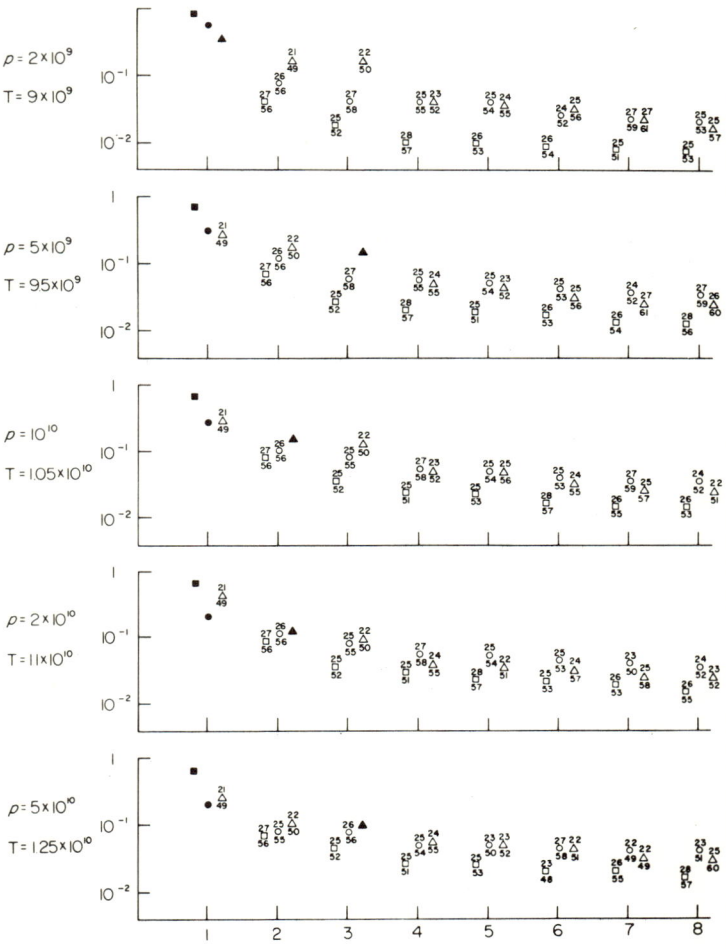

Figure 3
The contribution of the electron capture rate of the eight most important "electron capture nuclei" to the total electron capture rate for each of a sample of densities (ρ, in g/cm^3), temperatures (T, in °K), and total neutron-proton ratios. The quantity Λ(i) (defined in the text) is specified by the ordinate. The nuclei, in order of their electron capture importance, are arranged along the abscissa. Circles, squares, and triangles correspond, respectively, to a total neutron-proton ratio of 1.0, 1.15, and 1.5. Filled symbols denote free protons.

veals two important points. (1) Electron capture on free protons is the dominant contributor to the net electron capture rate for a substantial portion of the postdetonation evolution. Since electron capture on free protons is uncomplicated by excited states and is essentially the inverse of the laboratory measured β-decay of neutrons, it is unlikely that the rate for this process will be substantially revised. (2) With the exception of Sc^{49} and Ti^{50}, at large neutron-proton ratios, no one nuclear specie (other than free protons) contributes more than 10% to the net electron capture rate. Thus, a revision upward by a factor of ten or more in the electron capture rate of a given nucleus is necessary in order that the net electron capture rate be revised upwards by a factor of two. We conclude that a substantial revision in the net of the electron capture rate, and therefore in the net of all the β-rates, is very unlikely.

Collapse Prior to Detonation

A recent study (Couch and Arnett 1973) of hydrostatic carbon burning in electron degenerate carbon-oxygen stellar cores indicated that the increase in μ_e (the mean molecular weight per electron) due to electron captures on the products of the burning caused the evolution of the core to become rapid, and perhaps ultimately dynamic, before the completion of the carbon burning. It is possible, then, that the detonation of a stellar core will be preceded by a dynamic collapse. We consider here the effect of such a predetonation collapse on the value of ρ_{crit}. Such models will be referred to as Prior Collapse Models.

The important parameter of a Prior Collapse Model is the timescale of the collapse. Without the availability of a detailed calculation this timescale can only be crudely estimated. As stated above, the initial instability is caused by electron capture on someof the products of the carbon burning, so the timescale of the initial collapse is set by the timescales of these electron captures. In Figure 1 we have indicated by the dashed lines the electron capture timescales (computed from Hansen's (1966) β-rates) of all elements expected to be synthesized innon-negligible amounts by the carbon burning. Each dashed line is the locus of points for which the electron capture timescale of the indicated element is 10^3 times the hydrodynamical timescale, τ_{HD}, given by Equation (3). It appears that neither the electron captures inducing the initial instability nor those which subsequently occur during the collapse proceed on a timescale much less than $10^3\tau_{HD}$. Therefore, an electron capture induced predetonation collapse will probably occur on a timescale of similar magnitude.

General relativistic effects will tend to reduce the timescale of the collapse. As the density of the collapsing core increases, the pressure gradient given by the TOV (Tolman, Oppenheimer, and Volkoff) equation of hydrostatic equilibrium (see, for example, Zeldovich and Novikov 1971, p. 257) necessary to maintain a hydrostatic balance exceeds more and more that given by the corresponding Newtonian equation. In Figure 1 we have indicated by the dot-dashed lines the density at which the pressure gradient given by the TOV equation exceeds by 0.1% and 0.2% that given by the Newtonian equation. As can be seen, it is unlikely that general relativistic effects will reduce the predetonation collapse time-scale appreciably below 10^3 τ_{HD}.

We have also considered the reduction in pressure due to Coulomb, Thomas-Fermi, and exchange interactions (Salpeter 1951, Kovetz et al. 1972). These effects are very small for the extreme electron degeneracies encountered here, however, and work in such a way as to reduce the pressure imbalance during the collapse. We conclude from all of the above that should a predetonation collapse occur, the timescale of the collapse is unlikely to be substantially less than about 10^3 that of a full (zero pressure) hydrodynamic collapse.

To place an upper bound on the effect of a predetonation collapse on the value of ρ_{crit}, we took a 6×10^9 g/cm^3 central density Static Model and instantaneously increased the quantity μ_e in the innter 78% (by mass) of the core from its initial value of 2.003, to a value of 2.05. This highly schematic and arbitrary procedure was intended to simulate an upper bound of a very rapid and extensive electron capture in the core. (It is true that an instantaneous electron capture of the products of carbon burning to β-stable elements would result in a value of μ_e of the order of 2.25 for densities ~3×10^{10} g/cm^3, but this would require a comparatively long time.) Upon increasing μ_e the core began an immediate collapse, the timescale of which was about 40 τ_{HD} when the central density was between 10^{10} and 1.5×10^{10} g/cm^3. The results of detonation-postdetonation evolutionary calculations of these Prior Collapse Models are given in the eighth and ninth row of Table 1, and are illustrated for the Prior Collapse Model detonating at $\rho_c = 1.5 \times 10^{10}$ g/cm^3 by the broken line in Figure 2. It is seen that a reasonably dynamical predetonation collapse has an appreciable effect on ρ_{crit}, in this case reducing it to a value between 10^{10} and 1.5×10^{10} g/cm^3. The efficacy of a dynamical predetonation collapse in reducing ρ_{crit} is due in part to the maintenance of a high postdetonation temperature (and therefore a correspondingly large net electron capture rate) by the compression of the core (see Figure 2). The passage of the detonation wave only momentarily reversed the core compression.

A predetonation collapse more in accord with the timescale estimates discussed at the beginning of this section was brought about by instantaneously increasing μ_e from 2.003 to 2.01 in the inner 80% (by mass) of a 1.5×10^{10} g/cm^3 central density Static Model. The timescale of this induced collapse was about 130 τ_{HD} when the central density reached 2×10^{10} g/cm^3, and results of detonation-postdetonation evolutionary calculations are given in the tenth row of Table 1. The efficacy of this more mild predetonation collapse on reducing the value of ρ_{crit} is seen to be much smaller.

A further calculation was performed to test the effect of an extensive quasi-static electron capture occuring prior to a predetonation collapse. The occurrence of the quasi-static electron capture was represented by a 1.5×10^{10} central density Static Model with μ_e = 2.242. This value of μ_e corresponds to the electron capture at $\rho \sim 10^{10}$ g/cm^3 to β-stable elements of Ne^{20}, Na^{23}, and Mg^{24}, which were present in roughly the proportions expected from quasi-static carbon burning. A predetontion collapse was then induced by increasing μ_e instantaneously to 2.250 throughout most of the core. The timescale of the collapse was similar to that of the previous model, but detonation had a much more devastating effect--completely disrupting this model (see the eleventh row of Table 1). The obvious reason for this is that the occurrence of a quasi-static phase of electron capture prior to detonation increases the timescale of the post-detonation electron captures, and therefore reduces their effect on the hydrodynamics.

Off-Center Detonation

The possibility that detonation might be initiated in a region of the stellar core other than the center (i.e., off-center) has recently been suggested and discussed (Colgate 1972, Iben 1972, Buchler 1973). Spherical symmetry would be required in remarkable degree if detonation were to be initiated in a spherical shell with such a close approximation to simultaneity that each given region of the shell detonated before being perturbed by the detonation of other regions of the shell. In fact, such an occurrence, the complete simultaneous idealization of which would result in the formation of spherically symmetric inwardly and outwardly propagating detonation waves, would be fortuitous, as the individual events constituting the initiation of detonation would not be causally connected (the speed of sound being the appropriate signal velocity). Much more probable would be the occurence of an off-center detonation initiated at a point. The details of such an event would demand considerable computing time to follow numerically, as it would require a hydrosynamic computer code incorporating more than one spatial dimension.

We have performed hydrodynamic calculations of stellar cores undergoing a simultanous off-center detonation in a spherically symmetric shell (this extremely unrealistic model being the only one treatable with our computer code) to determine the effect on ρ_{crit}. In these calculations, spherically symmetric inwardly and outwardly propagating detonation waves were assumed to be formed as a result of the detonation. The calculations were made for models in which the spherically symmetric region of detonation enclosed 36% of the mass of the core, and the entire mass of the core, respectively, and the results are given in the twelfth to fourteenth rows of Table 1. The main conclusion to be drawn is that such idealized off-center detonations tend to <u>increase</u> ρ_{crit}. This is because the outer layers of these models are expelled before the outer layers of corresponding models undergoing central detonation, so that the inner regions of the core more quickly experience a reduced pressure. The effect of the inwardly propagating detonation wave is minimal due to the strong rarefaction wave which follows immediately behind it.

Summary

The detonation-postdetonation evolution of a variety of degenerate, predetonation stellar cores has been examined with the principal aim of determining the effect of some of the input physics and model parameters on the value of ρ_{crit}. We draw the following conclusions.

1) The value of ρ_{crit} for a quasi-static, degenerate stellar core with $X_{c12} \geqslant 0.07$ (and possibly also with $X_{c12} < 0.07$) is too high to be reached by quasi-static evolution. (This assumes, of course, that the input physics is accurate).

2) The value of ρ_{crit} is not a very sensitive function of the overall β-rates.

3) It is unlikely that revisions in the β-rates of individual nuclei will substantially affect the overall β-rates.

4) A collapse prior to detonation can substantially reduce the value of ρ_{crit} provided the collapse is reasonably dynamic (collapse timescale $\leqslant 40\ \tau_{HD}$).

5) A crude examination of the factors inducing and sustaining a predetonation collapse indicates that such a collapse would occur on a timescale $\sim 10^3 \, \tau_{HD}$.

6) Increasing the quantity μ_e by a prior quasi-static phase of electron capture increases the value of ρ_{crit}.

7) The occurrence of a spherically symmetric off-center detonation increases the value of ρ_{crit}.

In Figure 1 we have represented by the dotted line the evolutionary track computed by Couch and Arnett (1973) of the center of a degenerate stellar core undergoing quasi-static carbon burning. The large dot indicates the point at which the evolution became "rapid" and the calculation terminated, and the dashed line extending upward and to the right represents the path of an adiabatic collapse. From this figure and the above summary, we conclude that whether or not the ensuing detonation will lead to the formation of a neutron star depends critically on the final value of X_{c12}, the rate of the predetonation collapse, and the amount of cooling that occurs during the collapse. These factors should be examined carefully in future calculations.

It is a pleasure to thank the staff of the Florida Atlantic University Computer Center for providing me with the requisite computer time for the calculations.

REFERENCES

Arnett, W.D. 1969, Ap. and Space Science, 5, 180.
Arnett, W.D. 1971, Ap.J., 169, 113.
Arnett, W.D. and Truran, J.W. 1969, Ap.J., 160, 181.
Barkat, Z., Wheeler, J.C., and Buchler, J.R. 1972, Ap.J. 171, 651.
Bruenn, S.W. 1972a, Ap.J. Suppl., 24, 283.
Bruenn, S.W. 1972b, Ap.J., 177, 459.
Bruenn, S.W. 1973, (submitted to Ap.J.).
Couch, R.G. and Arnett, W.D. 1973, Ap.J. Lett., 180, L101.
DeWitt, H.E., Graboske, H.C., and Cooper, M.S. 1973 (preprint).
Fowler, W.A. and Hoyle, F. 1964, Ap.J. Suppl., 9, 201.
Graboske, H.C., DeWitt, H.E., Grossman, A.S., and Cooper, M.S. 1973 (preprint).
Hansen, C.J. 1966, unpublished Ph.D. thesis, Yale University.
Hansen, C.J. 1968, Ap. and Space Science, 1, 499.
Iben, I. Jr. 1972, Ap.J., 178, 433.
Kovetz, A. Lamb, D.Q., and Van Horn, H.M. 1972, Ap.J., 174, 109.
Paczynski, B. 1972, (preprint).
Paczynski, B. 1973, (preprint).

Salpeter, E.E. 1961, Ap.J., 134, 669.
Truran, J.W. and Arnett, W.D. 1970, Ap.J., 160, 181.
Zel'dovich, Ya. B. and Novikov, I.D. 1971, Relativistic Astrophysics, Vol. I (Chicago; University of Chicago Press).

OFF-CENTER DETONATION SUPERNOVAE

Jean-Robert Buchler

Belfer Graduate School of Science
Yeshiva University
New York, New York

Evolutionary calculations of stars in the mass range 4 - 10 $M_\odot$ (Rose 1969, Arnett 1969, Paczynski 1970, Barkat 1971, Barkat et al.1972) indicate that these stars gradually develop identical degenerate carbon (- oxygen) cores of ~1.4 $M_\odot$. The temperature profile of these presupernova models was found to be highest in the center and to decrease monotonically outwards. These cores were believed to ignite carbon violently in the center when the central density reached about 3×10^9 g/cm^3. It seemed plausible that this violent ignition would generate a detonation wave. Hydrodynamic calculations showed that such a detonation wave runs all the way outwards and completely disrupts the star (Arnett 1969, Wheeler et al.1970, Mazurek 1972). An investigation of the various uncertainties in the carbon content, the rate of mass addition at the helium burning shell, and of the nuclear screening factors showed that the ignition density could be pushed as high as 6×10^9 g/cm^3 (Barkat et al.1972). The hope had been that at these higher desnities electron capture in the detonated material would be sufficiently effective to induce a reimplosion of part of the core (Colgate 1970). A study of the capture rates (Barkat et al.1971) indicated that the central density would have to be as high as 6×10^9 g/cm^3 for the captures to operate on a dynamic time scale. A subsequent hydrodynamic calculation by Bruenn (1972) indicated that central densities as high as 2×10^{10} g/cm^3 were needed for reimplosion, a density which was way above the value thought possible on evolutionary grounds.

This left the carbon detonation model as a very unlikely candidate for producing a supernova together with a neutron star (pulsar) remnant until Paczynski (1972) pointed out that Urca shells in the presence of convection could be extremely efficient in regulating the burning of carbon. However, in view of the complexity of the evolutionary behavior of convectively-driven Urca shells together with the uncertainty in the abundance of Urca seed-nuclei, one can, at present, only guess what the evolution will be.

One of the possibilities, which we shall follow up, is that the convective core stays cool while burning the carbon quietly and gives rise to a temperature inversion. It is quite plausible that the ensuing off-center ignition still generates a detonation wave. It has been suggested by various authors that such an off-center detonation, if it propagates both inwards and outwards, could give both the necessary inwards momentum to make the core collapse to neutron star densities and the outwards kick to expel the outer part of the core. Iben (1973) has recently analyzed the uncertainties involved in the evolutionary calculations of the presupernova cores and suggested that it is quite possible that a temperature inversion develops naturally, independently of convective Urca losses, giving also rise to an off-center detonation.

In view of these recent developments we have considered it worthwhile to investigate the aftermath of an off-center detonation.

Our pre-detonation models all have a mass close to the Chandrasekhar limit and are composed of half carbon, half oxygen. They are divided into three groups according to their central density:

1) $\rho_c = 8.6 \times 10^9$ g/cm^3
2) $\rho_c = 1.6 \times 10^{10}$ g/cm^3
3) $\rho_c = 2 \times 10^{10}$ g/cm^3

The models, initially in hydrostatic equilibrium, were rendered unstable by uniformly adding 2 percent to the mass. Model (1) contained 37 zones and models (2) and (3) contained both 47 zones of equal mass. The presence of a detonation was simulated by giving a velocity of 5×10^8 cm/sec to the boundary of the zone in which a nuclear runaway (flash) occurred first. This artificial kicking of the detonation wave was found to be necessary and can be justified on the ground that it is the coarseness of the zone width (>>> actual shock width) which prevents the numerical formation of a true detonation wave.

Since the question whether off-center ignition generates a detonation wave moving both outwards and inwards is still open, although a double detonation seems more plausible, we ran model (1) by (a) giving only the outer boundary of the first flashing zone an outwards velocity, and (b) giving its boundaries both an inwards and an outwards velocity. The final result is a complete disruption of the star in both cases. In model (a) the inner zones first expanded as the outwards moving detonation reduced the outside pressure. Subsequently an inwards moving compression wave due to electron capture in the detonated zone caused the first non-detonated inner zone to flash and generated an inwards moving detonation wave. The ease with which this detonation wave formed makes case (b) a more likely alternative from the start. The ultimate reason for complete disruption of model (1) is that electron captures at these densities are too slow to prevent a bounce and subsequent expansion, just as in the central detonation models (Bruenn 1972).

We thought that a comparison of the post-detonation behavior of central and off-center ignition would be interesting. Consequently models (2) and (3) were constructed with an initial temperature profile decreasing monotonically outwards, whereas models (2') and (3') were given a temperature inversion at a mass fraction of about 0.4.

After a transient phase, the corresponding central and off-center ignition models behaved essentially identically. Models (2) and (2') expelled the outer core and formed an extended (iron-nickel) white dwarf of about 1 $M_\odot$. Models (3) and (3'), on the other hand, completely collapsed to neutron star densities except for the last zone. In cases (2') and (3') we initiated both an inwards and outwards detonation wave.

For the purpose of completeness we also followed the behavior of model (3') with a detonation wave propagating outwards only. In this case the result was an expulsion of the outer core and the formation of a white dwarf remnant, just as in the lower central density model (2). The reason is that the inner zones have time to expand because of the reduction of outside pressure and, although these zones ultimately flash, electron captures are not as efficient at the lower densities.

Our central detonation models behaved similarly to those of Bruenn (1972), except that the initial infall caused our $\rho_c = 2 \times 10^{10}$ g/cm^3 models to completely collapse to neutron star densities while Bruenn's initially static models collapsed only for $\rho_c \geq 3 \times 10^{10}$ g/cm^3.

We have also compared the behavior of our models with those of Wheeler et al. (WBB 1973). WBB found that an initial isothermal (2.5 $\times$ 10^8 °K) $\rho_c = 10^{10}$ g/cm^3 model, rendered unstable by a small uniform mass addition, collapses to $\rho_c = 2.7 \times 10^{10}$, until it ignited in the center. A subsequent outwards running detonation wave expelled ~ 0.1 $M_\odot$ (4 mass zones) while electron captures caused the inner core to collapse to neutron star densities. The major difference is that WBB used mass zones decreasing outwards by 4 percent per zone, whereas the models reported here had equal mass zones. While the latter zoning gives a better overall numerical accuracy and energy conservation, it is clearly inadequate to describe the exact behavior of the outer part of the core.

Even with our crude zoning at the core boundary we are able to estimate that in the $\rho_c = 2 \times 10^{10}$ g/cm^3 model the detonation wave should deposit in excess of 10^{50} ergs into the surrounding envelope which is of the order of the binding energy of the latter. The detonation wave could therefore, at worst, be able to hold the envelope sufficiently long to have the core collapse to a neutron star (pulsar), give the radiation time to decouple from the plasma and allow the Ostriker-Gunn (1972) mechanism time to develop and drive off the envelope.

We should also point out that if the collapse lasts sufficiently long before thermal runaway occurs, the center may flash first due to adiabatic or viscous heating in spite of a temperature inversion, or it may ignite before the off-center detonation wave has had time to spread very far.

The off-center ignition most likely does not occur in a spherical shell to which we have constrained our dynamic models. However, on the basis of the results presented here, we believe that it is fairly irrelevant for the final result of these explosions, where exactly ignition occurs. We do not have to worry too much either about the difficult question of the evolutionary pre-detonation temperature profile. The present results are therefore encouraging to future work on exploding carbon cores.

The author would like to acknowledge the support of the National Science Foundation, the hospitality of the Goddard Institute for Space Studies in New York, and a grant by the Luxembourg Ministère des Affaires Culturelles.

REFERENCES

Arnett, W. D. 1969, Astrophys. Space Sci., 5, 180.
Barkat, Z. 1971, Astrophys. J., 163, 433.
Barkat, Z., Buchler, J. R., and Wheeler, J. C. 1971, Ap. J. Lett., 8, 21.
——————. 1972, Ap. J., 171, 651.
Bruenn, S. W. 1972, Ap. J. Suppl., 24, 283.
Colgate, S. A. 1971, Ap. J., 163, 221.
Iben, I. 1973, Ap. J., 178, 433.
Mazurek, T. J. 1972, thesis, Yeshiva University (unpublished).
Ostriker, J. P., and Gunn, J. E. 1972, Ap. J. Lett., 164, L95.
Paczynski, B. E. 1970, Acta Astr., 20, 47.
——————. 1972, Ap. Lett., 11, 53.
Rose, W. K. 1969, Ap. J., 155, 491.

Wheeler, J. C., Barkat, Z., and Buchler, J. R. 1970, Ap. J. Lett., 162, L129.
Wheeler, J. C., Buchler, J. R., and Barkat, Z. 1973, to be published.

CHAPTER V. Massive Stars, Deuterium and Gamma Rays

If our ideas about the evolution of stars of mass $4 \leqslant M/M_\odot \leqslant 8$ are even vaguely correct, then we must look elsewhere for the site of almost all stellar nucleosynthesis and for possible progenitors of massive black holes. The evolution of stars more massive than $8M_\odot$ is not well understood and has not been explored in any sort of detailed and comprehensive way. Arnett gives a progress report on his attempts to make such a survey.

As a supernova shock wave travels outward through ever more tenuous material, the matter and radiation begin to decouple, allowing a high temperature precursor to occur. Colgate discusses the physics of such phenomena and its implications for the possible noncosmological production of deuterium.

Explosive nucleosynthesis predicts that many stable nuclei were formed as unstable progenitors. If such debris is rapidly ejected into space, the concurrent expansion could allow gamma rays from discrete nuclear transitions to escape without being degraded in energy. Clayton discusses the fundamental importance of the observations of such lines for astrophysics.

References for Chapter V

Zeldovich, Ya. B. and Novikov, I. 1971, <u>Relativistic Astrophysics, Vol. I.</u> (Chicago:Univ. of Chicago Press).

Zeldovich, Ya. B. and Raizer, Yu. P. 1969, <u>Physics of Shockwaves and High-Temperature Hydrodynamic Phenomena.</u> (New York: Academic Press).

SOME QUANTITATIVE CALCULATIONS OF FINAL
STAGES OF STELLAR EVOLUTION

W. David Arnett

The University of Texas at Austin

I. Introduction

If a star is sufficiently massive ($M \geq 8\ M_\odot$) it will ignite $^{12}C + ^{12}C$ in a nondegenerate manner. Even if uncertainties in the ^{12}C-detonation model could be solved, there are several reasons to consider the evolution of these more massive stars.

(1) Stars of mass $4 \leq M/M_\odot \leq 8\ M_\odot$ are not good prospects for the site of nucleosynthesis. Because of the high degeneracy, ^{12}C and ^{16}O cannot be shock ejected without being processed to such high temperatures that they are destroyed. If processed matter is ejected it will be in the form of iron group nuclei, or neutron-rich iron group nuclei. In the solar system these comprise less than 7% and 1%, respectively, of the matter heavier than helium. If every star of $4 \leq M/M_\odot \leq 8$ were to eject $1.4\ M_\odot$ of iron group nuclei, then galactic models develop an awkward and pronounced propensity for making too much iron.

(2) The difficulty of making neutron stars from the mass range $4 \leq M/M_\odot \leq 8$ has been discussed earlier. Since the identification (Gunn and Ostriker 1970) of the pulsar progenerators with this mass range is unreliable to the accuracy needed, an examination of other possible modes of pulsar formation is desirable. In particular, the death rates of stars in the ranges $4 \leq M/M_\odot \leq 8$ and $8 \leq M/M_\odot \leq 60$ are about the same.

(3) Theory does not now suggest any plausible way that stars of $4 \leq M/M_\odot \leq 8$ can form massive black holes. What about more massive stars? The calculation of the evolution of a massive nonrotating single star is at least a first step toward a solution of the problem of massive multiple systems. Recent x-ray observations emphasize the importance of such systems for the black hole question.

II. Status of the Author's Stellar Evolutionary Calculations

During the past two years I have attempted to follow the evolution of massive stars from a well-understood evolutionary stage to their final state. For computational efficiency hydrogen burning was neglected: the evolution of helium cores of various masses ($M_\alpha/M_\odot$ = 2, 4, 8, 16, 32, 64 and 100) during helium burning was calculated, compared and calibrated to standard calculations from the hydrogen burning main sequence. See Arnett (1972 a,b) for details. Hydrostatic stages were done with a standard Henyey-type relaxation scheme with automatic mass rezoning; hydrodynamic stages were treated both implicitly or explicitly as the situation warranted. The different modes of computation are completely consistent, using the same mass zoning, equation of state, convection algorithms, nuclear rates, etc. Rezoning was suppressed during hydrodynamic stages. Energy conservation was carefully monitored throughout all the calculations; numerical errors in energy conservation were completely negligible. In the nuclear reaction sequences nucleon conservation was valid to the limit of the computer (14 significant figures). An integral test for dynamic instability was continually performed; it agreed with direct numerical results to about one part in 10^4 or 10^5.

Table 1 contains a list of the relevant thermonuclear burning stages and list of references upon which the nuclear evolution subroutines were based. Detailed comparison with realistic nuclear reaction network calculations suggests that the approximations for energy generation are superb through oxygen burning, and reasonably good beyond. The group of nuclei from Si to Ca were lumped together (they would be in a common quasi-equilibrium pool) as were the iron-group nuclei. The result of all this analysis, taken with new experimental results on $3\alpha \rightarrow {}^{12}C$, ${}^{12}C(\alpha,\gamma)$, ${}^{12}C + {}^{12}C$ and ${}^{16}O + {}^{16}O$, is a much more realistic treatment of the thermonuclear evolution.

Table 2 summarizes the status of the computations. The massive stars have simpler structure and therefore have been easier to evolve numerically. The approximate total mass M corresponding to a given helium core mass M_α is given as is the approximate (hydrogen burning) main sequence spectral type. It is not clear that stars of M ~ 120 and 170 $M_\odot$ (M_α = 64 and 100 $M_\odot$) actually exist; the computation of their evolution may just be an academic exercise. The lower masses are more representative of observed stars. The check marks in Table 2 indicate that the relevant masses have been evolved into (or through) that burning stage.

The evolutionary calculations for the 64 and the 100 $M_\odot$ helium cores have been performed through their explosive disruption. They become dynamically unstable due to electron-

TABLE 1. NUCLEAR BURNING STAGES

FUEL	REFERENCES
H	
He	Arnett (1972) Ap. J., 176, 681. Couch and Arnett (1972) Ap. J., 178, 771.
C	Arnett (1972) Ap. J., 176, 699. Arnett (1973) Ap. J., 179, 249.
Ne	Reeves (1965) in *Stellar Structure*, (U. Chicago Press: Chicago), p. 175. Fowler, Caughlan and Zimmerman (1972), private communication
O	Spinka and Winkler (1972) Ap.J., 174, 455. Woosley, Arnett and Clayton (1972) Ap. J., 175, 731.
Si and beyond	Bodansky, Clayton and Fowler (1968) Ap. J., Suppl., No. 148, 16, 299. Woosley, Arnett and Clayton, in press. Arnett, unpublished

TABLE 2. STATUS OF EVOLUTIONARY SEQUENCES (APRIL 1, 1973)

		4	8	16	32	64	100
$M_\alpha/M_\odot$		4	8	16	32	64	100
$M/M_\odot$		15	22	36	70	120	170
M.SEQ.Sp		B0	09.5	09	06		
He	core	✓	✓	✓	✓	✓	✓
	shell	✓	✓	✓	✓	✓	✓
C	core	✓	✓	✓	✓	✓	✓
	shell	✓	✓	✓	✓	✓	✓
Ne	core		✓	✓	✓	✓	✓
	shell		✓	✓	✓	✓	✓
O	core		✓	✓	✓	EXPLOSION, 2.2 $M_\odot$ REMNANT	EXPLOSION, NO REMNANT
	shell			✓	✓		
Si	core			WEAK Si FLASH, PROBABLY LIKE 32 $M_\odot$	DYNAMIC INFALL, 1.4 $M_\odot$ CORE		
	shell				✓		

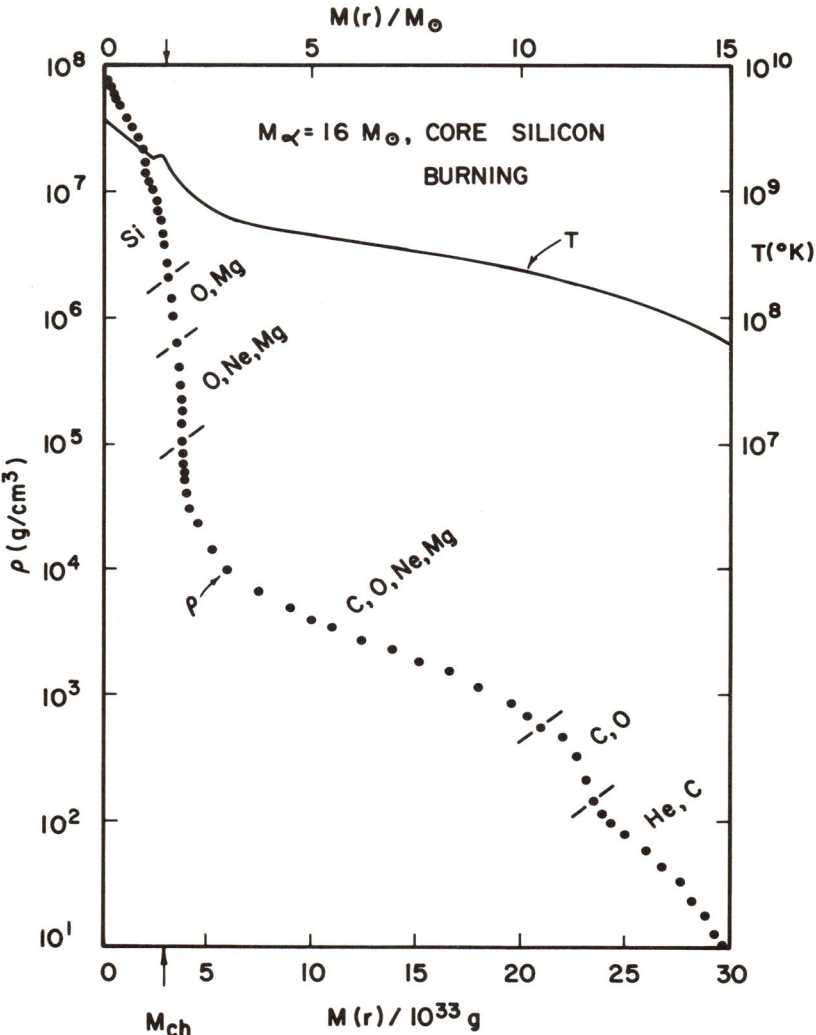

Fig. 1. Temperature and density structure for core mass $M_\alpha = 16\ M_\odot$ during core silicon burning. Each dot on the density curve represents a mass zone. Compositions of various regions are indicated. The last solar mass in the ^{4}He envelope is not shown; M_{ch} refers to the Chandrasekhar limit $M_{ch} \simeq 1.4\ M_\odot$. Evolution at this point is mildly hydrodynamic.

positron pair formation, and proceed to explode and disrupt in much the way suggested by Barkat, Rakavy and Sack (1967) and by Fraley (1968) . The present calculations have the advantages that He and C shells were included, the nuclear physics was more carefully treated, the total mass of the star was determined, and the mass of any remnant was very carefully examined. Table 3 exhibits the composition of ejected matter, normalized to ^{16}O. The last column gives solar system abundances for comparison. The 100 $M_\odot$ core overproduces "Si" and "Fe"; He, C and Ne are underproduced in both masses. The 64 $M_\odot$ leaves a 2.2 $M_\odot$ Si core at low density ($\sim 10^{-3}$ g/cm^3). The evolution of this remnant has not yet been followed; it may be a likely prospect for the formation of a small black hole (?). More massive stars will encounter a more extreme $e^{\pm}$ instability; to the extent that the 100 $M_\odot$ evolution is representative, no remnants will be left. Both objects released a considerable amount of energy; the hydrodynamics was followed until the internal and potential energy of escaping zones was negligible. The last entry in Table 3 gives velocity at the surface of the helium core. These values could be modified by interaction with a hydrogen envelope.

Figure 1 displays the run of temperature and density as a function of lagrangian mass variable for M_α = 16 $M_\odot$ just after a mildly hydrodynamic core silicon flash. Each dot on the density curve represents a mass zone. Note the steep density gradient outside the silicon core. This is characteristic of all the lower mass cores ($M_\alpha/M_\odot$ = 4, 8, 16 and 32). This gradient occurs at the Chandrasekhar mass $M(r) \approx 1.4 M_\odot$ due to the effect of electron degeneracy and neutrino cooling. See Arnett (1972 c) for an analysis of this phenomenon. Figure 1 contains a crude indication of the composition structure; much of the potentially explosive fuel (C, O, Ne) is at relatively low density.

The 32 $M_\odot$ helium core begins an accelerating contraction after core silicon burning. Figure 2 displays the run of density at two points in the evolution. The lower density curve corresponds to the early stage of contraction after Si exhaustion. The central, high density region soon goes into a subsonic hydrodynamic collapse. The evolution was followed to the curve shown, where $\rho_c \approx 9 \times 10^{10}$ g/cm^3. The lower density "mantle" contracts very little while this happens. The iron-rich core is already above its Chandrasekhar mass. For the first time we have a consistent evolutionary calculation of a reasonably plausible astronomical object which seems to lead to the formation of (at least) a neutron star. The subsequent evolution should be interesting! A summary of characteristics of the most advanced model is given in Table 4.

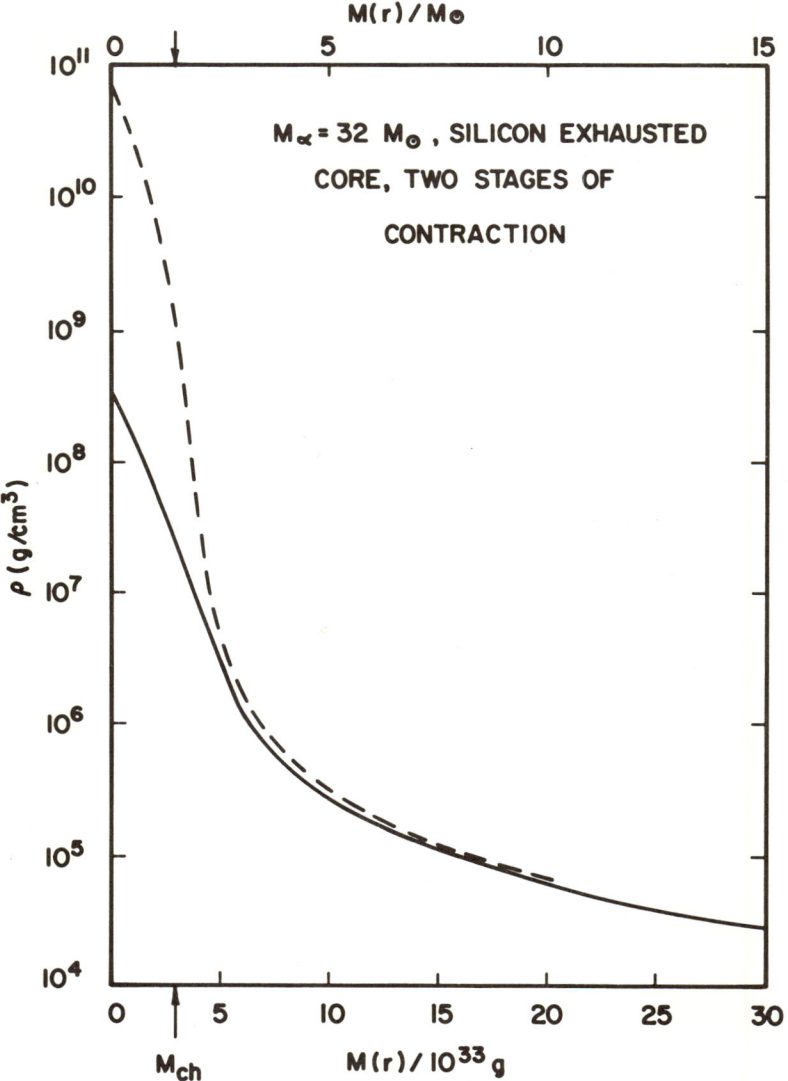

Fig. 2. Density structure for core mass M_α = 32 $M_\odot$ at two points during core collapse. The dashed curve is the most evolved case; a subsonic collapse is occurring. Notice how the central regions "run away" from the relatively low density "mantle". Most of the mantle is not shown; the mass scale is the same as in Figure 1 for comparison.

TABLE 3. ELECTRON-PAIR INSTABILITY "SUPERNOVAE"

$M_\alpha/M_\odot$	64	100	$\odot$
He/O	.120	.158	22.2
C/O	.0698	.0389	.423
Ne/O	.0135	.00711	.114
Mg/O	.0556	.0439	.0524
Si/O	.0957	.718	.126
Fe/O	0.	.247	.139
TOTAL ENERGY	4.50×10^{51} ergs	3.83×10^{52} ergs	------
MASS OF REMNANT	2.2 $M_\odot$ (Si)	0.	------
"SURFACE VELOCITIES"	9,500 Km/sec	15,000 Km/sec	

TABLE 4. SOME CHARACTERISTICS OF AN EVOLVED HELIUM CORE

QUANTITY	VALUE
Core mass (M_α)	32 $M_\odot$
time since core helium ignition	8.52×10^{12} sec.
ρ_c	8.46×10^{10} g/cm^3
T_c	15.5×10^9 °K
Binding Energy/M_α	6.515×10^{16} erg/g
Kinetic Energy/M_α	1.110×10^{16} erg/g
Potential Energy/M_α	6.184×10^{17} erg/g
$\Gamma_{effective}$	1.3350
sound travel time through core	5.00×10^2 sec
collapse time = $(\rho/\dot{\rho})$	1.19×10^{-2} sec
photon luminosity	5.93×10^{39} erg/sec
neutrino luminosity	4.72×10^{49} erg/sec
nuclear energy release rate	2.76×10^{50} erg/sec

TABLE 5. COMPOSITIONS FOR LOWER MASS CORES

$M_\alpha/M_\odot$	4	8	16	32	$\odot$
He/O	2.80	1.19	.652	.408	22.2
C/O	.450	.329	.209	.174	.423
Ne/O	1.02	.400	.332	.0908	.114
Mg/O	.211	.153	.104	.0466	.0524
Si/O	-----	.0179	.178	.0729	.126
Fe/O	-----	-----	.0815	.135	.139
STAGE	PRE-Ne IGNITION IN CORE	CORE O BURNING	Si FLASH	CORE COLLAPSE	-----

For the moment let us defer consideration of the gravitational collapse problem, and consider what would happen if some of that outer mantle were ejected. We must be cautious; some shock processing might occur and some species, especially in the core, could be buried in a remnant. With these caveats, consider Table 5 which gives compositions (relative to ^{16}O) for the most advanced model for each core mass. Solar system values are quoted for comparison, and a brief comment on the evolutionary stage is given. Clearly ^{4}He will be underproduced; this was expected. Otherwise it seems quite possible (and perhaps even likely) that a roughly solar-like abundance distribution will result! Taken with the work reported by Talbot and by Woosley at this conference, as well as that of Arnett (1971) and Arnett and Clayton (1970), a coherent picture of nucleosynthesis, galactic and stellar evolution begins to emerge.

III. Gravitational Collapse

What about the collapsing core? While we wait for the evolution to be continued it is interesting to review the state of gravitational collapse calculations. I suggested (Arnett 1967, 1968) that massive collapsing cores would form hot neutron star configurations and, with further infall of matter and neutrino cooling, evolve to a gravitational singularity. That is, a black hole would be formed. Lower mass cores exploded, ejecting mass. Previously Colgate and White (1966) had found that neutrino energy transport would blow off the infalling matter before a massive remnant formed; they always got a violent explosion and a neutron star, and no massive black holes.

Imshennik, Nadezhin and their collaborators (see Zel'dovich and Novikov 1971 for a review of their extensive work and a list of references) performed a similar but independent analysis; in view of the differing formulations their results agree surprisingly well with crude work of Arnett (1967). It appears that the infalling central regions of the core heat up and the finite neutrino opacity inhibits nonadiabatic effects. They found a reflected shock arising from the bounce of the core. This occurs at less than neutron star densities because the nucleons are "hot" and have thermally contributed pressure as suggested by Arnett (1967). Ivanova, Imshennik and Nadezhin (1967) found that an oxygen detonation could enhance the explosion due to the reflected shock, the energy release going from 3×10^{50} to 3×10^{51} ergs.

Wilson (1971) has examined the problem again, but with a better mathematical treatment of the neutrino transport. For large masses his results agree with Arnett (1967) but for lower masses he finds a much weaker explosion. Only his 1.25 $M_\odot$ star (just above the Chandrasekhar mass $M = 1.2$ $M_\odot$

for his equation of state) actually could eject mass, and then only about 0.05 $M_\odot$. This is more consistent with the Russian work which found weaker explosions at the lower masses also. Wilson (1971) attributed this difference to an inconsistency in Arnett's treatment of the region where neutrinos decouple from the matter. Actually this was the reason that the Colgate-White models all exploded. Since Arnett's more massive cores, which had a larger neutrino emission, did not explode, the correct explanation is clearly more complex.

Part of the answer is the following. As emphasized in the original paper, Arnett (1967) intentionally used a rather optimistic estimate of neutrino transparency in the critical density range $10^{11} \leqslant \rho \leqslant 10^{13}$ g/cm^3. This tended to maximize the probability of explosion; even so the more massive cores would not explode. The velocity of a neutrino diffusion wave across a region of width ΔR is

$$v(\nu) \approx \lambda c/(3\Delta R)$$

where λ is the mean-free-path and c the velocity of light. For $\lambda > 3\Delta R$ superlight velocities are implied; this sort of gibberish occurred for some parts of the lower mass cores. Arnett (1967) used neutrino opacities about 0.3 to 0.1 of Wilson's in the critical regime. More realistic opacities and an emissivity limit would probably remove much of the discrepancy between the low mass results.

In order to understand gravitational collapse better we must at least:

(1) include neutronization properly,
(2) use realistic pre-collapse structure, and
(3) properly treat nuclear effects.

We should try to understand to what extent Wilson's results depend upon his complex coding of the problem, and to what extent upon his physical assumptions.

A horrible and fascinating consideration is that of rotation. A glance at the work of Hoyle (1946), Ostriker and Gunn (1969 for example), and Leblanc and Wilson (1970) is enough to intrigue us as to the possibilities.

The current embarrassment of a very wide range of possible rotation parameters may remain with us until we begin to understand magnetohydrodynamics, rotation and convection in slowly evolving stars.

A final point: the considerations above emphasized non-rotating single stars. Suppose that a real star in a multiple system managed to remain well mixed through all its nuclear evolution, or that finally nuclear fuel regions were lost. Apparently a large object of this type would collapse to the black hole state without in the process changing the orbital character of the system; a massive iron core collapses with little (if any) mass ejection. This speculation might be of interest in interpreting x-ray source observations.

Clearly research on these topics is at a very interesting stage.

This work was supported in part by NSF Grant GP-32051.

REFERENCES

Arnett, W. D. 1967. Can. J. Phys., 45, 161.
———. 1968. Ap. J., 153, 341.
———. 1971. Ibid., 166, 153.
———. 1972a. Ibid., 176, 681.
———. 1972b. Ibid., 176, 699.
———. 1972c. Ibid., 173, 393.
Arnett, W. D. and Clayton, D. C. 1970. Nature, 227, 780.
Barkat, Z., Rakavy, G. and Sack, N. 1967. Phys. Rev. Lett., 18, 379.
Colgate, S. A. and White, R. 1966. Ap. J., 143, 626.
Fraley, G. S. 1968. Ap. Space Sci., 2, 96.
Gunn, J. E. and Ostriker, J. P. 1970. Ap. J., 160, 979.
Hoyle, F. 1946. M.N.R.A.S., 106, 343.
Ivanova, L. N., Imshennik, V. S. and Nadezhin, D. K. 1967. preprint, subsequently published in Sci. Inf. Astr. Council USSR Acad. Sci., 13 (1969).
Leblanc, J. M. and Wilson, J. R. 1970. Ap. J., 161, 541.
Ostriker, J. P. and Gunn, J. E. 1969. Ap. J., 157, 1395.
Wilson, J. R. 1971. Ap. J., 163, 209.
Zel'dovich, Ya. B. and Novikov, I. D. 1971. Relativistic Astrophysics (Chicago: Univ. of Chicago Press).

SUPERNOVA SHOCK WAVES

Stirling A. Colgate
New Mexico Institute of Mining and Technology
Socorro, New Mexico 87801

Introduction

The shock ejection of matter from supernova envelopes at progressively higher energy as the shock progressed to sequentially lower density (and external mass fraction) was first proposed by Colgate and Johnson (1960) as a possible origin of cosmic rays. In a subsequent paper, Colgate and White (1966) discussed both one possible origin of supernova associated with the dynamical formation of a neutron star and, in addition, the nonrelativistic speed-up of the explosion shock wave in the stellar envelope. Subsequent theoretical work on supernova has emphasized the problem of formation presumably recognizing that the speed-up of the shock wave should be applicable to any mechanism of supernova explosion. Recently, Kirshner (1972) has concluded that spectroscopic observations of the hydrogen and Ca II lines of the Type II supernova in M101 are consistent with the theoretical prediction of Colgate and White; namely, that the shock speeds up according to the law of ejected matter velocity $U_s \propto F^{-1/5}$, where F is the external mass fraction.

The application of this theory to the acceleration of cosmic rays has concentrated on the difficulty of preserving the higher atomic number nuclear species from spallation in the shock transition that may reach many GeV per nucleon in strength Ginzburg and Syrovatskii (1964) and Kinsey (1969). Colgate (1970) then proposed a possible solution where the shock acceleration took place in matter that was sufficiently dense such that the shocked fluid temperature was high enough $\geq mc^2$ that the electron-positron pair density exceeded the original electron density by many orders of magnitude and hence increased the dynamic friction sufficiently to preserve the heavier nucleon species through the shock transition. At this time, Weaver and Chapline (1973) at Lawrence Livermore Laboratory undertook an extensive and not-yet-completed numerical analysis of the shock structure in both the pair dominated and ordinary radiation limit. One of their early conclusions was the possible presence of a high ion temperature precursor to the shock when pairs were *not* important, i.e., for shocks in low density stellar envelopes. A major consequence of such a high ion temperature precursor is the production of some of the light elements.

The distinction between high and low density supernova envelopes is presumably the distinction between Type I and Type II supernova.

Type I supernovae are presumably initially highly evolved and highly condensed objects.

The observation by Oke and Kirshner (1972) that less than 10% hydrogen can be identified in the ejecta is strong observational evidence of a highly evolved star. Theoretically, the production of a neutron star from a carbon core requires a pre-collapse density $\geq 2 \times 10^{10}$ gm cm^{-3} (Bruenn 1971) and a unique evolutionary path to this point has been proposed by Paczynski (1972). The number of pulsars discovered and the Crab indicate that at least some Type I supernovae leave a neutron star remnant and therefore at least some, and most likely all, Type I supernovae are initially highly condensed objects. Type II supernovae, on the other hand, are accepted to be far more massive with extensive hydrogen envelopes. The Balmer lines are dominantly pronounced during the major fraction of the light curve and the associated population I stars are massive and young.

We then recognize two extreme limits of supernova shock waves, the one high density (Type I's) where the dynamic friction due to pairs is large enough to possibly preserve the various high atomic number nuclear species and the second limit of low density Type II's in a red giant-like envelope where the dynamic friction is small enough such that a high temperature precursor forms that gives rise to major degree of spallation. In particular, the spallation of ^{4}He leads to free neutrons which are subsequently captured by protons to give deuterium. Other light elements are formed when ^{13}C and ^{14}N are spalled at a lower energy.

Limiting Shock Strengths For Type I and II Supernovae

The limiting shock strength is determined by both the stellar structure and the initial specific energy release. We can measure this "strength" and specific energy release in various units, but it is convenient to use the energy per nucleon because of the close association between this kinetic energy, nuclear binding energies and cosmic ray energies. In addition, we note that since the specific kinetic energy of fluid motion behind a strong shock is equal to the specific internal energy, the shock strength measured in energy per nucleon is both the relative kinetic energy between an incoming nucleon and the moving matter behind the shock and also the specific internal energy per nucleon behind the shock.

The average specific kinetic energy released (Minkowski 1968) in a supernova is apparently somewhat higher for a Type I supernova about 2-4 MeV/nucleon (4 to 8×10^{51} ergs for 1 M$_\odot$

ejected) compared to 1/2 to 1 MeV/nucleon for a Type II (1 to 2 x 10^{52} ergs for 10 $M_\odot$ ejected). The limiting shock strength is determined not by the density, but by the external mass fraction of that external layer that corresponds to the shock thickness. In general, this thickness will depend upon the details of the dynamic friction between the nucleons and the predominant component of the internal energy. For reasons to be explained later, the low density limiting thickness is some 10 gms cm^{-2} for the precursor ion-ion collisions thermalization and somewhat less than 1 gm cm^{-2} for positron-electron pair equilibrium in the high density limit. It is then readily apparent that the external mass fractions for the two limits are vastly different.

Type I - It is presumed that carbon-burning has somehow proceeded without detonation and then dynamical collapse is initiated by sufficiently rapid neutrino emission. This occurs at a central density $\simeq$ 2 to 4 x 10^{10} gm/cm^2 and a radius r $\simeq$ 5 x 10^7 cm. The external mass fraction, F, at shock breakout then becomes

$$F = \frac{4\pi r^2 \delta}{M}.$$

For δ = 1 gm cm^{-2}, M = $M_\odot$, F = 1.5 x 10^{-17}. This is so small that the shock strength becomes extremely relativistic and numerical calculations (Colgate, McKee, and Blevins 1972; Colgate and McKee 1973) are required to determine the scaling laws bridging nonrelativistic to relativistic behavior. The <u>ejected</u> matter becomes relativistic at F $\simeq$ 3 x 10^{-4}, the shock strength becomes relativistic at F $\simeq$ 10^{-6} and the shock strength at breakout becomes $\simeq$ 100 c^2 or 10^5 MeV per nucleon.

Type II - It is presumed that the presupernova star has an extended red giant envelope typically r $\simeq$ 2 x 10^{13} cm, M = 10 $M_\odot$, in which case, for δ = 10 g cm^{-2}, F = 2.5 x 10^{-6}, and the shock strength at breakout scaling from a mean energy of 1 MeV per nucleon becomes

$$S = E_o F_s^{-1/2.5} \simeq 180 \text{ MeV per nucleon.}$$

In the nonrelativistic limit, the kinetic energy after expansion of the shocked material increases by roughly x4, whereas, in the relativistic limit, the increase is a power ($\simeq$ 2.5) of the relativistic energy factor. Hence, the Type II supernova ejects nonrelativistic matter of the order of a GeV per nucleon, Type I's may eject matter of 10 $\gamma_s^{2.5}$ $\simeq$ 10^6 GeV per nucleon in the purely hydrodynamic limit.

The total energy of the ejected matter may be comparable in the two cases, but the energy spectrum of ejected matter would be truncated, in the case of Type II's, at less than a GeV per nucleon. It should be pointed out that if the hydrogen envelope of a Type II is more compact, $r \simeq 10^{12}$, somewhat greater than main sequence, but less than a red giant, then the limiting shock strength (and ejected energy) would increase only as $r^{-0.8}$ nonrelativistically and with slightly greater power relativistically.

We envisage two extreme conditions for supernova shock waves, the one at low density where we expect spallation due to a high ion temperature precursor and the other at higher density where we expect sufficient positron-electron pairs such that small spallation occurs.

The optimum condition for spallation occurs for the following conditions:

1. Relatively high counterstreaming velocities (energy $\geq$ 30 MeV) among the nuclear species in the shock.

2. Low enough temperature behind the shock such that a) electron-positron pair dynamic friction is small, $T \ll 500$ keV and b) the thermonuclear burn-back of the important light elements is small, $T \leq 1$ to 10 keV.

3. Hopefully, a high enough density such that spallation produced neutrons will capture on hydrogen and add to the deuterium production.

Strong Shocks and Stellar Structure Limits

The internal energy density behind a very strong shock is equal to the fluid kinetic energy density and because we desire a small equilibrium temperature (where $kT \ll$ fluid kinetic energy per nucleon) then the internal energy density must reside almost entirely in radiation so that

$$\rho_s U_f^2/2 = aT^4. \qquad (1)$$

ρ_s = post shock mass density; U_f = post shock fluid velocity. If we use an a posteriori estimate of the maximum temperature $T < 3$ keV (3×10^7 deg) and $U_f^2/2 \simeq 10$ to 100 MeV/nucleon, then the maximum density behind the shock becomes 10^{-3} to 10^{-4} g/cm^3 and 1/7 this ahead of the shock. (The compression ratio of a strong radiation dominated shock = 7.) The external mass fraction at which this strength shock should occur is 10^{-2} to 10^{-5} (Colgate and White 1966). (100 MeV per nucleon shock fluid energy corresponds to 400 MeV/nucleon post-expansion ejection energy.) This range of mass fraction and density falls midway between a pre-supernova red giant structure and a massive star just before the occurrence of the pair instability. The stellar radius of such a mass fraction and density will be $10^{12} < r < 10^{13}$ cm, so that the time available for neutron capture and/or thermonuclear burn-back will be of the order $\tau \simeq r/2U_f$ or 50 to 500 seconds. In order that the hydrogen capture the spallation neutrons

Colgate   252

($\overline{\sigma v} \simeq 7 \times 10^{-20}$ sec^{-1} for T ≤ 3 keV) $\rho_s > 10^{-6}$ to 10^{-7}gm/cm^3 and a factor of 7 less for the pre-shock density. The lower limiting density is consistent with red giant structure and the higher density with a massive star at the pair instability. Finally, if about 20% of the helium and heavier nuclei are spalled in the shock (to be justified later), then for normal abundance, the post-shock deuterium number density becomes about 10% of the post-shock helium or $n_D < 5 \times 10^{17}$cm^3. The temperature such that the d-d reaction results in $\overline{\sigma v}\, n_d\, \tau < 1/2$ with $\tau = 50$ seconds (the higher density case) is T ≤ 3 keV. Since ^{3}H is produced in the spallation of He at roughly 1/6 the deuterium (D) production (Reeves 1971) for 2 ≤ T ≤ 3 k3V the ^{3}H would be thermonuclearly burned to ^{4}He and a neutron. After subsequent neutron capture on ^{1}H, the D is regenerated and ^{3}H becomes D and not ^{3}He. Similarly, the ^{3}He will capture neutrons with 1.5×10^4 the cross section of hydrogen, producing ^{3}H which may either burn back to give a free neutron and subsequently D, or at kT < 2 keV survive and decay back to ^{3}He having removed one available neutron.

Shock Structure

We have chosen shock conditions that are nonrelativistic in the fluid velocities, but where the photon energy density exceeds the particle thermal energy density by a factor of $\simeq 10^4$. The dominant stress far behind the shock must therefore be the gradient in the photon energy density. The phenomenological description of the shock structure consists of a description of how this stress is transferred to the heavy component, the nuclei, of the fluid. The photons of the thermal Planck radiation interact only weakly with the nuclei and so there is an overwhelming tendency for the gradient of the radiation energy density to transfer this stress to the electrons of the unshocked fluid. The electrons in turn must then transfer their resultant motion to the ions. If the dynamic friction between ions and electrons is "large" so that their relative displacement is "small" then there is apparently a small problem in finding a self-consistent shock structure solution in which:

1. The thickness δ of the shock transition is dominated by the diffusion gradient of the Planck radiation such that $\delta \simeq \lambda_c (c/3)(1/V_s)$, i.e., the self-consistent diffusion solution where λ_c = Compton mean-free-path and V_s = shock velocity, U_f = fluid velocity. See Colgate (1972) for a numerical simulation of this case.

2. The ion-electron dynamic friction is large enough such that ions are accelerated to the fluid velocity in a distance small compared with δ. This distance is analogous to the classical range of a high energy ion traversing an electron gas.

3. The dynamic friction between ions and electrons heats the electrons by just the change in kinetic energy of the ions. This is necessarily so for a totally inelastic collision.

4. The electrons exchange energy with the photons of the Planck radiation sufficiently rapidly such that quasi-thermal radiation equilibrium is maintained.

We have purposely chosen conditions for the strong supernova shock that violates some of these assumptions. There are two approximations that are violated and that will lead to counter-streaming. We discuss these in some depth because they lead to rather different spallation conditions and a hierarchy of shock models that depart from the simplistic "standard" collisional shock.

The violation of 2 occurs for shocks where $U_f^2/2 \geq 100$ MeV/nucleon. In this limit, $V_s \simeq c/2$ and the diffusion thickness of the shock will be roughly 2/3 Compton mean-free-paths (1.5 g/cm^2). The "range" of a 100 MeV proton in a plasma of kT = 3 keV will also be of the order of 3 g/cm^2 so that the electrons and ions are not tightly coupled by dynamic friction, but instead by an electric field of charge separation. If the shock is viewed in that moving frame such that the shock is stationary, the ions and electrons enter the radiation front at $-V_s$. The radiation gradient exerts a stress tending to stop the in-coming electrons. If the electrons were to stop, the resulting electric field would be very large compared with what is required to drag the electrons through the radiation against the radiation stress. (The Debye length for kT = 100 MeV, $\rho = 10^{-7}$ is $\lambda_D \simeq 2 \times 10^{-2}$ cm compared with $\lambda_c \simeq 10^7$ cm.) As a result, an electric field decelerates the ions rather than electron-ion dynamic friction. In the high energy limit where the electron dynamic friction can be neglected, the electrostatic potential difference is just that necessary to conserve energy and momentum of the Hugoniot relations. The equations would be complete if the friction between the electrons being dragged through the photon gas were sufficient to recreate the Planck spectrum. In this limit we will assume that the electrons recreate the Planck radiation by scattering in the strong electric field, $E^2/8\pi \simeq aT^4 = \rho_s U_f^2/2$, and that the electrostatic potential is just such as to cause an average momentum change of the ions of $\rho_s U_s$. In the supernova shock there are two different kinds of ions Z/A = 1, and 2 of roughly 2/3 and 1/3 mass fraction. It is obvious that a traversal of an average electrostatic potential will result in different velocities. In the shock frame, the potential is static and the change in energy per nucleon is double for protons as for the heavier nucleii. If one demands momentum conservation, it is elementary to show that the electrostatic potential is such that the relative counterstreaming energy is $U_f^2/2$, but the corresponding internal energy of the counterstreaming is only $0.22\ U_f^2/2$. This is composed of $(2/3)(0.11)(U_f^2/2)$ in the protons and an oppositely directed $(1/3)(0.44)(U_f^2/2)$ in the heavies each measured relative to the shocked fluid center of mass and/or the electrons (Colgate 1959). The interpenetrating range of the protons through the heavy nuclei is roughly 7 gm/cm^2 (4.5 barns) at a mean energy of 70 MeV. If we assume a density such that the post shock temperature is in the range 2 to 3 keV such that all the ^{3}H is burned back to n, etc., then from Table II-2b of Reeves (1971) about 0.09 ^{3}He and 0.13 D will be produced per ^{4}He in the stellar mass fraction. Since the mass fraction after acceleration will be ejected with roughly 400 MeV/nucleon, it represents a stellar mass fraction (for massive stars or red giants) that cannot be larger than 3×10^{-5} (Colgate and White 1966) and for most models is much smaller. The number fraction of deuterium produced per supernova would then be less than 3×10^{-6} in even the most extremely energetic supernova. What is more, the effect of post shock acceleration which doubles the post shock fluid velocity strongly favors spallation _after_ ejection. The mass fraction of ejected matter that has the same energy after ejection as the mass fraction at the shock with counterstreaming spallation is larger by the factor $2^6 = 64$ fold nonrelativistically and greater relativistically. Unless some cooperative deceleration mechanism (other than ionization loss) is particularly effective in the interstellar medium, one is forced to the conclusion that shock front spallation in a _presumed_ radiation equilibrium shock will _not_ give rise to the major contributions of the light elements.

Non-Radiative Equilibrium Shocks

We presumed, in shock assumption, 4 that quasi-radiation equilibrium was maintained and noted later the necessary assumption of photon creation in the shock transition. Remember that radiation equilibrium implies a photon number density $\gtrsim 10^4$ the electron density so that a very large number of photons per electron must be created. We now assert that no physical processes exist that can recreate a quasi-thermal Planck photon distribution within a radiative diffusion shock thickness $\delta \simeq \lambda_c$. The more obvious mechanisms are:

1. Bremsstrahlung
2. Bound-Bound transitions on C, N, O.
3. Radiation emission from electrons accelerated in the strong electrostatic field of charge separation.
4. Double Compton scattering.
5. Pair annihilation.

1. Bremsstrahlung alone requires many Compton mean-free-paths to generate the energy content in photons and will be discussed in greater detail later.

2. It is well recognized that in radiation-transparent high temperature plasmas whose composition is similar to normal abundances that the bound-bound radiation emission can greatly exceed bremsstrahlung (Cox and Tucker 1969). However, as discussed by Post (1961) these "impurity" ions will be stripped of bound electrons by the high temperature electron plasma at an energy cost in photon emission comparable to their binding energies. This is a trivial contribution to the net required photon energy. Also, the contribution to bremsstrahlung by the enhanced $\Sigma\, n_i\, Z_i^2$ is also negligible.

3. Finally, electron radiation due to acceleration in the electrostatic field of charge separation should be small unless this field is comparable to the critical field $E_c \simeq 10^{16}$ esu where classical radiation reaction becomes important. The actual charge separation field is limited to $E^2/8\pi \le aT^4$; $E \le 5 \times 10^8$ esu.

4. Double Compton scattering (Heitler 1949) produces one extra photon in the photon-electron scattering process and hence adds to the radiation field. Since it is a second order process, it is smaller by the factor $1/\alpha = 1/137$ than single scattering, but so also is bremsstrahlung and so double Compton scattering may significantly add to the radiation field. However, when the relative energies are of the order mc^2, the cross section for the double process to emit the energy of the electron in 2nd photons is about 10% of the bremsstrahlung cross section so that the high energy photon energy density must become x10 the electron energy density before the processes compete. In the case at hand, by the time this occurs, the ions are effectively cooled so that double Compton scattering is only a small correction to the life time of the high ion temperature precursor. However, double Compton scattering does become important in the final process of thermalization to a Planck spectrum.

5. Pair annihilation requires the creation of pairs in the first place. Pair creation becomes comparable to bremsstrahlung only when the photon or electron energy $\ge 10\, mc^2$ which is much higher than the effective temperatures of electrons or photons during the cooling of the high ion temperature precursor.

Therefore, the remaining mechanism for photon creation is bremsstrahlung. Let us first examine the shock structure consequences before making a more detailed estimate of the bremsstrahlung thermalization time.

If there were no radiation at all, then we would naturally visualize a standard collisional shock where the specific internal energy $U_f^2/2$ behind the shock was distributed among the various particle degrees of freedom. In particular, if $U_f^2/2$ were of the order of 10 MeV per nucleon, we would imagine a hot plasma where $kT \simeq 3$ MeV where we have included the specific heat of one electron per ion. The shock thickness would be determined predominantly by the collision mean-free-path of the ions recognizing that the ion-electron equilibrium time is somewhat longer than the ion-ion collision time. (We have neglected the possibility of "collision free" shocks that in this case would depend upon electrostatic turbulence. Regardless of the possible existence of an electrostatic precursor, its effect would be merely to steepen the initial transition.) As we permit radiation to take place, the temperature will decrease as more energy appears in photons and a slight re-adjustment in density will take place at near-constant-pressure corresponding to the different ratio of specific heats.

Such a shock structure with a high temperature precursor is well known in the theory of shocks in molecular gases where the binary collisions among the molecules represent the high temperature precursor and the relaxation into the various molecular modes of vibration and rotation in the analogue of the radiation specific heat. See page 438 of Bond, Watson, and Welch (1965) for an extensive analysis of such molecular relaxation shocks and page 416 for examples of shocks with radiation where a non-radiation equilibrium high temperature excursion is analyzed.

Precursor Conditions for Spallation

In our case of searching for the maximum spallation in supernova shocks, we look for the lowest temperature (and hence largest mass fraction) of the hot precursor plasma at which spallation can occur. We therefore require a relatively high temperature plasma 1 to 10 MeV such that a relatively truncated Maxwell tail (of opposed collisions) will result in the spallation of N^{14}, C^{13} at 4 MeV or ^{4}He at 25 MeV. If either ^{4}He, ^{12}C, ^{16}O can be spalled, then all the light nuclei products can be spalled with one order of magnitude larger cross section and at 1/10 the temperature and hence for larger times during radiation cooling.

Summary of Electron Photon Thermalization of the Ions

The radiation thermalization process is too complicated to present in detail in this paper, but can be summarized in the following.

The electrons are heated by a presumed quasi-thermal distribution of ions, $1 \leq T_i \leq 10$ MeV. The electrons emit bremsstrahlung radiation that accumulates until the inverse Compton scattering cools the electrons at the rate they are being heated by ions. This occurs in 10 $T_i^{-1/42}$ Compton periods and causes a temperature maximum at $\tilde{T}_e \simeq 1.0\, T_i^{4/21} mc^2$. Thereafter the inverse Compton cooling occurs more rapidly than ion electron heating, so that the electron temperature falls as $\tilde{T}_e \simeq 2.5\, T_i^{19/105}$ $(t/\tau_T)^{-2/5} mc^2$. During this electron temperature decrease, the ions are cooled more rapidly with an energy loss $\Delta W_i = -6.2 \times 10^{-3}$ $(t/\tau_T)^{8/5} T_i^{8/35} mc^2$ per period, so that the total time required to cool the ions to half value becomes $\simeq 100\, T_i^{1/2}$ Thompson scattering periods.

Subsequent cooling to equilibrium at $kT_e \simeq 1$ to 3 keV again depends upon the Compton heating of photons primarily emitted as bremsstrahlung. The double Compton process is reduced by the factor $(h\nu/mc^2)^2$ and so is not important. The time required to emit all the energy in bremsstrahlung at $kT_e \simeq 2$ keV is approximately $7 \times 10^3\, T_i$ periods but the e-folding time of the photon energy by inverse Compton scattering is $1/T_e = 160$ periods so that enough low energy photons ($h\nu \simeq 1/10 \cdot T_e$) should be produced in 1/10 the bremsstrahlung time (700 T_i periods) to be pumped up in energy x10 to fill the Planck spectrum at $kT_e = 2$ keV in a time of roughly 1000 T_i Thompson periods.

Shock Precursor Ion Collisions

The transport cross section for proton-proton and proton-helium ions for isotropizing the incoming ion fluxes is comprised of both a coulomb and nuclear part. The nuclear elastic scattering is the largest part of the transport cross section for $E \simeq 10$ MeV, and the combined cross section is approximately 1 barn, 10^{-24} cm^2, for p-p, α-p, and α-α collisions.

The result is that roughly 60 to 100 ion-ion elastic collisions occur during the radiation cooling. This number is not large, but is still sufficient to:

1. justify the existence of a non-equilibrium high ion temperature shock precursor,

2. give rise to an isotropic pressure,

3. build up part of a Maxwellian "tail".

The incoming α particles contain $4\, kT_i$ of energy relative to the more rapidly thermalized protons so that opposed collision between proton and α's can occur at $4\, kT$ initially as compared with $(1.5)^2\, kT$ for a mono-energetic isotropic distribution. Hence, we feel it is a reasonable approximation to assume that the effective collision energy distribution is near Maxwellian up to roughly $4\, kT_i$ and partially truncated above this energy.

Spallation to Free Neutrons and Protons

The threshold for He spallation by protons $\simeq 2.1$ MeV rising steeply to a near constant 0.1 barns above 35 MeV. The total spallation becomes

$$\int^{t_{max}} \eta(E)\sigma(E) V \, dt$$

and the fraction of the collision occurring in the internal 2 to 3 kT_i is $V/V_o e^{-2.5} \simeq 0.15$ of those occurring at kT_i so that the total spallation of He by proton at kT_i = 10 MeV becomes from 23

$$F_{sp} = 1 - \exp\left[-8 \bar{\sigma}_{sp} T_i \frac{V}{V_o} e^{-2.5}\right] \simeq 1 - 1/e \simeq 65\%,$$

where $\sigma_{sp} \simeq 0.08$ barns. The subsequent spallation of d, $He^3 T$ should also go to completion to free neutrons and protons because the thresholds are an order of magnitude less and the cross sections 3 to 5 times larger (Reeves 1971).

Neutron Capture During Cooling

The capture rate of neutrons on protons is a near constant 7.2×10^4 barns cm sec^{-1} compared with the cooling rate to kT = 3 keV of $\simeq 10000 \, \tau_m$ or 6×10^6 barns cm sec^{-1}. We therefore expect less than 1% neutron capture during cooling and therefore no burn-back of the deuterium formed during cooling.

Mass Fraction of 1 to 10 MeV Shock

For the range of supernova specific energies discussed in the introduction $\simeq$ 1 MeV/nucleon, $F_{10 \, meV} \simeq 3 \times 10^{-3}$ to 10^{-4}. If the number fraction of helium is the usual 12%, then the number fraction of deuterium formed per supernova of Type II will be of the order 1 to 5×10^{-4}, which is significantly more than the current galactic gas estimates of a few $\times 10^{-5}$. It then requires roughly 1/10 the mass of the Galaxy to have been cycled through supernova explosions during its history. This is not an unlikely value in view of the metalacity and requires a rate of supernova in the early history of the galaxy some 10 times the current value.

B^{11}, B^{10}, Li^7 Production

If a significant fraction of N^{14} and C^{13} exist in the preshock fluid, then a precursor ion temperature of $T_i \simeq$ 1-2 MeV is sufficient to result in the production of C^{11} and B^{10} by the (p,α) reactions on N^{14} and C^{13}. Neutrons will be produced by $C^{13}(p,n)N^{13}$ so that some of the C^{11} will be destroyed by capture to C^{12} or to $Be^8 \rightarrow 2\alpha$, but some will create Li^7 by B^{10} (n,α) Li^7.

High Density Supernova Type I Shocks

In the opposite extreme of Type I supernova, the shock becomes relativistic $(\gamma-1) \simeq 1$ at a mass fraction $\simeq 10^{-6}$

(Colgate, McKee, and Blevins 1972). If we scale a radiative zero solution, for the envelope, then $\rho \propto T^{1/3} \propto F^{3/4}$ for the outer layers of near constant radius, then

$$\rho_{\bar{6}} = \left(\frac{F_{\bar{6}}}{F_s}\right)^{3/4} \rho_s,$$

where $\rho_{\bar{6}}$ is the density at the mass fraction $F = 10^{-6} = F_{\bar{6}}$, and ρ_s and F_s in the respective density and mass fraction of the surface. We have already pointed out that $F_s \simeq 1.5 \times 10^{-17}$ and for a surface temperature $T_s \simeq 10^5$ deg, the surface scale height $h_s \simeq 300$ cm so that $\rho_s \simeq 3 \times 10^{-3}$ gm cm^{-3}. Then $\rho_{\bar{6}} \simeq 3 \times 10^5$. The shock compression is roughly 10-fold (for a shock strngth $(\gamma-1)c^2 \simeq c^2$) so that the energy density behind the shock is of the order $10 \, \rho_{\bar{6}} c^2 = 3 \times 10^{27}$ gm cm^{-3}. When the electron pair specific heat is included, the temperature is determined by

$$\tfrac{11}{4} aT^4 = \varepsilon = 3 \times 10^{27}$$

or $T \simeq 3.5 \, mc^2$.

The lepton (electron + positron) number density, $n_\pm$, is then very much larger than the nucleon or original electron, n_n number density. At $kT \geq mc^2$, where the mean energy of a pair particle $\cong (3kT + mc^2) = 4kT$, the ratio of lepton number density to nucleon number density becomes

$$\frac{n_\pm}{n_n} \simeq \frac{\tfrac{7}{11} \mu_s \rho_s c^2}{4 \, kT \, n_n} \simeq 260 \, \mu_s,$$

where $\rho_s = n_n M$, and $\mu_s = (\gamma_s - 1)$ the shock energy factor.

This large lepton density then becomes the determining feature of the shock structure. It greatly enhances the dynamic friction between the at-rest and moving fluid and similarly the resulting large photon opacity serves to confine the photon gas as well as drastically reduce the thermalization times. To the extent that the temperature remains constant behind the shock, the lepton and photon number density remains constant and only the nuclear density decreases.

If we use the similarity solution that $\gamma \propto F^{-0.178}$ in the relativistic limit and $\rho \sim F^{3/4}$, then $T \propto (\rho \gamma^2)^{1/4} \propto F^{1/10}$. Since F decreases by 10^{10} before breakout, the temperature falls sensibly below mc^2 only near breakout. The pair density is still adequately high for the following argument to apply at $T > (1/3)mc^2$.

Shock Structure with Pairs

If the pair density is some 300 times the normal electron density, we can then ask the question whether the dynamic friction is large enough such that

1. the incoming (to the shock) high energy ions can give up their energy to the pair fluid before spallation or gamma-nucleon destruction,

Colgate 260

2. whether the pair number density can be recreated within the slowing down length of the ions, or before any other collision process takes place.

In general, we consider these questions in the relativistic limit $\mu_s \simeq 1$ because of their application to cosmic ray production and the preservation of nuclear species.

If the pair density is formed in equilibrium, then the slowing down dynamic friction should be larger by the pair number density ratio; namely, x300.

For an unionized gas the stopping power for a high energy particle is well known. In an ionized gas the stopping power is slightly larger than in neutral matter but at very high densities, as in the current case, it decreases. It is given by Ginzburg and Syrovatski (1964)

$$-\frac{dE}{dt} = \frac{2\pi e^4 z^2 n_\pm}{mc} \left[\ln \frac{m^2 c^2 E}{4e^2 n h^2} - \frac{3}{4} \right]$$

or

$$-\frac{dE}{dx} = 2.78 \times 10^{-28} n_\pm z^2 \left[\ln \frac{E}{mc^2} - \ln n + 73.4 \right] \text{ Mc}^2 \text{ per cm}$$

for $v \to c$.

Using $n_\pm/n$ at $kT = mc^2$, and assuming the composition of the outer layers of a pre-supernova star is carbon, $z^2/A = 3$ and further that the mean density during slowing down of a nucleus is $1/2\rho_s$ as compared to ρ_s behind the shock, then 35 becomes

$$-\frac{dE}{dx} = 1.4 \times 10^{-1} \gamma_s \left[\ln \gamma_s + 3.4 \right] \text{ Mc}^2 \text{ per nucleon per gram}$$

of rest mass cm^{-2}

Since the relative kinetic energy between moving matter and nuclei at rest is $\gamma_s Mc^2$ per nucleon, then the range Δx of an incident nucleus becomes

$$\Delta x \simeq \frac{\gamma_s Mc^2}{\frac{dE}{dx} Mc^2} \simeq \frac{7}{\ln \gamma_s + 3.4} \text{ gm cm}^{-2}$$

The range is measured in units of grams per cm^2 of rest mass nucleons. Since this unit is independent of frame, it can be related to the mass fraction of the shock transition layer and, in particular, to the mass fraction where the shock no longer propagates due to insufficient matter.

When $\gamma_s \simeq 100$ at the surface mass fraction, $\Delta x \simeq 1$ gm cm^{-2} which justifies the shock breakout condition derived earlier. In addition, Δx varies between 1 to 2 gm cm^{-2} so that nucleon-nucleon collisions will be unimportant in the shock transition as opposed to the several orders of magnitude greater thickness in the low density case of the high ion temperature precursor.

In addition, we must justify the self-consistency of the rapid pair formation within the slowing down length Δx of the ions.

The simplest way to visualize the pair multiplication during the period of the shock traversal of its own thickness is to realize that a relativistic ion traverses 300 μ_s $\Delta x A$ (A = Avogadro's number) pairs at the relative velocity c/2 in the same time. Each pair should then interpenetrate a number of pairs or photons (they have roughly equal number density) equal to $n_\pm/$ shock=$7 \times 10^{26} \mu_s$ at velocity $\simeq$ c during the passage of the shock. The pair annihilation cross section (Heitler 1949) is

$$\sigma_{an} \simeq \pi r_o^2 \frac{mc^2}{E} (\log \frac{2E}{mc^2} - 1),$$

so that at temperatures $\simeq 3$ mc^2, $\langle E \rangle \sim 10$ mc^2 and $\langle \sigma_{an} \rangle \simeq 4 \times 10^{-26}$ cm^2. By detailed balance, the pair creation processes must be equal in rate, so that the photons and pairs must recreate each other at least $\langle \sigma_{an} \rangle \times n_\pm/$shock $\simeq$ 30 μ_s fold. This is amply large to maintain thermal equilibrium. We therefore envisage a shock in which the photons and pairs are in thermal equilibrium and are created by the heat of the slowing down ions.

Photon Spallation

The gamma ray or photon spallation occurring during the ion traversal of the shock can be treated approximately in the same fashion. Hence, we are concerned with the Doppler shifted photons of the Planck distribution seen in the moving (slowing down) frame of the ions. The major fraction of the ion history in the shock occurs at the relative ion energy γ_s Mc^2, so that the mean photon energy will be γ_s (3 kT). Since $T \propto F^{1/10}$ and $\gamma_s \propto F^{-0.178}$, $T \propto \gamma_s^{-0.56}$ and the Doppler shifted photon energy becomes

$$E_r \simeq 5 \gamma_s^{0.44} \text{ MeV}.$$

The cross section for carbon photonuclear peaks at $(h\nu) \simeq 25$ MeV and so $\gamma_s \simeq 40$.

The cross section at the energy is roughly 0.9×10^{-25} MeV cm^2 so that the spallation probability during traversal becomes

$$n_\pm /\text{shock} \quad \sigma_{\text{spallation}} \quad 2/\gamma_s \simeq 130 \text{ fold}.$$

The integration over the Planck spectrum and slowing down function reduces this to 30 fold, but nevertheless gamma nucleon reactions should take place extensively for $\gamma_s \geq 40$. However, the energy of the matter ejected from such a shock is approximately $10\gamma_s^{2.5} = 10^5$ GeV. Whether the high atomic number nuclear species continues to be a significant fraction of cosmic rays above this energy is problematical. On the other hand, when heavy matter is spalled to free neutrons and protons at $T_9 \simeq 5$, the burnback and re-synthesis of heavy elements will be extremely rapid. This problem has yet to be evaluated.

ACKNOWLEDGMENT

I am intially indebted to Al Cameron for the continuing encouragement to pursue this problem and to discussions with Tom Weaver, George Chapline, and Edward Teller of Lawrence Livermore Laboratory, Livermore, who imprinted the original possibility of high temperature precursor -- to the detriment of cosmic rays -- and finally, to Al Petschek without whom the simplifying concepts of the radiation process would not have emerged. In addition, Montgomery Johnson and Marshall Rosenbluth have contributed significantly to the high density case.

REFERENCES

Bond, J., Watson, K., and Welch, J. 1965, Atomic Theory of Gas Dynamics (New York: Addison-Wesley).

Bruenn, S. 1971, Ap.J., 168, 203.

Colgate, S.A. 1959, Phys. Fluids, 2, 485.

Colgate, S.A. 1969, "Cosmic Rays From Supernova Ejecta", OG4, Proc. 11th International Conf. on Cosmic Rays, Budapest, p.23, Acta Physica Academiae Scientiarum Hungaricae, 29, Suppl. 1, pp. 353-359, 1970.

Colgate, S.A. 1972, Ap.J., 174, 377.

Colgate, S.A., and Johnson, M.H. 1960, Phys. Rev. Letters, 5, 235.

Colgate, S.A., and White, R.H. 1966, Ap.J., 143, 626.

Colgate, S.A., and McKee, C.R. 1973, Ap.J., in press.

Colgate, S.A., McKee, C.R., and Blevins, B. 1972, Ap.J., 173, L87-L91.

Cox, D.P., and Tucker, W.H. 1969, Ap.J., 157, 1157.

Fraley, G. 1968, Astrophys. Space Sci., 2, 96.

Ginzburg, V.L., and Syrovatskii, S.I. 1964, The Origin of Cosmic Rays (New York: Pergamon Press).

Heitler, W. 1949, The Quantum Theory of Radiation (New York: Oxford University Press).

Kirshner, L. 1972, Supernova Theory and Rapid Nucleosynthesis, Winter Workshop, February 21-25, Kitt Peak National Observatory, Tucson, Ariz.

Kinsey, J.H. 1969, Ap.J., 158, 295.

Minkowski, R.L. 1968, in Stars and Stellar Systems, Vol.7, ed.B.M. Middlehurst and L.H. Aller (Chicago: U. of Chicago Press), chap. 11.

Oke, B., and Kirshner, L. 1972, Supernova Theory and Rapid Nucleosynthesis, Winter Workshop, February 21-25, 1972, Kitt Peak National Observatory, Tucson, Arizona.

Paczynski, B. 1972, Ap.J.Lett., 11, 52.

Post, R.F. 1961, Plasma Physics (J. of Nuclear Energy Part C), 3, p.273.

Reeves, H. 1971, Nuclear Reactions in Stellar Surfaces and their Relation with Stellar Evolution (New York: Gordon and Breach Science Publishers, Inc.)

Weaver, T., and Chapline, G. 1973, thesis to be published.

CONFIRMING EXPLOSIVE NUCLEOSYNTHESIS
WITH GAMMA-RAY TELESCOPES

Donald D. Clayton

Rice University, Houston 77001

The idea that the common intermediate-mass nuclei are synthesized during their explosive ejection from stars, rather than before it, has one extremely important observational consequence. Several abundant nuclei are ejected in the form of radioactive progenitors, and their decay outside the star can clarify many unproven hypotheses concerning nucleosynthesis. Specifically, if the gamma-ray lines from radioactivity in supernova ejecta and in the accumulated background of the universe can be detected (and the anticipated fluxes are promising) it will be possible to accomplish the following:

(1) Prove supernovae eject new nuclei and measure the supernova yield;
(2) Prove nucleosynthesis occurs during the explosion rather than prior to it;
(3) Measure the supernova structure by the profiles of the lines and their Compton tails;
(4) Demonstrate that nucleosynthesis is occurring today in the universe and measure its average rate in the isotropic background;
(5) Determine whether the average rate of nucleosynthesis has been relatively constant or peaks in the distant past;
(6) Gain additional information about the average density in the universe;
(7) Decide between evolving and steady-state cosmologies.
That is a lot to promise; if it is correct, these observations will be as entertaining and profound as other great experiments in astronomy, such as the solar neutrino experiment and the microwave background experiment, for example. My object will be to outline these possibilities and to estimate the chances of successful detection.

The Radioactive Species

The most abundant species having a radioactive progenitor is ^{56}Fe. Bodansky, Clayton and Fowler (1968) showed that

Table 1. Average Supernova Yield (1.7×10^9 SN)

Nucleus	$X_\odot$	Progenitor	$\tau_{1/2}$	Y_{SN}	E_γ (%) MeV
^{56}Fe	1.3×10^{-3}	^{56}Co	77d	3.0×10^{54}	0.84(100),1.24(67),2.60(17),1.03(16), 1.76(14),3.26(13),2.02(11),e^+(20)
^{56}Co	1.3×10^{-3}	^{56}Ni	6.1d	3.0×10^{54}	0.812(85),0.748(51),0.472(34), 1.56(15)
^{48}Ti	2.3×10^{-6}	^{48}Cr→^{48}V	16d	6.2×10^{51}	0.983(100),1.31(97),e^+(50)
^{44}Ca	1.9×10^{-6}	^{44}Ti→^{44}Sc	48yr	5.6×10^{51}	1.156(100),e^+(94)
^{60}Ni	2.0×10^{-5}	^{60}Fe(%p$_{60}$) ^{60}Co(%p'$_{60}$)	3×10^5 yr 5.26yr	4.4×10^{52}	1.17(100),1.33(100)
^{238}U (example)	1.3×10^{-10}	r-process — (example)	4.5×10^9 yr	1.3×10^{47}	Transuranic plus daughters (many weak possibilities)

ejecta in the process of silicon burning resemble the solar abundances between A=28 and A=57 if they contain roughly equal amounts of ^{28}Si and ^{56}Ni. This result suggested that several prominent nuclei, primarily ^{44}Ca, ^{48}Ti and ^{56}Fe were ejected as radioactive ^{44}Ti, ^{48}Cr and ^{56}Ni respectively. Clayton and Woosley (1969) strengthened that result by showing that if the silicon burning had occurred slowly enough for beta decays to raise the neutron excess to a value for which ^{56}Fe itself could be ejected during silicon burning, implausible overabundances of key species would result. They further strengthened the case for ^{56}Ni by showing that something similar to an e-process centered on ^{56}Ni would also synthesize otherwise troublesome ^{58}Ni, especially if the free-particle densities were somewhat in excess of their equilibrium values. Clayton, Colgate and Fishman (1969) seized on these results to make the first estimates of the importance of ^{44}Ti, ^{48}Cr and ^{56}Ni to the gamma-ray astronomy of young supernova remnants. Because of their importance to gamma-ray astronomy, Hainebach, Arnett, Woosley and Clayton (1973) have pursued the arguments for ^{56}Ni even further. They show that two-or-three component e-processes with differing neutron enrichments and with freezeout corrections overwhelmingly select ^{56}Ni production when asked to produce the solar abundances by superposition. I think the evidence now makes it virtually certain that ^{56}Fe was ejected dynamically from the synthesizing events as ^{56}Ni. The preference for low-η solutions [Arnett and Clayton (1970); Arnett (1971); Woosley, Arnett and Clayton (1973)] in explosive burning of carbon, oxygen, and silicon and continuity arguments strongly suggest that ^{44}Ca and ^{48}Ti were also ejected as ^{44}Ti and ^{48}Cr. The solar mass fractions of these species, their half lives, and the prominent gamma-ray lines emitted during their decay are included in Table 1. The ^{56}Co→^{56}Fe decay should be, because of its rich spectrum, high abundance, and 77 day half-life, the single most important radioactive decay for gamma-ray astronomy. It remains possible, however, that a less abundant product may prove to be easier to detect.

Clayton (1971) discovered that a significant fraction of ^{60}Ni may be made as radioactive ^{60}Fe, with $\tau_{1/2} = 3 \times 10^5$yr, or perhaps as ^{60}Co, with $\tau_{1/2}$=5.26yr. In either case gamma rays of 1.17 MeV and 1.33 MeV are subsequently emitted. The arguments for and against ^{60}Fe synthesis are complex and by no means certain. About 1% of ^{60}Ni could be synthesized by arresting about half of the Cr seed at ^{60}Cr (which decays to ^{60}Fe) in the rapid neutron-induced reactions on seed nuclei during explosive carbon burning (Howard, Arnett, Clayton and Woosley 1971, 1972). Several-to-fifty percent of ^{60}Ni may have been synthesized as ^{60}Fe directly from ^{56}Fe seed nuclei in the same event. Clayton (1971) has made the intriguing observation in this regard that only ^{60}Ni is abundant enough to have absorbed the ^{56}Fe seed in explosive carbon burning, thereby suggesting that much of the iron seed is arrested at

^{60}Fe. Because of the strong (p,n) flows during high temperature carbon burning, it also seems plausible that a percent or so of the ^{60}Ni is due to ^{60}Co nuclei ejected in the explosion. Without going into the matter further here, I let P_{60} be the percentage (i.e. fraction x 100) of ^{60}Ni nuclei synthesized as ^{60}Fe nuclei and p'_{60} be the percentage synthesized as ^{60}Co, and I expect

$$1 < P_{60} (\%) < 50$$

$$0.1 < p'_{60} (\%) < 5$$

I note here that Clayton (1971) did not explicitly include ^{60}Co in his considerations. However, there do appear to be circumstances in which the gamma rays due to ^{60}Co synthesis could, for many years, exceed those due to synthesis of all other nuclei.

The r-process synthesizes many heavy radioactive nuclei, which are expected to have unfortunately small yields. Clayton and Craddock (1965) considered the flux expected from supernova remnants if the r-process yield were great enough for the "californium hypothesis" of Type I light curves to be correct. In particular they calculated the expectations of the Crab nebula in that regard. There is a large range of half lives present in initial transbismuth debris, however, so their conclusions on the 920 year-old Crab (that the strongest line should be no greater than 10^{-4} cm^{-2} sec^{-1}) would require recalculation for remnants having different ages and distances. The main problems with this idea would seem to be that it requires the r-process to be concentrated in relatively rare events in order that these nuclei not be greatly overproduced and that there seems to be no compelling reason to associate the Type I light curves with radioactivity. I therefore currently hold little hope for this gamma-ray source, although additional clarifying remarks will be made later.

Typical Supernova Yield

In the absence of more certain knowledge, I take a simple model of galactic nucleosynthesis in supernovae. Arnett and Clayton (1970) and, more specifically, Arnett (1971) have described the conceptual framework more accurately; however, my aim is only to extract typical numbers for the typical supernova event. Let the explosively synthesized nuclei be coproduced in the same abundance ratios that we find in the solar system in identical supernova events occurring at the Galactic rate

$$\dot{N}_{SN} = R e^{-t/T_R} \quad . \tag{1}$$

Fowler (1972) finds that $T_R \simeq 4 \times 10^9$ yr and Galactic age $A_G = 12 \times 10^9$ yr are not unreasonable caricatures of r-process nucleosynthesis (which I take here to characterize all explosive nucleosynthesis). Taking a current supernova rate $\dot{N}_{SN}^{\odot}$ (today) = 0.025 yr^{-1} then gives R=0.5 yr^{-1}. The initial supernova rate would, with these parametric values, have been twenty times greater.

Let the average yield of the typical event be such that its product with the total number of events prior to the birth of the sun shall have produced a galactic mass having solar composition. The total number of such events is

$$N_{SN} = \int_0^{t_\odot} \dot{N}_{SN} dt = \dot{N}_{SN}^{\odot} T_R \left[1 - e^{-t_\odot/T_R} \right] e^{A_G/T_R} \qquad (2)$$

where $t_\odot$ is the time of solar formation (approximately 7×10^9 yrs). The number of events is nearly exponential in A_G/T_R if $T_R < t_\odot$, as seems likely. With the specific choice of parameter values taken above the number of events would have been $N_{SN} = 1.7 \times 10^9$.

If the mass of the Galaxy is 1.8×10^{11} $M_\odot$ (Schmidt 1965) and the mass fraction of iron in the sun is $X_\odot = 1.3 \times 10^{-3}$ (Cameron 1968), and if the average composition of the galaxy at that time was solar, the galaxy would have contained 2.3×10^8 $M_\odot$ of ^{56}Fe. The average yield for each of the 1.7×10^9 contributing events would have been

$$M_{SN}(^{56}Fe) = \frac{2.3 \times 10^8 M_\odot \text{ of } ^{56}Fe}{1.7 \times 10^9 \text{ SN events}} = 0.14 M_\odot/SN. \qquad (3)$$

The corresponding number of ^{56}Fe atoms per event is

$$Y_{SN}(^{56}Fe) = \frac{0.14(2.0 \times 10^{33})(6.0 \times 10^{23})}{56} = 3.0 \times 10^{54} \qquad (4)$$

which would have been ejected initially as ^{56}Ni atoms. These numbers for several interesting abundances formed explosively as radioactive progenitors are shown in Table 1.

It is not difficult to question the appropriateness of many of the assumptions leading to this estimate. However, my point of view is that the simplest reasonable argument is the most appropriate one for gearing our expectations.

Table 1 shows the total yield of ^{60}Ni to be $Y_{SN}(^{60}Ni) = 4.4 \times 10^{52}$ atoms persupernova. According to the earlier discussion, the yields of ^{60}Fe and ^{60}Co are evaluated as

$$Y_{SN}(^{60}Fe) = 4.4 \times 10^{50} \, P_{60}$$
$$Y_{SN}(^{60}Co) = 4.4 \times 10^{50} \, P'_{60} \qquad (5)$$

The yield of ^{238}U under these assumptions is listed in Table 1 only as an example of transbismuth r-process yield rather than as a nucleus of particular importance for gamma-ray astronomy. Indeed, Clayton and Craddock found that the most important nuclei for the Crab were likely to be ^{249}Cf and ^{214}Bi. Nonetheless it is instructive to note that this "typical ^{238}U yield" is about four orders of mangitude too small for that required for the californium hypothesis of the light curve. If the latter hypothesis is correct, the r-process will have to have occurred in events about 10^4 times less numerous than the typical supernovae we are considering in this section. Whereas this is possible, it suggests that all Type I events are not r-process events, in which case the original hypothesis loses its raison d'etre.

Typical Line Fluxes

If species Z decays with mean lifetime $\tau(Z) = 1/\lambda_Z$, and if each decay is accompanied by g_i photons of type i, then the flux of those gamma rays at the earth due to nearby supernovae is

$$F_i = g_i \frac{\lambda_z Y_{SN}(Z)}{4\pi R^2} e^{-\lambda_z t} \qquad (6)$$

where R is the distance to the supernova and t is the time since its detonation. This formula neglects attenuation due to absorption or scattering in the source and will, therefore, be correct only for times long enough that the expanding remnant has become transparent to gamma rays.

Using information from Table I one obtains

$$F_i(^{56}Ni) = g_i \frac{3.3 \times 10^4}{R^2 (kpc)} e^{-(t/8.8d)} \quad cm^{-2} s^{-1} \qquad (7)$$

$$F_i(^{56}Co) = g_i \frac{2.6 \times 10^3}{R^2 (kpc)} e^{-(t/111d)} \quad cm^{-2} s^{-1} \qquad (8)$$

$$F_i(^{48}V) = g_i \frac{26}{R^2(\text{kpc})} e^{-(t/23d)} \quad \text{cm}^{-2}\text{s}^{-1} \tag{9}$$

$$F_i(^{44}Ti) = g_i \frac{2.1 \times 10^{-2}}{R^2(\text{kpc})} e^{-(t/69\text{yr})} \quad \text{cm}^{-2}\text{s}^{-1} \tag{10}$$

$$F_i(^{60}Fe) = g_i \frac{2.7 \times 10^{-7} P_{60}}{R^2(\text{kpc})} e^{-(t/4.3 \times 10^5 \text{yr})} \quad \text{cm}^{-2}\text{s}^{-1} \tag{11}$$

$$F_i(^{60}Co) = g_i \frac{1.6 \times 10^{-2} P'_{60}}{R^2(\text{kpc})} e^{-(t/7.6\text{yr})} \quad \text{cm}^{-2}\text{s}^{-1} \tag{12}$$

Several of these fluxes are shown in Figure 1 as a function of time. The supernova itself has been placed at R = 10^3kpc to emphasize that the A = 56 lines may even be observable from supernovae in other galaxies. A supernova in M31, for example, would present a ^{56}Co line flux above the detectable level of about 10^{-4}cm^{-2}sec^{-1} for more than a year. These lines show a rise time, rather than a pure exponential decay, because a specific model was adopted by Clayton et al. (1969) for the transparency of the expanding supernova. They took a rather optimistic (in light of recent nucleosynthesis theory) model--a 0.5 M$_\odot$ ball of iron expanding at 1.7×10^9 cm/sec so that the product of mean density times radius is

$$\bar{\rho}(t) R(t) \approx 8 \times 10^{13} t^{-2} \quad \text{gm cm}^{-2} \tag{13}$$

which falls below 10 gm cm^{-2} (a rough estimate of the optical depth for gamma rays) for t > 3×10^6 sec. Thus at best the lines will be poorly visible for the first month. Even then it is clear that Compton scattering will have a serious effect on the gamma-ray spectrum near those times when they begin to emerge. Brown (1973) has calculated this effect for some special cases similar to those considered by Clayton et al. (1969). Figure 2 shows one of his results, when 3.5 MeV and 1.25 MeV lines are emitted isotropically from a depth of 18.6 gm cm^{-2} within an iron sphere of radius 37.2 gm cm^{-2}. The total mass of such a sphere depends upon its metric radius, of course, so with R(t) = 1.7×10^9t we find that the mass whose line spectrum corresponds to Figure 2 is

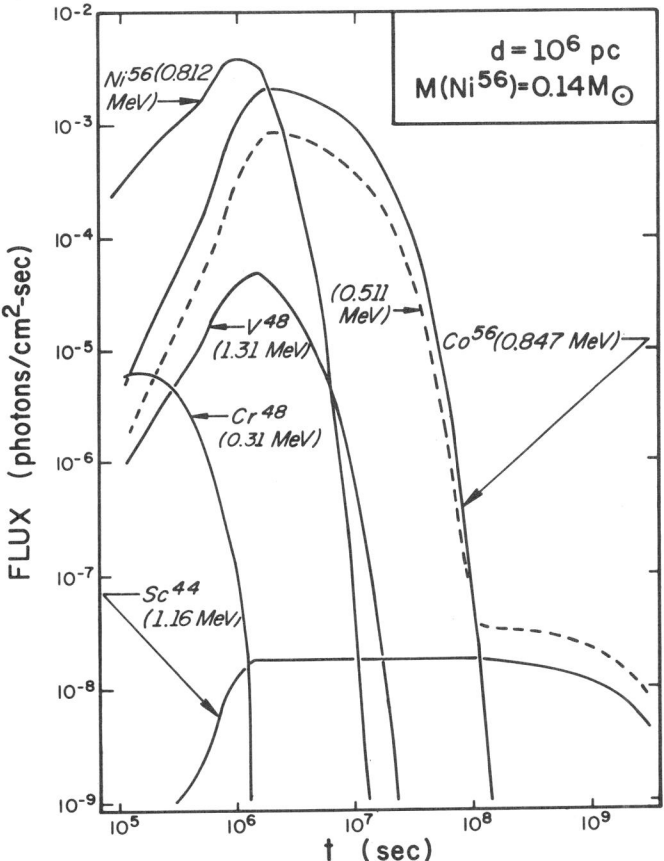

Figure 1. Prominent medium-lifetime gamma-ray line fluxes as a function of time from a distant ($d=10^6$ pc) supernova ejecting $0.14 M_\odot$ of ^{56}Ni and $2.0 \times 10^{-4} M_\odot$ of ^{44}Ti. The early growth reflects the increasing transparency of an expanding model (Clayton et al 1969)

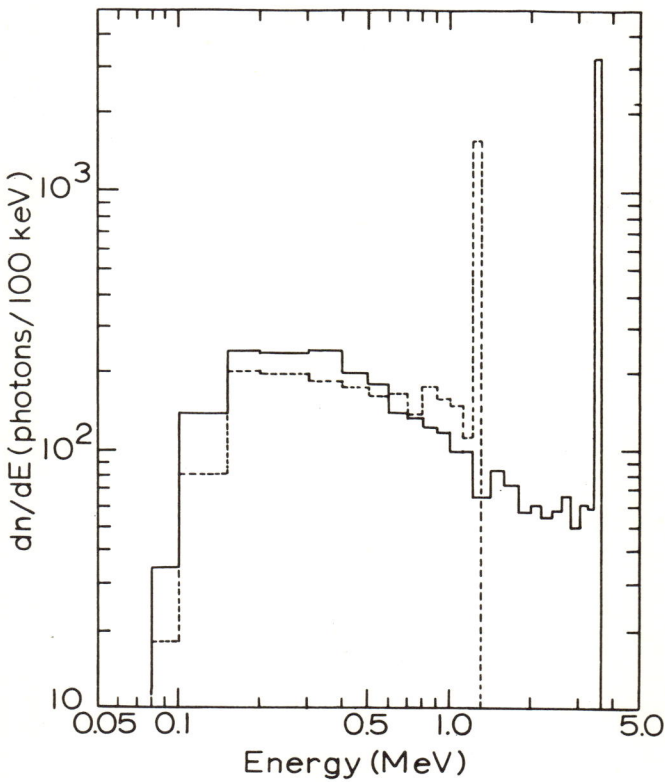

Fig.2. Effect of Compton scattering on 3.5 MeV lines (solid histogram) and 1.25 MeV lines (dashed histogram) emitted isotropically from a point at a depth of 18.6 gm cm^{-2} from the surface of an Fe sphere of radius 37.2 gm cm^{-2} (Brown 1973).

$$M(Fig.2) = \left(\frac{t}{2\times 10^6 \text{sec}}\right)^2 M_\odot \qquad (14)$$

Therefore the total mass of that example could be any reasonable multiple of the solar mass at time of order a few months. The question of the mass of layers overlying the CO core at the time of detonation is even more uncertain, but it will clearly be worthwhile to evaluate dynamic models of gamma-ray opacity for exploding massive evolved stars. For the time being I wish only to emphasize that whether the ^{56}Ni lines emerge at all (they did in Figure 1) depends on the structure and dynamics of the exploding object. Ideally we may one day watch these and the ^{56}Co lines rise to peak intensity before beginning their decay, and the rise time of these fluxes will be a crucial measure of the structure of the exploding object. This astronomy will allow us to measure that structure somewhat analogously to the way neutrino astronomy has allowed us to measure the interior of the sun -- and probably with all the surprises!

The 1.16 MeV line emitted subsequent to the decay of the ^{44}Ti could be quite strong in several present Galactic remnants, and will surely emerge even if the A=56 lines should happen not to get out. In this sense the ^{44}Ti synthesis may prove to be extremely important. The real need, of course, is for the Galaxy to arrange a visible supernova, preferably after (if ever) instruments like HEAOB are operational. The A=48 lines, on the other hand, seem likely to be of no special importance, because they are both weaker and shorter lived than the ^{56}Co lines.

The ^{60}Co lines have not been entered on Figure 1, but a comparison of Eqs. (12) and (10) show that they are comparable to those of ^{44}Ti for about 10 years if p'_{60} is around unite (i.e., about 1% of ^{60}Ni is due to synthesis of ^{60}Co, which requires about 2% of ^{56}Fe seed to reside at ^{60}Co at completion of explosive carbon burning). Remnants throughout the Galaxy (R<20 kpc) should ultimately prove detectable for a decade.

The ^{60}Fe lines (actually the same as the ^{60}Co lines but with a much longer halflife) are also not shown in Figure 1. They are a special case due to the long 3×10^5yr halflife, which insures that many radiating remnants exist but they may have large angular size due to the long time available for dispersal. For the flux to exceed a detectable 10^{-4} cm^{-2} sec^{-1} requires R≤160pc if 10% of ^{60}Ni is due to synthesis of ^{60}Fe. A circle of 160pc radius constitutes about 10^{-4} of the area of the Galactic disk, and should thus contain one of the approximately 10^4 supernovae that should have occurred during the

Clayton

lifetime of ^{60}Fe. The size of a remnant 10^5 years old might cover a significant fraction (even half!) the sky for an event about 100 pc away, however, so simple on-source-off-source differences will have to be measured with this in mind. The radiation from such sources seems more likely to appear as a general galactic background. The general flux from a wide angle containing the galactic center might be

$$F_{60} \text{ (Galactic)} \simeq 3 \times 10^{-4} \text{ cm}^{-2} \text{ sec}^{-1}$$

if p_{60} is about 10%. This is also about the same as the average flux from the Galaxy due to the ^{44}Ti lines (Clayton 1971).

The Universal Background

One need only appreciate that the average Galactic luminosity due to radioactive gamma rays has been 3×10^{40} erg sec^{-1} to realize that their contribution to the isotropic background radiation may be significant. The cosmological principle allows us to estimate their flux very easily. Taking $H_0 = 55$ km sec^{-1} Mpc^{-1} (Sandage 1972) the observed universal density of matter is $\rho = 1.7 \times 10^{-31}$ gm cm^{-3} (Oort 1958). If the average mass fraction of ^{56}Fe is $X_\odot(^{56}\text{Fe}) = 1.3 \times 10^{-3}$, this corresponds to 2.2×10^{-34} gm cm^{-3} of ^{56}Fe. Consequently, the average iron number density in the observed universe is

$$n(^{56}\text{Fe}) = 2.3 \times 10^{-12} \text{ cm}^{-3} \qquad (15)$$

The flux of these gamma rays per steradian is (Clayton and Silk 1969)

$$\frac{\partial F}{\partial \Omega} = \frac{c}{4\pi} g_\gamma \, n(^{56}\text{Fe}) \qquad (16)$$

where g_γ is the number of gamma rays emitted per ^{56}Fe nucleus synthesized. The value of g_γ is 2.8 for only ^{56}Co decays and $g_\gamma = 4.9$ if both ^{56}Ni and ^{56}Co decays are used. Taking the latter value yields

$$\frac{\partial F}{\partial \Omega} = 2.7 \times 10^{-2} \text{ cm}^{-2} \text{ ster}^{-1} \text{ sec}^{-1}$$

To emphasize the size of this flux, Clayton and Silk pointed out that it is as large as the total integrated universal background at photon energies in excess of 300 keV. Because this estimate is based on the observed mass density, it will be proportionately greater if the universe contains "hidden matter" that has also synthesized ^{56}Fe.

The simple density argument does not determine the frequency distribution of the photons comprising the flux in eq. (16). For those ^{56}Fe nuclei synthesized early in the universe, the associated gamma rays will now have been considerably redshifted. It is just this feature that allows the spectrum to carry a wholly new astrophysical datum; i.e., the redshift distribution in the gamma-ray spectrum measures the distribution of the ages of ^{56}Fe nuclei. Hidden in it is the chronological account of the rate of nucleosynthesis.

Let f(t) be the rate per unit of cosmic time at which ^{56}Fe nuclei were (and are being) synthesized. Let it be normalized such that

$$\int_0^{t_o} f(t) dt = 1 \tag{17}$$

so that f(t)dt is the fraction of all ^{56}Fe nuclei that were synthesized at cosmic time t in the interval dt (t_o = cosmic time today). It follows that f(t)dt is also the fraction of A=56 gamma rays whose travel times are t_o-t in the interval dt. In any standard cosmological model the travel time t_o-t is some function of the redshift z. Thus f(t)=f(z) and f(t) (dt/dz) dz becomes the fraction of the photons having redshift z in the interval dz. The gamma-ray source function per unit ^{56}Fe nucleus per unit energy interval is just

$$P(E,t) = \sum_i P_i(E,t) = \sum_i g_i \, \delta(E-E_i) f(t) \tag{18}$$

where the sum is over the lines of type i emitted with rest energy E_i at the rate of g_i gammas per ^{56}Fe nucleus synthesized. The differential flux today due to gamma rays of type i is

$$\frac{\partial^2 F_i}{\partial E \partial \Omega} = \frac{c}{4\pi} n(^{56}Fe) \int_0^{t_o} \frac{R(t_o)}{R(t)} P_i\left[\frac{R(t_o)}{R(t)} E, t\right] dt$$

$$= \frac{c}{4\pi} n(^{56}Fe) \int_0^{\infty} (1+z) P_i\left[(1+z)E, t\right] \frac{dt}{dz} dz \tag{19}$$

where E is the energy today of the photon and $R(t)$ is the scale factor of the universe. [e.g., McVittie 1965]. Because

$$\frac{E}{E_i} = \frac{R(t)}{R(t_o)} = \frac{1}{1+z} \qquad (20)$$

the integral over cosmic emission time can also be expressed as an integral over received energies:

$$dt = \frac{R(t_o)}{\dot{R}(t)} \frac{dE}{E_i} \quad. \qquad (21)$$

This integral is easily done due to the δ-function nature of P_i to give

$$\frac{\partial^2 F_i}{\partial E \partial \Omega} = \frac{c}{4\pi} \frac{g_i \, n(^{56}Fe)}{E_i} \frac{R(t_o)}{\dot{R}(t_E)} f(t_E) \qquad (22)$$

where the time t_E is the solution of Eq. (20). In the Friedman dust models one has [e.g., Weinberg 1972]

$$\frac{\dot{R}(t_E)}{R(t_E)} = \frac{E_i}{E} H_o \left[1 - 2q_o + 2q_o \frac{E_i}{E}\right]^{\frac{1}{2}} \qquad (23)$$

so that Eq. (22) reads

$$\frac{\partial^2 F_i}{\partial E \partial \Omega} = \frac{c}{4\pi} \frac{g_i \, n(^{56}Fe)}{E_i \, H_o} f(t_E) \left[1 - 2q_o + 2q_o \frac{E_i}{E}\right]^{-\frac{1}{2}}. \qquad (24)$$

Clayton and Silk (1969) evaluated the flux in a simpler form for the two cases where

$$R(t) \propto t^{1/\gamma} \quad. \qquad (25)$$

They are the low-density universe ($q_o \simeq 0$, $\gamma \simeq 1$) and the Einstein-de Sitter universe ($q_o = 1/2$, $\gamma = 3/2$). In those

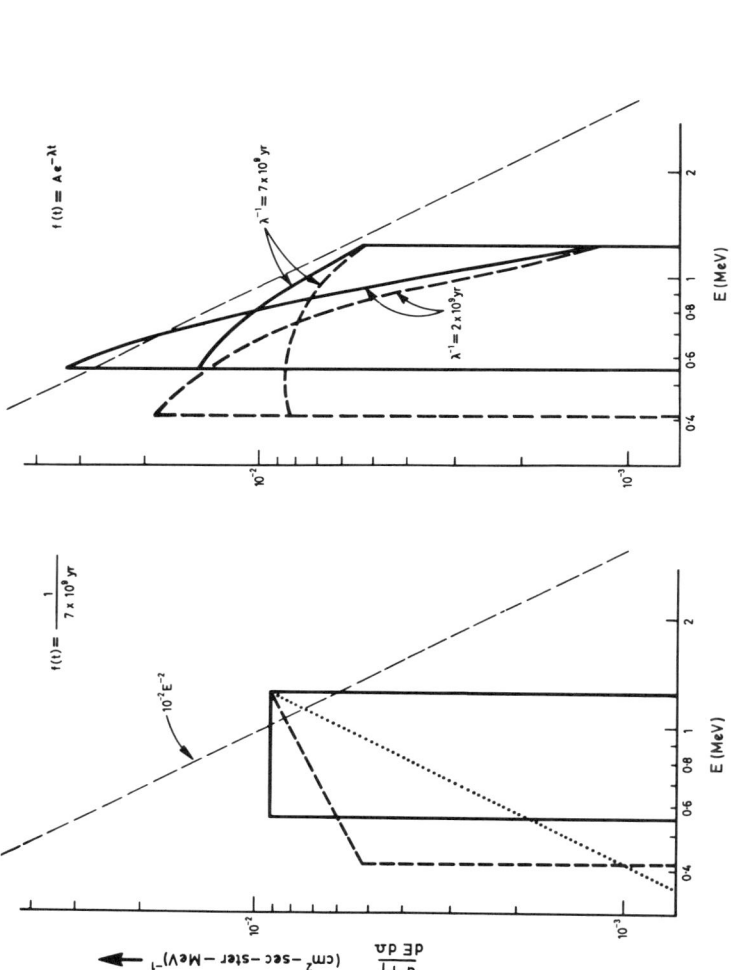

Figure 3. Differential flux due to a single line (^{56}Co at 1.24 MeV). Models of constant galactic synthesis of ^{56}Fe over a period of 7×10^9 yr are shown on the left, and models of exponentially decreasing nucleosynthesis are shown on the right. This rather short duration of galactic nucleosynthesis was chosen only for ease of comparison, so that it could fit in the age of the Einstein-de Sitter universe with $H_o = 75$ km/sec/Mpc. The low density universe is shown as a solid line and the Einstein-de Sitter universe as a dashed line. The steady-state-universe line profile is dotted on the left figure. The line $10^{-2} E^{-2}$ is also shown to indicate the approximate level of the observed diffuse background. (Clayton and Silk 1969)

cases Eq. (20) can be explicitly solved for t_E and, furthermore, the factor involving q_o simplifies:

$$\frac{\partial^2 F_i}{\partial E \partial \Omega} = \frac{c}{4\pi} \frac{g_i n(^{56}Fe)}{E_i H_o} f\left[t_o\left(\frac{E}{E_i}\right)^\gamma\right] \left(\frac{E}{E_i}\right)^{\gamma-1} \qquad (26)$$

It is straightforward to confirm with Eq. (24) or Eq. (26) that

$$\int_o^{E_i} \frac{\partial^2 F_i}{\partial E \partial \Omega} dE = \frac{c}{4\pi} g_i n(^{56}Fe) \qquad (27)$$

as required by photon conservation.

The spectrum due to each line is characterized by a step at the rest energy

$$\Delta\left(\frac{\partial^2 F_i}{\partial E \partial \Omega}\right)_{E=E_i} = \frac{c}{4\pi} \frac{g_i n(^{56}Fe)}{E_i H_o} f(t_o) \qquad (28)$$

that is directly proportional to the average rate of nucleosynthesis today in the universe. Detection of the series of correlated rest edges will confirm that nucleosynthesis is still occurring and measure its rate. Each rest edge is followed at immediately lower energies by identical redshifted continua, whose shape and extent depend upon the cosmological model and the history $f(t)$ of galactic nucleosynthesis.

Some simple profiles for the ^{56}Co line of 1.24 MeV are shown in Figure 3. If nucleosynthesis has occurred within galaxies at a constant rate up to the present time t_o since it began at some time t_*, then $f(t)=(t_o-t_*)^{-1}$ between t_* and t_o and is zero elsewhere. The left half of Figure 3 shows that case from Clayton and Silk (1969), who took $t_o-t_* = 7 \times 10^9$yrs so that it could fit within the Einstein-de Sitter universe based on $H_o=75$ km/sec/Mpc. The right half of Figure 3 shows this line profile for exponential nucleosynthesis $\dot{f}(t)=A \exp[-\lambda(t-t_*)]$, where A is a normalization constant and $\lambda=1/T_R$ from Eq. (1). The two choices of λ shown there give different relative strengths to present-day nucleosynthesis in comparison with the initial galactic rates. The rest edges are still detectable here, but smaller than for the case of constant nucleosynthesis.

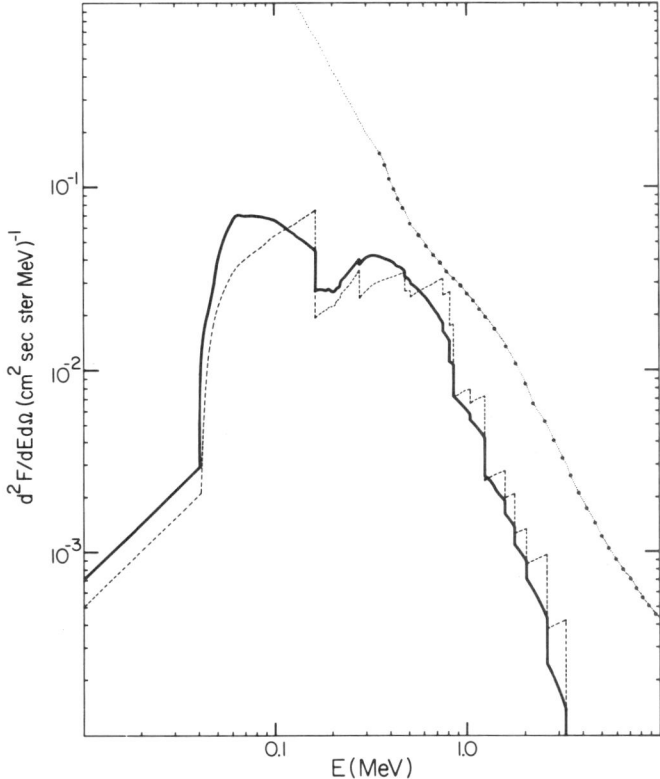

Figure 4. The composite $^{56}Ni \rightarrow {}^{56}Co \rightarrow {}^{56}Fe$ gamma-ray spectrum in a specific Einstein-de Sitter universe. The age of this universe is 11.8×10^9 years, and nucleosynthesis in galaxies began at $t = 2 \times 10^9$ years, corresponding to $z = 2.5$. The solid line represents an exponentially decreasing rate of nucleosynthesis per galaxy which is today e^{-2} of the initial rate, whereas the dashed line represents a constant rate of nucleosynthesis per galaxy. The spectrum has series of rest-frequency edges and redshifted continua. The rest edges, which are calculated without compton scattering in the source, are smallest for nucleosynthesis peaked in the early galactic history. The dotted line, shown for comparison, is the background observed on the Apollo 15 scapecraft by Trombka et al.-- the heavy solid dots being their data points. The number of photons in the radioactivity background is obviously significant, but higher-energy-resolution observations will be needed to extract the presence of detailed structure. Although the density required for $q_o = 1/2$ with $H_o = 55$ km/sec Mpc is $\rho_c = 5.9 \times 10^{-30}$ gm cm^{-3}, this figure assumes that only the galaxies, with density $\rho_G = 0.028 \rho_c$, contain ^{56}Fe. This calculation (Clayton and Ward) is thus a lower limit to the anticipated gamma-ray density.

It is worth noting here that these figures are applicable to Sandage's (1972) value $H_o=55$ km/sec/Mpc if one only increases t_o-t_* by the factor 75/55, giving the more reasonable $t_o-t_* = 9.6 \times 10^9$ yr, and if the value of the flux is reduced by the factor $(55/75)^2$. The latter comes about because $(^{56}Fe) \propto H_o^2$ and $f(t) \propto H_o$ if we require $t_o-t_* \propto H_o^{-1}$. It is clear that the flux at this rest edge may well be comparable to the isotropic background, whose approximate value is shown for comparison. The model $T_R = 4 \times 10^9$ yr and $A_G = 12 \times 10^9$ yr used in estimating the typical supernova yield resembles the curve labeled $\lambda = (2 \times 10^9 yr)^{-1}$ in Figure 3. Its rest edge is the smallest shown -- about 15 percent of the observed background.

The steady-state universe, shown as a dotted line in Figure 3, affords a somewhat different problem. The number of objects of age A is proportional to $\exp(-3H_o A)$, and if ^{56}Fe nuclei are taken to be the objects, we see that the redshift distribution is already fixed. The line profile is proportional to $(E/E_i)^2$ independent of the time history of nucleosynthesis within individual galaxies. On the other hand, the size of the rest edge does depend upon the rate of galactic nucleosynthesis, because the average galactic age must be 1/3 of the Hubble time -- about 6×10^9 yrs. Therefore our Galaxy is twice as old as the average galaxy, so our iron content should be somewhat higher than average -- especially if nucleosynthesis were relatively constant rather than peaked early. Since the rate of nucleosynthesis in the large must be such as to replace the expanding iron density in the large, the normalization of the steady state curve depends upon the degree to which our Galactic iron concentration is applicable to galaxies half our age. The normalization of Figure 3 is therefore arbitrary, although it is <u>not</u> <u>unreasonable</u>, since $(3H_o)^{-1}$ is not greatly different than the average age of iron nuclei within our Galaxy.

Figure 4 illustrates the entire A=56 spectrum for Einstein-de Sitter case. Two points need be made: (1) the rest edges are clearly more prominent in the case of constant galactic nucleosynthesis than they are in the e^{-2} - exponential case, but (2) the general shape of the continuum feature produced is quite similar for the two cases. The fascinating thing about the Apollo 15 points of Trombka, Metzger, Arnold, Matteson, Reedy and Peterson (1973) is the way they show a positive curvature (excess) around one MeV just after negative curvatures near 400 keV had suggested the continuum was beginning to fall more steeply. This suggests a multi-source spectrum, and it is quite conceivable that the radioactivity spectrum may be significant in the overall shape. Certainly the changes of second derivative will, if they remain after further experimental scrutiny, be important keys to the origins of this spectrum. The radioactivity spectrum may be less visible if the exploding source remains opaque for several months. Compton scattering as extensive as that in Figure 2 would remove at least half of the photons from the rest frequency at the source and redistribute them at energies of 0.5 MeV or so. If this source function were employed in Figure 4, the rest edges would be smaller by a factor of two or so, and the whole high-energy slope would be diminished in importance. At present no firm conclusion can be made, because the NaI(Tl) scintillator aboard Apollo 15 had not the energy resolution to detect structure like that in Figure 4.

Conclusion

It is within scientific grasp to learn many or all of the questions enumerated in the introduction. What is needed is a gamma-ray telescope with high energy resolution, moderately good angular resolution, and long operation times outside the earth's atmosphere while responsive to ground command. Of primary importance is energy resolution of a few percent or better to extract lines from continua and to detect rest edges in the universal continuum. The angular resolution is needed to identify specific radiating objects (supernovae). As far as I know, the best type of instrument for accomplishing these two needs would be one like I described in the NASA x-and-γ-Ray Committee Study of November 1965 -- a honeycomb of parallel holes drilled through actively collimating CsI or NaI with solid-state (say Li-drifted Ge) gamma detectors at the bottom of each hole. Operation outside the earth's atmosphere is necessary to reduce to emission background of the earth's atmosphere and its opacity. Ground command will be necessary for viewing different objects and for extracting the isotropic component. Last but by no means least, we need nature's cooperation in presenting us with a new Galactic supernova, preferably a visible one, although an invisible one could be immediately recognized by a large increase of the A=56 lines [See Eqs. (7) and (8)]. Good observation of at least one supernova is needed to measure what fraction of the gamma lines emerge unscattered from their source, for without this calibration the interpretation of the universal background will remain insecure. With a little bit of luck, the entire science of explosive nucleosynthesis will gain a firm observational footing from these very special photons. Like all photons, they tell us that an electromagnetic de-excitation occurred; unlike any other photons, they alone tell us that a new nucleus was just born.

Because it does not have an immediately obvious relationship to the science of explosive nucleosynthesis, I have not tried to discuss nuclear gamma rays produced in cosmic-ray collisions. Nonetheless it is worth noting that the first positive detection of such events may have occurred (Fishman and Clayton 1972). We may expect contributions to cosmic-ray science from the same instruments we need for nucleosynthesis.

This research was partially supported by the National Science Foundation under GP-18335. I wish to explicitly thank my past coauthors, who have taught me so much about the various aspects of gamma-ray astronomy.

REFERENCES

Arnett, W. D. and Clayton, D. D. 1970, Nature, 227, 780.
Arnett, W. D. 1971, Ap. J., 166, 153.
Bodansky, D., Clayton, D. D., and Fowler, W. A. 1968, Ap. J. Suppl. 16, 299.
Brown, R. T. 1973, Ap. J., 179, 607.
Cameron, A. G. W. 1968 in Origin and Distribution of the Elements, L. H. Ahrens, ed., (New York: Pergamon Press) p. 125.
Clayton, D. D. 1971, Nature, 234, 291.
Clayton, D. D. and Craddock, W. 1965, Ap. J., 142, 189.
Clayton, D. D., Colgate, S., and Fishman, G. J. 1969, Ap. J., 155, 75.
Clayton, D. D. and Silk, J. 1969, Ap. J., 158, L43.
Clayton, D. D. and Woosley, S. E. 1969, Ap. J., 157, 1381.
Fishman, G. J. and Clayton, D. D. 1972, Ap. J., 178, 337.
Fowler, W. A. 1972, in Cosmology, Fusion, and Other Matters, ed. F. Reines, (Boulder: Univ. of Colorado Press) p. 67.
Hainebach, K., Arnett, W. D., Woosley, S. E. and Clayton, D. D. 1973, in preparation.
Howard, W. M., Arnett, W. D., Clayton, D. D. and Woosley, S. E. 1971, Phys. Rev. Letters, 27, 1607; 1972, Ap. J., 175, 201.
McVittie, G. C. 1965, General Relativity and Cosmology (Urbana: Univ. Illinois Press).
Oort, J. H. 1958, in Solvay Conference on Structure and Evolution of the Universe (Brussells: R. Stoops).
Sandage, A. 1972, Ap. J., 178, 1.
Schmidt, M. 1965, in Galactic Structure, A. Blauw and M. Schmidt, eds. p. 528 (Chicago: University of Chicago Press).
Trombka, J. I., Metzger, A. E., Arnold, J. R., Matteson, J. L., Reedy, R. C. and Peterson, L. E. 1973 "The Cosmic γ-Ray Spectrum Between 0.3 and 27 MeV Measured on the Apollo 15", preprint and private communication.
Weinberg, S. 1972, Gravitation and Cosmology (New York: John Wiley) p. 495.

CHAPTER VI. Nuclear Reaction Rates: The Basis of the Subject

The empirical foundations of explosive nucleosynthesis were laid in nuclear physics laboratories. Our understanding of the subject is only as good as the nuclear data and theory which we use. One of the central questions currently being pursued is: how can we determine, experimentally if possible and semi-empirically if necessary, those reaction rates needed for nucleosynthesis calculations? Michaud and Fowler discuss the current progress and problems involved in estimating nuclear reaction rates, and provide comparisons with recent experimental results. Fowler emphasizes the importance of excited states in astrophysical situations.

References for Chapter VI

Textbooks

    Blatt, J. M. and Weisskopf, V. F. 1952, <u>Theoretical Nuclear Physics</u>. (New York: John Wiley & Sons)
    McCarthy, L. E. 1968, <u>Introduction to Nuclear Theory</u> (New York: John Wiley & Sons)

Original Papers

    Michaud G. and Fowler, W. A. 1970, <u>Phys. Rev. C.</u>, 2, 2041.
    Truran, J. W., Cameron, A. G. W. and Gilbert, A. 1966, <u>Can. J. Phys.</u>, 44, 463.

PROBLEMS WITH NUCLEAR REACTION RATES

Georges Michaud

Departement de Physique, Universite de Montreal
Montreal, Canada

I. Introduction

Large numbers of nuclear reaction rates are needed in "Explosive Nucleosynthesis" calculations. (Truran, Arnett and Cameron 1967; Arnett 1969; Truran and Arnett 1970; Howard, Arnett and Clayton 1971). Those involved in the later stages of the cooling are specially important for the relative abundances of the elements that come out of the star (Howard, Arnett, Clayton and Woosley 1972; Arnett, Truran and Woosley 1971). Few of the needed reaction cross sections have been measured in the energy range of importance to calculate the reaction rates (Fowler, Caughlan and Zimmerman 1967; Fowler 1972). The others must be evaluated from a theoretical model (Reeves 1966; Fowler and Hoyle 1964; Truran, Hansen, Cameron and Gilbert 1966; Wagoner, Fowler and Hoyle 1967; Michaud and Fowler 1970; Michaud 1972; Truran 1973). Here we will analyze the uncertainties in the calculated rates. In our analysis we will use the optical model and the statistical theory and we will try to explain the difficulties met in extrapolating the $^{12}C + ^{12}C$ cross sections, in determining the radius to use in charged particle reactions and in evaluating the contribution of radiative channels.

We will use mainly the optical model with Woods-Saxon shape. It is reasonably well established for nuclear channels. The use of a specific model puts constraints on the calculations and so restricts the error bars. We will study what would be the effect of relaxing some of the constraints imposed by the optical model specially for alpha-particle or heavy ion channels.

II. The Optical Model and Statistical Theory

The calculations of cross sections start from the results of the evaporation theory averaged over the resonances of the compound nucleus (Hauser and Feshbach 1952; Vogt 1968).

$$\bar{\sigma}_{\alpha\alpha'} = \frac{\pi}{k_\alpha^2} \sum_{J^\pi} \frac{2J+1}{(2I+1)(2i+1)} \left[\sum_\ell T_\ell(\alpha) \right]$$

$$\times \left[\sum_{\ell'} T_{\ell'}(\alpha') \Big/ \sum_{\alpha''\ell''} T_{\ell''}(\alpha') \right], \quad (1)$$

where α represents a pair of particles and their state of excitation, I and i are the intrinsic spins of the initial pair of interacting particles, s is the channel spin, ℓ the orbital angular momentum of the pair, and J^π is the angular momentum and parity of the compound nucleus. The vector sums $\vec{s} = \vec{I} + \vec{i}$ and $\vec{J} = \vec{\ell} + \vec{s}$ must be consistent with the conservation laws for angular momentum and parity. The unprimed quantities refer to the incoming channel, the primed quantities refer to the outgoing channel, and the double-primed quantities should be summed over all open channels in the compound nucleus, including excited states of the residual nuclei.

In what follows we will not discuss the effect of the excited states of the target and residual nuclei. It should however be born in mind that in the laboratory target nuclei are all in their ground state but not so in stars. Bahcall and Fowler (1969) found that the rates were not changed by the inclusion of the excited states. However Michaud and Fowler (1970) pointed out that it was not necessarily so if there was a large spin difference between the target and the residual nuclei. Arnould (1972) found an important reaction where this changed the reaction rate by a factor of 10^3. So, when there is a large spin difference the effect of excited states should be studied carefully. It affects not only the calculated but also the measured rates!

Once the spins and parities are known all the needed nuclear physics in Eq. (1) appears in the transmission functions, T_ℓ. The transmission functions for γ-ray channels will be studied in section V. In sections III and IV we will essentially study the transmission functions for particle channels, for heavy ions in section III and for nucleons and α-particles in section IV.

To analyze the sources of errors in calculated transmission functions, we will use a nuclear model which is more general than the one generally used in the calculations and we will see how the "neglected" nuclear physics could influence the results. We will calculate transmission functions with Saxon-Woods optical model potentials:

$$V = -(V_o + iW_o) / \left[1 + \exp\left((r - R_o)/a\right)\right] + V_{Coulomb} \quad (2)$$

where V_o and W_o are the strengths of the real and imaginary parts of potentials, R_o is the radius and $\underline{a}$ the surface thickness. Using it in the Schnodinger equation one then calculates transmission functions (Michaud 1973). For all cases of interest:

$$10.0 \lesssim V_o \lesssim 100.0 \text{ MeV} \quad (3)$$

$$1.0 \lesssim W_o \lesssim 10. \text{ MeV} \quad (4)$$

In all calculations of reaction rates (Truran 1973; Michaud and Fowler 1970), V_o and W_o are assumed large enough that:
1) there is no resonance effect in the well ($W_o \lesssim 10.0$ MeV), and
2) the well is large enough that the exact position of the bottom of the well does not affect the results ($V_o \gtrsim 50.0$ MeV). Those two assumptions lead to the use of the so-called "black nucleus" model. We will lift those two assumptions. We will also discuss the effect of adding a repulsive core to the Saxon Woods potential.

Finally, we will try to specify over what temperature range and over what mass range, optical potentials may be used to calculate reaction rates. Specifically, the experimentally verified Wigner distribution.

$$P(s > s_o) = \exp - \pi s_o^2 / (4 D^2) \quad (5)$$

gives the probability that the spacing between two neighboring levels (of a given spin and parity) be larger than s_o when the average level spacing is D. The probability that the distance between levels be larger than twice the average distance is 4% and it is 0.0004% that it be four times the level spacing. The optical potential of Eq. (2) averages over such compound nucleus resonances. Rates calculated with the optical potential will be reasonable approximations if, and only if, a few compound nucleus resonances contribute to the rate at the temperature of interest. The most important energy of the cross section to calculate a reaction rate at temperature T_9 ($T_9 = T \times 10^{-9}$°K) is approximately (Fowler and Hoyle 1964)

$$E_o = 0.122 \left(z_1^2 z_o^2 A T_9^2\right)^{1/3} \text{ MeV} \quad (6)$$

where Z_0 and Z_1 are the charges of the target and residual nuclei respectively and A is the reduced mass in AMU. The full range of important energies around E_0 is:

$$\Delta E_0 = 0.237 \left(Z_1^2 Z_0^2 A T_9^5 \right)^{1/6} \text{ MeV} \tag{7}$$

Requesting

$$\Delta E_0 = 4D \tag{8}$$

where we will take D to be the distance between two levels whatever be its spin or parity will give us a reasonable rule to check the statistical applicability of the optical model. It should guarantee the presence of a few resonances in the important energy range.

III. Extrapolation of the ^{12}C + ^{12}C Total Cross-Section Measurements

The ^{12}C + ^{12}C reaction has been carefully studied in the laboratory, most recently by Mazarakis and Stephens (1972) and Patterson et al. (1969). However the lowest energy measurements have been made at 2.5 MeV, whereas to calculate stellar reaction rates the cross-section is needed down to 1.0 MeV. The extrapolation is made more difficult by giant and intermediate structure resonances which appear in the experimental results (see Figure 1b). We will briefly discuss a few possible models.

First one might assume that the apparent giant nuclear structure between 2.5 and 6.0 MeV is due to statistical fluctuations in the density of intermediate structure resonances. It appears plausible since the average spacing is around 0.4 MeV if one considers levels of all spins, but is about 1.0 MeV between levels of the same spin. Eq. (5) applies to levels of a given spin. Fluctuations over intervals of 2 MeV are expected to occur for levels of a given spin. If the levels with the different spins conspired they could all be grouped around 6.0 MeV and around 2.5 MeV. The levels around 2.5 MeV would not have been resolved in the ^{12}C + ^{12}C reaction but could correspond, as suggested by Fowler (1972), to levels of ^{24}Mg which have been observed in the ^{16}O(^{12}C, α)^{24}Mg reaction by Middleton et al. (1970).

The most reasonable extrapolation is then that of Reeves (1966b). In Figure 1b we used it to normalize all measurements and calculations. The experimental cross-sections are equal to Reeves' at 2.5 and 5.5 MeV, they are larger above 5.5 MeV but lower between 2.5 and 6.0 MeV. The cross-section calculated by Reeves (1966b) then averages over the statistical fluctuations. However it is easy to see that the

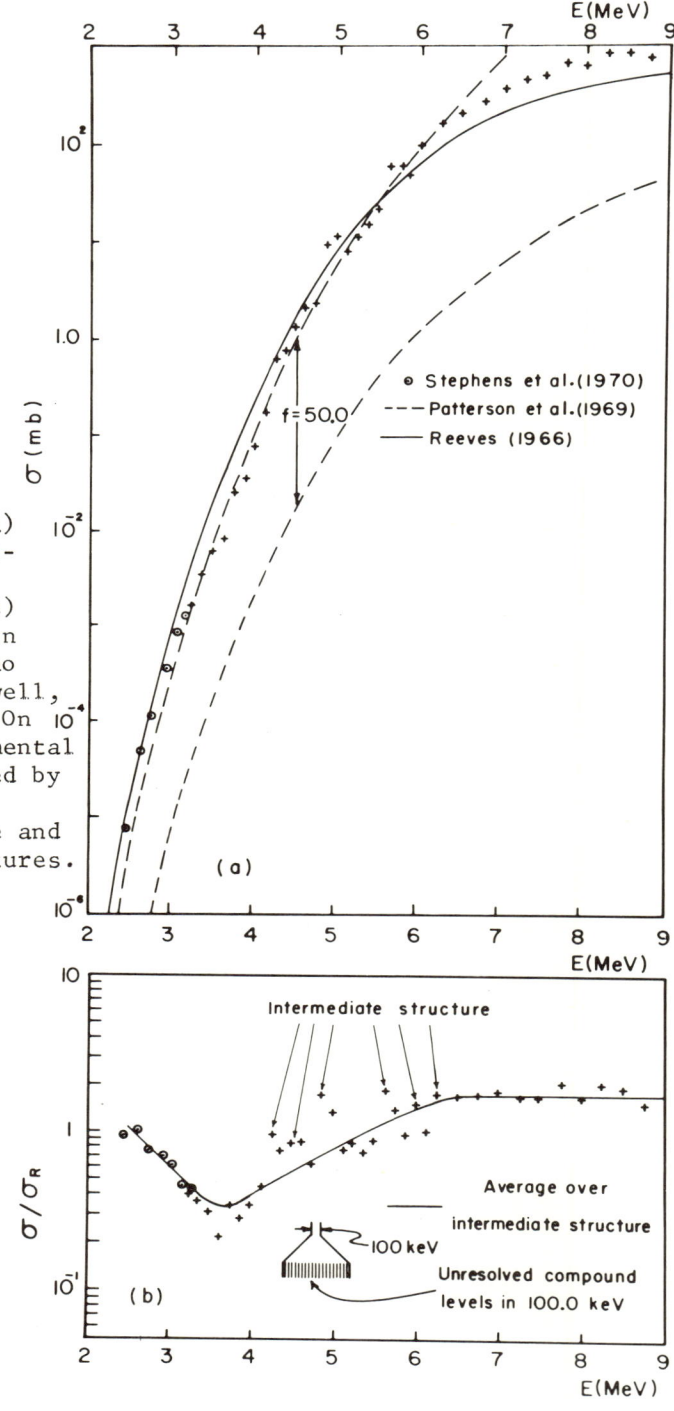

Figure 1. On part a) are shown the experimental measurements (crosses and circles) and the extrapolation of Reeves (1966), who used a Saxon-Woods well, with $W_0 = 1.0$ MeV. On part b) the experimental data has been divided by Reeves' values thus showing intermediate and giant nuclear structures.

extrapolation is very uncertain. The width of the energy range of importance is (from Eq. 6) around 0.6 MeV at $T_9 = 0.5$. If there are statistical fluctuations over intervals of 2.0 MeV, there could be many intermediate structure resonances just as there could be none in the energy range of importance. This leads to an uncertainty of at least a factor of ten either way around Reeves (1966b) extrapolation. The extrapolation of Patterson et al. (1969) follows approximately the straight lines between 3.2 and 6.0 MeV on Figure 1b and is a reasonable lower limit in this case.

We have two objections to the proceeding model: first the three resonances in ^{24}Mg which are needed to explain the rise below 3.0 MeV have been shown to have $J^\pi = 8^+$ (Zurmuhle et al. 1972). This would affect strongly the predicted ratio of α to p emission in the $^{12}C + ^{12}C$ reaction (see Figure 2b). The experimental results contradict such a prediction. Our second objection comes from the study of the $^{12}C + ^{16}O$ and $^{16}O + ^{16}O$ reactions. Both show giant resonance structure of a similar nature (Michaud 1972, 1973). One could perhaps accept that there be a statistical fluctuation mocking giant resonance structure in one of those three reactions but that there be in all of them seems most unlikely. So we prefer to base the extrapolation on giant resonance structure in the $^{12}C + ^{12}C$ reaction and not to assume a statistical conspiracy.

Nuclear structure puts constraints on the possible extrapolations and so allows us to reduce the probable error from a factor of 10 to a factor of 4 at $T_9 \simeq 0.5$ (see Figure 2a). If the measured oscillations between 2.5 and 6.0 MeV are of a statistical nature, as assumed in the first model, they do not allow us to predict what will happen below 2.5 MeV, but if they are to be explained by giant resonances they do put limits on the possible extrapolations. We will not here discuss the details of the models that have the appropriate structure. They can be found elsewhere (Michaud 1972, 1973). The measured giant resonance structure led us to potentials with: a) small imaginary parts ($W_0 \simeq 1.0$ MeV); b) small real parts ($V_0 \simeq 10$ MeV). The 6.0 MeV resonance is the first one in the well; c) soft repulsive cores. It was found necessary to add to Saxon-Woods potentials (Eq. 2), a repulsive part of the form:

$$V_R = U \exp(-cr^2) \tag{9}$$

where U and c are positive constants. The constants are such that V_R becomes negligible beyond $r \simeq 0.6\ R_0$.

Finally, in our model most of the reactions (more precisely the absorption) at $E \simeq 2.0$ MeV, occur at $R \simeq 2\ R_0$, under the Coulomb barrier. The $^{12}C + ^{12}C$ reaction is the first reaction where the cross-section measurements suggest such a phenomenon and before it is considered as well

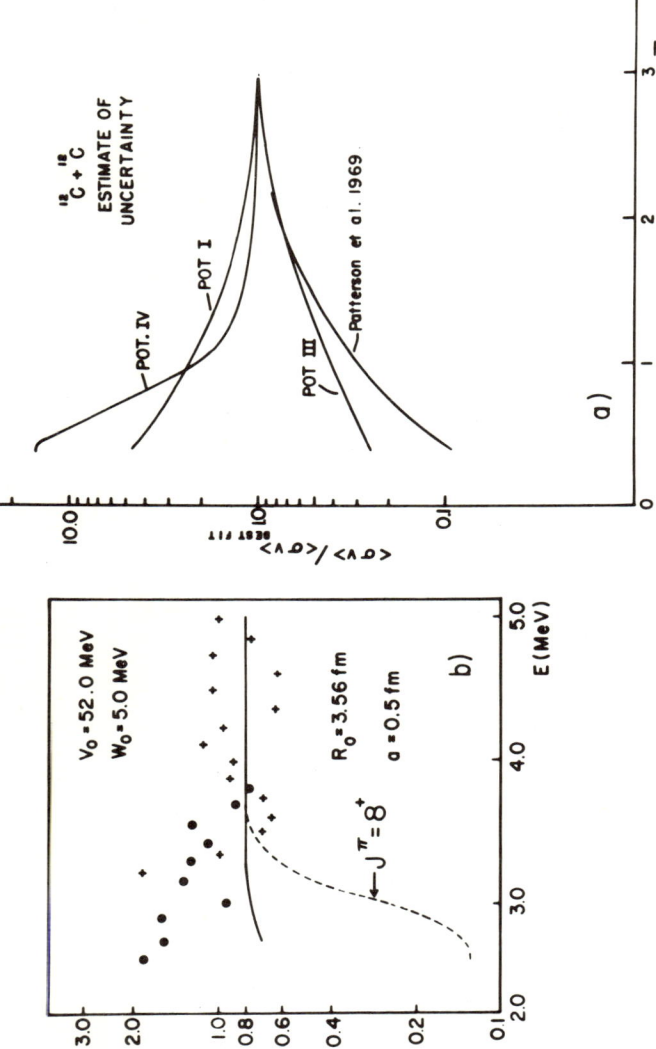

Fig. 2. On part a) are found the calculated reaction rates for a number of possible extrapolations, devided by the "Best Fit" reaction rate of Michaud (1972) where formulae for the rates may be found. At $T_9 = 0.5$ the extrapolation of Reeves (not shown) is a factor of 1.8 below our best fit. On part b) are the relative yields of the proton and alpha channels. The experimental results are inconsistent with a high spin for the partial waves involved.

established, its presence should be confirmed experimentally in other heavy ions reactions.

IV. Nucleon and Alpha-Particle Reactions

The most evident indication of a problem with nucleon and alpha particle reactions comes from the comparison of the measured and calculated (p, α) and (α, p) reaction rates. Truran (1973) first determined the radius of the particle channels by fitting low energy (p, n) and (α, n) reaction involving nuclei of relatively high mass. He so determined:

$$R_o = 1.35 \, A_o^{1/3} \text{ fm} \tag{10}$$

for proton channels and

$$R_o = 1.23 \, A_o^{1/3} + 1.6 \text{ fm} \tag{11}$$

for alpha-particle channels. He then used those radii to calculate the (α p) and (p α) cross-sections of lower mass nuclei and compared them to experimental values (Fowler 1972). One would have expected agreement within a factor of two. But the calculated rates are on the average larger than the measured ones by a factor of 1.5 at $T_9 = 1$ and between $T_9 = 1$ and 3 the disagreement increases to a factor of 10.1, and of 30 at $T_9 = 5$. In all eight cases compared the disagreement increases as the temperature increases. As suggested by Truran (1973) part, but not all, of the disagreement can be explained by missed resonances. This could be verified by checking whether or not the measured resonances satisfy the Porter-Thomas distribution (see for instance Preston 1962). But most of the disagreement nearly certainly comes from the theory. We will in turn discuss how uncertain the proton and alpha-particle channel transmission functions are.

From Eq. (1) the (α p) or (p α) cross sections are proportional to a sum of terms of the form

$$T_\ell (\alpha) \, T_\ell (p) / \sum_i T_\ell(i) \, .$$

where the sum in the denominator is to be taken over all open channels. In most reactions the denominator is dominated by one of the transmission functions of the numerator so that the errors in the incoming and outgoing channels should not be added. Here however, at high temperature, the denominator may often be dominated by the neutron channel or by inelastic scattering. We will add the errors in the proton and alpha

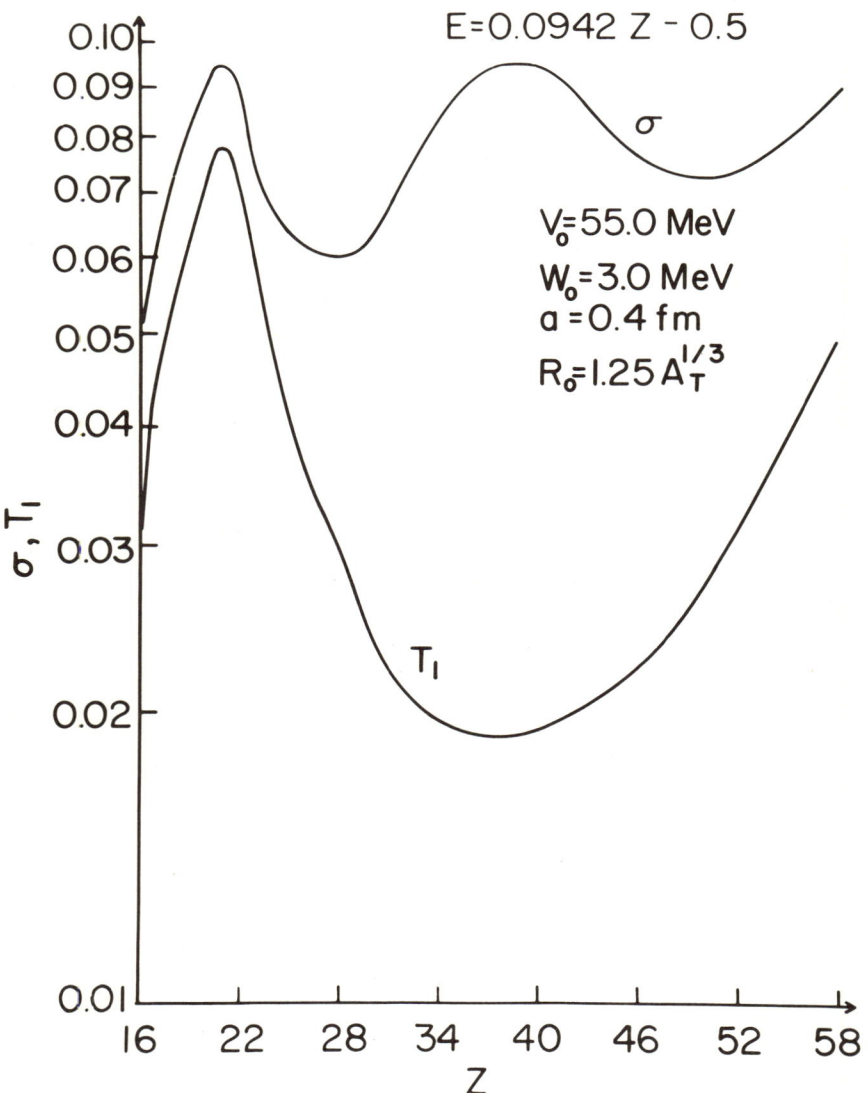

Fig. 3. Effect of the giant resonance in Saxon-Woods potentials without repulsive cores (σ in fm^2). The parameters used are appropriate to proton reactions on stable nuclei (identified by their atomic charge). The energy (in MeV) was adjusted so that the "black nucleus" cross-section would be the same for all nuclei.

channels but will neglect that in the neutron channel. We will not discuss the effect of inelastic scattering but it should be included in a detailed comparison. It would help to explain the discrepancy.

In Figure 3 are shown the total reaction cross-sections of protons with target nuclei of charge Z, as a function of the charge of the target nucleus. The energy of the reaction was chosen so that as Z increased, the cross-section remained the same except for giant resonance effects. The cross-section varies by 25% around the average value. The transmission function for $\ell = 1$, however, varies by more than a factor of 2 around the average value. In the total cross-section most of the variation in one partial wave is cancelled by that in the other. This will be the case so long as the wells used have the Saxon-Woods shape without repulsive core. For protons such wells are well justified since, except for the Coulomb field, protons are expected to react with the nuclei just like neutrons and the giant resonance structure of the $\ell = 0$ partial wave is well established for neutrons and well explained by Saxon-Woods wells with the same parameters as used here (see Vogt 1968 for a recent discussion).

In theoretical calculations (Truran 1973; Michaud and Fowler 1970) of reaction rates, to simplify calculations Saxon-Woods wells with $W_0 = 10$ MeV are used, so eliminating all resonance structure. From Figure 3 this is seen to introduce errors of 25% in the cross-section. It is much smaller than the measured errors. The only way to increase these fluctuations would be to assume that the $\ell = 0$ and $\ell = 1$ giant resonances occur at the same energy. This is not the case for Saxon-Woods wells nor for any similar well and is not believed to be the case for nucleons.

Apart from W_0 which governs giant resonance structure, two parameters have an important effect on the cross section, R_0 and a. The uncertainties in a can be related to uncertainties in R_0 and will not be further discussed (Michaud, Scherk and Vogt 1970). The radius used by Truran (1973) is uncertain by at least 10%, which leads to a factor of two variation in the cross-section, with the largest effect at the lowest energy where it is least needed. The correction is not large enough on the average, and would only make the relative behavior at $T_9 = 1$ and 5 worse.

Uncertainties in the proton channels apparently cannot explain the discrepancy most of which must then be explained in the alpha-channel. We will suggest two models of the alpha-particle channel which might explain the discrepancy. Many more experimental studies will however be needed to confirm our suggestion. In addition to being of interest in astrophysics such studies would bring new information on the average properties of nuclei which are of interest in nuclear physics.

Our first model tries to explain most of the discrepancy

by a giant resonance. The potential well would be of Saxon-Woods shape with a soft repulsive core (Equation 9) added. The repulsive core would be needed to have the even and odd partial waves resonate at the same energy and cause a strong giant resonance (Michaud 1973). The imaginary part of the potential, W_0, would be small (< 1.0 MeV) also to permit giant resonances. The radius would be given by Eq. (9). In this model, the maximum of the giant resonance would be, for target nuclei with $A \simeq 50$ where the discrepancy is seen, at low energy, ($E \simeq 2.0$ MeV) corresponding to $T_9 = 1$. At higher energy ($E \simeq 6.5$ MeV, $T_9 \simeq 5$) would be the valley between two resonances. This could lead to a reduction of the high temperature reaction rate by a factor of 5 or 10 depending on the value of W_0. The discrepancy between experimental and theoretical rates in (p α) reactions would then disappear. Finally the small value of V_0 in the Saxon-Woods well, would cause the equivalent square well to have

$$R_s = R_0 + \Delta R$$

with ΔR smaller than zero (see Michaud, Scherk and Vogt 1970, Fig. 6, with $KR_0/\pi = 1$ or 2). The low temperature reaction rates would be reduced compared to those of Truran (1973) who was doing calculations for large values of V_0 where $\Delta R > 0$.

Another possible model, would involve no repulsive core, a smaller R_0, a large V_0, a small W_0. Since the distance between resonances increases as V_0 increases, the giant resonance could there occur between an $\ell = 0$ and an $\ell = 1$ resonance. Both models are possible, both have their difficulties but both involve giant nuclear structure similar to that measured in the heavy ion reactions discussed earlier. Both calculations and measurements are needed to confirm or infirm the preceding suggestions.

The work of Truran (1973) indicates uncertainties of a factor of 10 when alpha particles are involved. It is an argument in favor of giant nuclear structure in the α-nucleus interaction. Until the systematics of this nuclear structure is understood, it will imply an uncertainty of a factor of 10 when alpha particles are involved.

V. γ-Ray Channels

The transmission functions which appear in Eq. (1) can be evaluated by using the optical model for particle channels, but not so for γ-ray channels. Their evaluation requires using a level density formula in one form or another. Level density formulas are needed only to evaluate γ-ray channels and to evaluate the correction due to inelastic scattering.

Transmission functions for γ-ray channels can be evaluated by assuming (Michaud and Fowler 1970):

$$T(\gamma) = c R^2 \int_0^B (B-E)^3/D(E) \, dE \qquad (12)$$

where $D(E)$ is the level spacing, B is the energy where the reaction occurs in the compound nucleus and c is an arbitrary constant. Using the level density formula for a Fermi gas, Michaud and Fowler (1970) then proceeded to determine the arbitrary parameters by comparing to experimental (nY) reaction rates. Corrections for shell effects were included in B. Surprisingly this formula agrees approximately as well with the experimental data as the more complete calculations of Truran et al. 1966. Discrepancies by factors of 2, between experiment and theory are frequent in both.

The reason for the discrepancy is unclear but is probably related to the energy dependence of the level density forumla and specially corrections made necessary by shell and pairing effects. No major progress is likely until the nuclear level density is more precisely corrected for these factors. The giant dipole resonance may also be involved. The parameterization of Michaud and Fowler (1970) may suggest ways to improve the level density formula but such a parameterization is limited since it neglects such effects as the spin dependence of the level density which probably leads to fluctuations of 30%.

REFERENCES

Arnett W. D. 1969, Ap. J., 157, 1369.
Arnett W. D., Truran J. W. and Woosley S. E. 1971, Ap. J., 165, 87.
Arnould M. 1972, Ay and Ap., 19, 92.
Bahcall N. A. and Fowler W. A. 1969, Ap. J., 157, 645.
Fowler W. A. 1972, private communication.
Fowler W. A., Caughlan G. R. and Zimmerman B. A. 1967, Ann. Rev. of Ay and Ap., 5, 525.
Fowler W. A. and Hoyle F. 1964, Ap. J. Suppl., 9, 201.
Hauser W. and Feshbach H. 1952, Phys. Rev., 87, 366.
Howard W. M., Arnett W. D. and Clayton D. D. 1971, Ap. J., 165, 495.
Howard W. M., Arnett W. D., Clayton D. D. and Woosley S. E. 1972, Ap. J., 175, 201.
Mazarakis M. G. and Stephens W. E. 1972, Ap. J. Lett., 171, L97.
Michaud G., Scherk L. and Vogt E. W. 1970, Phys. Rev. C., 1, 864.
Michaud G. 1972, Ap. J., 175, 751.
Michaud G. 1973, to be published.
Michaud G. and Fowler W. A. 1970, Phys. Rev. C., 2, 2041.

Middleton R., Garrett J. D. and Fortune H. T. 1970, Phys. Rev. Letters, 24, 1436.
Patterson J. R., Winkler H. and Zaidins C. A. 1969, Ap. J., 157, 367.
Preston M. A. 1962, Physics of the Nucleus (Reading, Mass.: Addison-Wesley).
Reeves H. 1966a, in Stellar Evolution, ed. R. R. Stein and A. G. W. Cameron (New York, Plenum Press).
Reeves H. 1966b, Ap. J., 146, 447.
Truran J. W. 1973, Ay and Ap., 00, 000.
Truran J. W., Arnett W. D. and Cameron A. G. W. 1967, Canadian J. of Phys., 45, 2315.
Truran J. W., Hansen C. J., Cameron A. G. W. and Gilbert A. 1966, Canadian J. of Phys., 44, 151.
Truran J. W. and Arnett W. D. 1970, Ap. J., 160, 181.
Vogt E. 1968, in Advances in Nuclear Physics, Vol.1, ed. M. Baranger and E. Vogt (New York: Plenum Press).
Wagoner R. V., Fowler W. A. and Hoyle F. 1967, Ap. J., 148, 3.
Zurmuhle R. W., Balamuth D. P., Fifield L. K. and Noe J. W. 1972, Phys. Rev. Letters, 29, 795.

EXPERIMENTS OF RELEVANCE TO EXPLOSIVE NUCLEOSYNTHESIS

Summary by the editors
of the talk given
by

William A. Fowler

California Institute of Technology
Pasadena, California

Fowler opened his talk by emphasizing that the basis of all nucleosynthetic calculations is the nuclear physics laboratory. With the exception of the quasi-equilibrium situations where only the Q-values are needed, astrophysicists need nuclear reaction rates. Basically the astrophysicist needs the Maxwell-Boltzman averaged cross section times velocity based on laboratory cross section measurements. Unfortunately not all cross sections of astrophysical importance can be measured experimentally. To determine these nonmeasurable cross sections one must have a good theory. This combination of theory and experiment will hopefully give analytic expressions for nuclear reaction rates over the large range in temperature used in astrophysics ($\sim 10^8$ °K to $\sim 10^{10}$ °K). Fowler described nuclear reaction rates as diagnostic probes since each reaction can help one learn about the temperature, density, time scale, and the presupernova conditions involved in each zone of a supernova explosion. In addition to the astrophysical implications the measurement of the appropriate cross sections can give the parameters of the nuclear optical model. Once the optical model parameters are determined then they can be used to calculate the cross sections for nonmeasurable reactions using the formalism described briefly later in this summary.

The method used by Fowler is based on the theory of Hauser and Feshbach (1952). This theory utilizes the concept of the formation of a compound nucleus; thus the problem of direct reactions remains for future research. In the Hauser - Feshbach theory the cross section can be expressed as follows:

$$\bar{\sigma}_{\alpha\alpha'} = \frac{\pi}{k_\alpha^2 (2I+1)(2i+1)} \sum_{J^\pi} \left\{ (2J+1) \frac{\left[\sum_{s,\ell} T_\ell(\alpha)\right]\left[\sum_{s',\ell'} T_{\ell'}(\alpha')\right]}{\sum_{\alpha''s''\ell''} T_{\ell''}(\alpha'')} \right\} \quad (1)$$

where α represents the incoming pair of particles and their state of excitation, α' represents the outgoing pair of particles and their state of excitation, I and i are the intrinsic spins of the incoming pair of interacting particles, s and s' are the channel spins of the incoming and outgoing states respectively, ℓ and ℓ' are the orbital angular momenta of the incoming and outgoing pair, J^π is the angular momentum and parity of the compound nucleus, and k_α is the de Broglie wave number for the incoming channel = $(2M_\alpha E_\alpha)^{1/2}/\hbar$ where M_α and E_α are the reduced mass and the center-of-momentum energy of the incoming channel respectively. Vector sums are carried out such that

$$\vec{s} = \vec{I} + \vec{i}$$
$$\vec{s}' = \vec{I}' + \vec{i}'$$ (1')

and

$$\vec{J} = \vec{\ell} + \vec{s} = \vec{\ell}' + \vec{s}'$$

where I' and i' refer to the intrinsic spins of the outgoing particles. The sums in equation (1') must be consistent with conservation laws for parity as well as for angular momentum. In general the primed quantities refer to the outgoing channel and the unprimed to the incoming channel. The sum over the double-primed quantities should include all open channels in the compound nucleus including excited states of residual nuclei. Note that this double-primed sum includes α and α'. This cross section in Equation 1 is an "averaged" cross section since it averages over excited states in the compound nucleus. It can be seen that the reaction is analyzed in two steps: one, the formation of the compound nucleus and two, the break up of the compound nucleus. The probabilities for these processes are calculated using the transmission functions, T_ℓ. The cross section is dependent on the competition between the various energetically allowed channels. Vogt (1968) and Michaud, Scherk and Vogt (1970) have shown that the transmission functions can be calculated using Woods-Saxon potentials

$$V_{WS}(r) = (V_0 + i W_0)\left(1 + e^{(r-R_0)/a}\right)$$ (2)

where a is the surface thickness, R_0 is the radius of the well, and V_0 and W_0 are the real and imaginary parts of the potential close to r = 0 for $R_0 \gg a$ as is usually the case. The basic details of how Fowler uses the optical model calculation are given by Michaud and Fowler (1970) and will not be repeated here. Basically the idea involves a blackbody absorber modified with a reflectivity factor, f.

In stars the target nucleus is not necessarily in its ground state but can occupy its excited states (Bahcall and Fowler 1969). The probability of occupation in an excited state, e, is

$$P_e = \frac{(2I_e + 1)\exp(-E_e/KT)}{\sum_{e'}(2I_{e'} + 1)\exp(-E_{e'}/KT)} \quad (3)$$

where the sum in the denominator is over all excited states e' including the ground state, $E_e' = 0$. Therefore, the averaged cross section at a temperature, T and Energy, E can be expressed as

$$\bar{\sigma}(T,E) = \frac{\sum_e\left[(2I_e + 1)\exp(-E_e/KT)\sum_{e'}\sigma_{\alpha\alpha'_{e'}}(E)\right]}{\sum_e(2I_e + 1)\exp(-E_e/KT)} \quad \text{(all states)} \quad (4)$$

where the cross section is in principle now averaged over all resonances in the compound nucleus and summed over all states in the nuclei involved. Notice that sum over all excited states, e', in the residual nucleus has also been made since all of these states will quickly decay down to the ground state of the residual nucleus. In a laboratory experiment the cross section naturally includes the contribution of each of the residual nucleus' excited states. This cross section can then be used to calculate the reaction rate at a given temperature

$$\langle\bar{\sigma}v\rangle = \frac{(8/\pi)^{1/2}}{M^{1/2}(KT)^{3/2}} \int_0^\infty \bar{\sigma}(T,E)E\exp\left(-\frac{E}{KT}\right)dE \quad (5)$$

where again M is the reduced mass of the incoming particles. The optical model parameters primarily used for calculating the transmission functions, T_ℓ, and thus ultimately determining the reaction rates are mostly from the work of Michaud, Scherk and Vogt (1970). It is found that the results are extremely sensitive to the radius parameter, R_o, and the surface thickness, a.

There is currently a program at Cal Tech to perform appropriate experiments and fit the optical model parameter to the experimental results. Fowler described several of these experiments in his talk but asked the editors to omit the results in this summary so that the experimentalists might have time to publish the results themselves.

In comparing theory with experiment quite good fits to the data were found for reactions where both the incoming and outgoing channels were particles rather than gamma rays. Some previous difficulties at fitting data were found to be caused by the improper treatment of barrier penetration at high temperatures. In comparing the experimental cross sections with the averaged cross sections of Equation 1 it is important to remember that individual resonances are not fit. The Maxwell-Boltzman distribution also averages over the closely spaced resonances of the nuclei being studied. Thus, missing individual resonances cause only minor discrepancies in the reaction rates. These discrepancies are on the order of a factor of 2. The fits to the data were usually good to this factor of 2 uncertainty.

Many of the reactions Fowler discussed were in the vicinity of mass 45 where there can be a "bottleneck" for silicon burning. Because nuclear experiments can be done on the dominant reactions in the relevant energy ranges for $1 \leq T_9 \leq 5$, the rates on that temperature region can be known fairly well. However at lower temperatures the cross section becomes smaller and only extrupolations can be used. Sometimes this problem can be avoided by measuring the reverse reaction and detailed balances.

Another point Fowler mentioned was that the effect of excited states in the target can be calculated if the inelastic scattering is measured at the same time. For example, if one were doing a (p,α) experiment, one should also do the (p,p') so that the effects of excited states in the target can be taken into account. With that data one can calculate the (p',α). It is important to notice that the good (factor of 2) agreement between theory and experiment that Fowler mentioned was only for reactions with both the incoming and outgoing channels being particles. The calculation of the transmission functions for the gamma channel is still very uncertain. Attempts at estimating $T(\alpha)$ (or equivalently $2\pi \Gamma_\gamma /D$) have been made by Michaud and Fowler (1970) and Truran, Cameron and Gilbert (1966). Fowler mentioned that it may be necessary to actually sum the individual level-transition probabilities rather than use the statistical treatments of Truran et al. (1966) and Michaud and Fowler (1970).

In the final few minutes of Fowler's talk the neutron capture cross section on ^{187}Os, $\sigma(187)$, was discussed. The ratio of this cross section to that of neutron capture on ^{186}Os, $\sigma(186)$, is the key to the ^{187}Re - ^{187}Os nucleocosmochronology (Clayton 1964). Fowler noted that ^{187}Os has a 9.8 KeV excited state with a 2J + 1 statistical factor of 4 compared to 2 for the ground state. Clayton (1964) estimated that the ratio $\sigma(186)/\sigma(187)$ was 0.4 ± 0.1. However he ignored the effect of this low lying excited state. Fowler showed that at the temperatures normally used for the s-process (30KeV) which produces ^{186}Os and all but the radiogenic part of ^{187}Os, the excited state would be sufficiently populated that $\sigma(186)/\sigma(187)$ would be increased by a factor of ~ 1.5 over the ratio determined when such occupation was neglected. Therefore the best estimate for $\sigma(186)/\sigma(187)$ is now $\sim 0.6\pm0.1$. This 1.5 correction is quite sensitive to temperature and may possibly be able to be used as a "thermometer" for the s-process. With the revised cross section ratio, the Re - Os chronology now yields time scales consistent with the Th - U time scales.

Fowler

From Fowler's talk it is clear that the ability to accurately determine cross sections is improving. However it is also clear that much more work is needed both from the nuclear experimentalist and from the nuclear theorist before all the reaction rates of importance to the nuclear astrophysicist are known with any high degree of confidence.

REFERENCES

Bahcall, N. and Fowler, W. A. 1969, Ap. J., 157, 695.
Clayton, O. D. 1964, Astrophys. J., 139, 637.
Hauser, W. and Feshbach, H. 1952, Phys. Rev., 87, 366.
Michaud, G. and Fowler, W. A. 1970, Phys. Rev. C, 2, 2041.
Michaud, G., Scherk, L. and Vogt, B. 1970, Phys. Rev. C, 1, 864.
Truran, J. W., Cameron, A. G. W. and Gilbert, A. 1966, Can. J. Phys., 44, 563.
Vogt, E. 1968, in Advances in Nuclear Physics Vol. I, ed. M. Baranger and E. Vogt (New York: Plenum Press, Inc.)

QB
450
C66
1973

~~Feb. 74~~

JUN 18 1974